Turing's Revolution

Birkhäuser

Giovanni Sommaruga • Thomas Strahm
Editors

Turing's Revolution

The Impact of His Ideas about Computability

Birkhäuser

Editors
Giovanni Sommaruga
Department of Humanities, Social and Political Sciences
ETH Zurich
Zurich, Switzerland

Thomas Strahm
Institute of Computer Science and Applied Mathematics
University of Bern
Bern, Switzerland

ISBN 978-3-319-22155-7 ISBN 978-3-319-22156-4 (eBook)
DOI 10.1007/978-3-319-22156-4

Library of Congress Control Number: 2015958070

Mathematics Subject Classification (2010): 03-XX, 03Axx, 03Dxx, 01Axx

Springer Cham Heidelberg New York Dordrecht London

Cover illustration: Image number: RS.7619, Credit: © Godrey Argent Studio
Cover design: deblik, Berlin

Printed on acid-free paper

Springer International Publishing AG Switzerland is part of Springer Science+Business Media (www.birkhauser-science.com)

Preface

June 23, 2012 was Alan Turing's 100th birthday. In the months preceding and following this date, there were widespread events commemorating his life and work. Those of us who had been writing about his work received more invitations to participate than we could accommodate. Speakers at one of the many Turing conferences were often invited to contribute to an associated volume of essays. This book developed from such a meeting, sponsored by the Swiss Logic Society, in Zurich in October 2012. The table of contents hints at the breadth of developments arising from Turing's insights. However, there was little public understanding of this at the time of Turing's tragic early death in 1954.

Thinking back on my own 65 years of involvement with Turing's contributions, I see my own dawning appreciation of the full significance of work I had been viewing in a narrow technical way, as part of a gradual shift in the understanding of the crucial role of his contributions by the general educated public. In 1951, learning to program the new ORDVAC computer shortly after I had taught a course on computability theory based on Turing machines, it became clear to me that the craft of making computer programs was the same as that of constructing Turing machines. But I was still far from realizing how deep the connection is. In the preface to my 1958 *Computability & Unsolvability*, I wrote:

> The existence of universal Turing machines confirms the belief ... that it is possible to construct a single "all purpose" digital computer on which can be programmed (subject of course to limitations of time and memory capacity) any problem that could be programmed for any conceivable deterministic digital computer.... I was very pleased when [it was] suggested the book be included in [the] Information Processing and Computers Series.

My next publication that commented explicitly on Turing's work and its implications was for a collection of essays on various aspects of current mathematical work that appeared in 1978. My essay [3] began:

> ... during the Second World War ... the British were able to systematically decipher [the German] secret codes. These codes were based on a special machine, the "Enigma" ... Alan Turing had designed a special machine for ... decoding messages enciphered using the Enigma.

> ...around 1936 [Turing gave] a cogent and complete logical analysis of the notion of "computation." ...[This] led to the conclusion that it should be possible to construct "universal" computers which could be programmed to carry out any possible computation.

The absolute secrecy surrounding information about the remarkable British effort to break the German codes had evidently been at least partially lifted. It began to be widely realized that Turing had made an important contribution to the war effort. For myself, I had come to understand that the relationship between Turing's mathematical construction that he called a "universal machine" and the actual computers being built in the postwar years was much more than a suggestive analogy. Rather it was the purely mathematical abstraction that suggested that, given appropriate underlying technology, an "all-purpose" computing device could be built.

The pieces began to come together when the paper [1], published in 1977, explained that Turing had been involved in a serious effort to build a working general purpose stored-program computer, his Automatic Computing Engine (ACE). Andrew Hodges's wonderful biography (1983) [5] filled in some of the gaps. Finally Turing's remarkably prescient view of the future role of computers became clear when, in [2], Turing's actual design for the ACE as well as the text of an address he delivered to the London Mathematical Society on the future role of digital computers were published. It was in this new heady atmosphere that I wrote my essay [4] in which I insisted that it was Turing's abstract universal machine, which itself had developed in the context of decades of theoretical research in mathematical logic, that had provided the underlying model for the all-purpose stored program digital computer. This was very much against the views then dominant in publications about the history of computers, at least in the USA. Meanwhile, noting that 1986 marked the 50th anniversary of Turing's universal machine, Rolf Herken organized a collection of essays in commemoration. There were 28 essays by mathematicians, physicists, computer scientists, and philosophers, reflecting the breadth and significance of Turing's work.

June 28, 2004, 5 days after Turing's 90th birthday, was Turing Day in Lausanne. Turing Day was celebrated with a conference at the Swiss Federal Institute of Technology organized by Christof Teuscher, at the time still a graduate student there. There were nine speakers at this one-day conference with over 200 attendees. Teuscher also edited the follow-up volume of essays [7] to which there were 25 contributors from remarkably diverse fields of interest. Of note were two essays on generally neglected aspects of Turing's work. One, written by Teuscher himself, was about Turing's work on neural nets written in 1948. The other recalled Turing's work during the final month's of his life on mathematical biology, particularly on morphogenesis: Turing had provided a model showing how patterns such as the markings on a cow's hide could be produced by the interactions of two chemicals. A more bizarre topic, discussed in a number of the essays, which had developed at the time under the name *hypercomputation*, sought ways to compute the uncomputable, in effect proposing to carry out infinitely many actions in a finite time. This was particularly ironic because computer scientists were finding that mere

Turing computability was insufficient for practical computation and were seeking to classify algorithms according to criteria for feasibility.

As Turing's centenary approached, there was a crescendo of events bringing Turing's importance to the attention of the public. An apology by the British government for his barbaric mistreatment along with a commemorative postage stamp was a belated attempt to rectify the unrectifiable. I myself spoke about Turing at various places in the USA and Britain and also in Mexico, Peru, Belgium, Switzerland, and Italy, crossing and recrossing the Atlantic Ocean several times in a few months' time. A high point was attending a banquet at King's College, Cambridge (Turing's college), with members of Turing's family present, celebrating his actual 100th birthday. I'm very fond of Zurich and was delighted to speak at the conference there from which this volume derives. The perhaps inevitable consequence of Alan Turing's developing fame was Hollywood providing in *The Imitation Game* its utterly distorted version of his life and work.

Considering the authors whose contributions the editors have managed to gather for this volume, I feel honored to be included among them. When it turned out that my good friend Wilfried Sieg and I were writing on very similar subjects, we agreed to join forces, providing for me a very enjoyable experience. Among the varied contributions, I particularly enjoyed reading the essay by Jack Copeland and Giovanni Sommaruga on Zuse's early work on computers. I had been properly chastised for omitting mention of Zuse in my *The Universal Computer*, and I did add a paragraph about him in the updated version of the book for Turing's centenary. However, I had never really studied his work and I have learned a good deal about it from this excellent essay. The writings of Sol Feferman can always be depended on to show deep and careful thought. I very much enjoyed hearing Stewart Shapiro's talk at the conference in Zurich, and I look forward to reading his essay. Barry Cooper has, of course, been a driving force in the 2012 celebrations of Alan Turing and in the "Computability in Europe" movement. Finally, I want to thank Giovanni Sommaruga and Thomas Strahm for bringing together these authors and editing this appealing volume.

Berkeley, CA, USA
February 2015

Martin Davis

References

1. B.E. Carpenter, R.W. Doran, The other Turing machine. Comput. J. **20**, 269–279 (1977)
2. B.E. Carpenter, R.W. Doran (eds.), *A.M. Turing's ACE Report of 1946 and Other Papers* (MIT Press, New York, 1986)
3. M. Davis, What is a computation, in *Mathematics Today: Twelve Informal Essays,* ed. by L.A. Steen (Springer, New York, 1978), pp. 241–267
4. M. Davis, Mathematical logic and the origin of modern computers, in *Studies in the History of Mathematics* (Mathematical Association of America, Washington D.C., 1987), pp. 137–165. Reprinted in [6], pp. 149–174

5. A. Hodges, *Alan Turing, The Enigma* (Simon and Schuster, New York, 1983)
6. R. Herken (ed.), *The Universal Turing Machine - A Half-Century Survey* (Verlag Kemmerer & Unverzagt/Oxford University Press, Hamburg, Berlin/Oxford, 1988)
7. C. Teuscher (ed.), *Alan Turing: Life and Legacy of a Great Thinker* (Springer, Berlin, 2004)

Introduction

During the beginning and about half of Turing's short life as a scientist and philosopher, the notions of computation and computability dazzled the mind of a lot of the most brilliant logicians and mathematicians. Some of the most prominent figures of this group of people are Alonzo Church, Kurt Gödel, Jacques Herbrand, Stephen Kleene, and Emil Post—and of course Alan Turing. The second quarter of the twentieth century was such a melting pot of ideas, conceptions, theses, and theories that it is worthwhile to come back to it and to dive into it over and over again. That's what this volume's first part is about.

What is the point of looking back, of turning back to the past? What is the good of looking at the history, e.g., of the notion of computability? The look backwards serves to better understand the present. One gets to understand the origin, the core ideas, or notions at the beginning of a whole development. The systematical value of the historical perspective may also lie in the insight that there is more to the past than being the origin of the present, that there is a potential there for alternative developments, that things could have developed differently, and even that things could still develop differently. The past and especially a look at the past may thereby become a source of inspiration for new developments. Looking at the past may contribute to planning the future, if one starts to wonder: Where do we come from? Where are we going to? And where do we want to go?

In the second half of the last century and up to this very day, these ideas, conceptions, etc. and in particular Turing's gave rise to a plethora of new logical and mathematical theories, fields of research, some of them extensions or variations, and others novel applications of Turing's conceptions and theories. This volume's second part aims at presenting a survey of a considerable part of subsequent and contemporary logical and mathematical research influenced and sparked off, directly or indirectly, by Turing's logical mathematical ideas, ideas concerning computability and computation.

These generalizations can take on different shapes and go in different directions. It is possible to generalize concepts, theses, or theories of Turing's (e.g., generalizing his thesis to other structures, or his machine concept to also account for the infinite). But it is equally possible to take Turing's fulcrum and by varying it think of new alternatives to it. Furthermore, seeds of notions or conceptions can be unfolded into full-blown theories (e.g., of Turing degrees). Or one can take a problem of Turing's and search for or discover solutions to it, new and different from Turing's own solution (e.g., transfinite rec. progressions). And what is the meaning of those generalizations in a broad sense? They all can be understood as explorations of Turing's own rich conceptual space (even still roughly restricted to the topic of computability). They all mean to further research into its wealth and its many dimensions, to analyze and transgress it, to widen and enrich it. And they all hint at the "fact" that there is yet more to be found in this mine of technical ideas.

Thus, Part I goes back to the roots of Turing's own as well as to the ideas of some of his contemporaries. Part II is dedicated to the utterly prolific outgrow of these ideas. Bringing Parts I and II together raises some questions such as: How was it possible that these original ideas, conceptions, and theories lead to such an explosion of new and highly innovative approaches in computability theory and theoretical computer science? How is this phenomenon to be understood? And what does it mean? Is there perhaps even more to this past which might be fruitful and of great interest in and for the future? A reflection on computability theory in general and Turing's in particular is what Part III of this volume is all about.

Turing's discoveries, important mathematical logical results and theses, etc. have also set off more penetrating, more daring, or more speculative questions reaching to the outskirts or to the core of mathematics. They raise questions concerning the computable and the incomputable, and concerning the meaning of computability and incomputability w.r.t. the real world. And the philosophical entanglements of Turing's thesis stir up questions such as: What is a proof? What is a theorem and what is a thesis? And different Turing ways lead to the same question, namely: What is mathematics? And what are its visions and its dreams? This needn't be idle speculation. If well and carefully done, it may in turn spark off new mathematical ideas and/or new mathematical research insights. But even if it doesn't, there can be plenty of meaning to it.

The spirit of the first part might be characterized as: There's a future to the past. The second part's spirit may be captured by: There's a future and a past to the present. And the third part's spirit might be put in the formula: Even if there isn't a path from the present to the future, there needn't by no means be nothing.

This volume makes no claim to completeness of coverage of the whole range of mathematical and logical research initiated or triggered off by Turing's respective ideas: Nothing is said or written about Turing and computational complexity theory, no mention is made of Turing computation and probability theory either, to list just a couple of examples. Even less so do Turing's ideas on number theory, cryptology, artificial intelligence, or mathematical biology and the offsprings of those ideas play any role in this volume.

Let's now turn to what actually can be found in this volume:

In 1936, two models of computation were proposed in England and in the USA, resp., which are virtually identical. Davis and Sieg claim that this is not a mere coincidence, but a natural consequence of the way in which Turing and

Post conceived of steps in mechanical procedures on finite strings. Davis and Sieg also claim that the unity of their approaches, Post's mathematical work on normal forms of production systems on the one side, and Turing's philosophical analysis of mechanical procedures on the other, is of deep significance for the theory of computability. Moreover, Davis and Sieg point out that the investigation by mathematical logicians like Hilbert, Bernays, Herbrand, Gödel, and others of mechanical procedures on syntactic configurations were following a line of research very different from the one of Post and Turing. While the former were concentrating on paradigmatic recursive procedures for calculating the values of number-theoretic functions, the latter were struck by the fundamental and powerful character of string manipulations. In what Davis and Sieg call "Turing's Central Thesis," Turing correlates the informal notion of "symbolic configuration" with the precise mathematical one of "finite string of symbols." This thesis has methodologically the problematic status of a statement somewhere between a definition and a theorem. Davis and Sieg finally raise the question whether there is a way of avoiding appeal to such a "Central Thesis," and answer it tentatively positively by referring to Sieg's representation theorem for computable discrete dynamical systems.

Thomas' paper focuses on a specific and also very prominent aspect of Turing's heritage, namely on his analysis of a machine or algorithm, that is "a process of symbolic computation fixed by an unambiguous and finite description."[1] Thomas outlines the history of algorithms starting with what he calls its prehistory and its most important figures, Al-Khwarizmi and Leibniz. The history proper may already start with Boole and Frege, but it decidedly starts with Hilbert and his Entscheidungsproblem, i.e. the problem whether there exists an algorithm for deciding the validity or satisfiability resp. of a given logical formula. Turing in 1936 solved this problem in the negative, and he did this by a radical reduction of the notion of algorithm to some very elementary steps of symbolic computation (later called a Turing machine). According to Thomas, Turing's work and also the one of his contemporaries Church, Kleene, Post, and others ended a centuries-old struggle for an understanding of the notion of algorithm and its range of applicability called computability. This success was made possible by merging two traditions in symbolic computation, namely arithmetic and logic. But for Thomas, 1936 not only marks a point of final achievement, but also a point of departure for the new and rapidly growing discipline of computer science, the starting point of what might be conceived of as posthistory. He claims that the subsequent rise of this new discipline changed the conception of algorithm in two ways: "First, algorithmic methods in computer science transcend the framework of symbolic manipulation inherent in Turing's analysis," and Thomas then offers an overview of such methods. Second, he emphasizes that algorithms are today understood in the context of highly complex software systems governing our life. "The span of orders of magnitude realized in huge data conglomerates and software systems" is so gigantic that diverse

[1] The quotations in this and the following sections are referring to the article which is summarized in the resp. section.

methodologies are required in computer science in order to study the rich landscape of computer systems.

After presenting a biographical sketch of Konrad Zuse, Copeland and Sommaruga "outline the early history of the stored-program concept in the UK and the US," and "compare and contrast Turing's and John von Neumann's contributions to the development of the concept." They go on to argue that, contrary to the recent prominent suggestions, the stored-program concept played a key role in computing's pioneering days, and they provide a logical analysis of the concept, distinguishing four different layers (or "onion skins") comprising the concept. This layered model allows them to classify the contributions made by Turing, von Neumann, Eckert and Mauchly, Clippinger, and others, and especially Zuse. Furthermore, Copeland and Sommaruga discuss whether Zuse developed a universal computer (as he himself claimed) or rather a general-purpose computer. In their concluding remarks, Copeland and Sommaruga "reprise the main events in the stored-program concept's early history." The history they relate begins in 1936, the date of publication of Turing's famous article "On Computable Numbers," and also of Zuse's first patent application for a calculating machine, and runs through to 1948, when the first electronic stored-program computer—the Manchester "Baby"—began working.

Feferman starts with a detailed summary and discussion of versions of the Church-Turing Thesis (CT) on concrete structures given by sets of finite symbolic configurations. He addresses the works of Gandy, Sieg as well as Dershovitz and Gurevich, which all have in common that they "proceed by isolating basic properties of the informal notion of effective calculability or computation in axiomatic form and proving that any function computed according to those axioms is Turing computable." Feferman continues to note that generalizations of CT to abstract structures must be considered as theses for algorithms. He starts by reviewing Friedman's approach for a general theory of computation on first order structures and the work of Tucker and Zucker on "While" schemata over abstract algebras. Special emphasis is put on the structure of the real numbers; in this connection, Feferman also discusses the Blum, Shub, and Smale model of computation. A notable proposal of a generalization of CT to abstract structures is the Tucker–Zucker thesis for algebraic computability. The general notion of algorithm it uses leads to the fundamental question "What is an algorithm?," which has been addressed under this title by Moschovakis and Gurevich in very different ways. Towards the end of his article, Feferman deals with a specific approach to generalized computability, which has its roots in Platek's thesis and uses a form of least fixed point recursion on abstract structures. Feferman concludes with a sensible proposal, the so-called "Recursion thesis," saying that recursion on abstract first order structures (with Booleans {T,F}) belongs to the class he calls Abstract Recursion Procedures, ARP (formerly Abstract Computation Procedures, ACP).

Tucker and Zucker survey their work over the last few decades on generalizing computability theory to various forms of abstract algebras. They start with the fundamental distinction between abstract and concrete computability theory and emphasize their working principle that "any computability theory should be focused equally on the data types and the algorithms." The first fundamental notion is the one of While computation on standard many sorted algebras, which is a high level imperative programming language applied to many sorted signatures. After a precise syntactical and semantical account of this language, Tucker and Zucker address notions of universality. An interesting question is to consider various possible definitions of semi-computability in the proposed setting. As it turns out, different characterizations, extensionally equivalent in basic computability theory over the natural numbers, are different in the Tucker and Zucker setting. A further emphasis in the paper is laid on data types with continuous operations like the reals, which leads to a general theory of many sorted partial topological algebras. Tucker and Zucker conclude by comparing their models with related abstract models of computation and by proposing various versions of a generalized Church–Turing thesis for algebraic computability.

Welch gives a broad survey of models of infinitary computation, many of which are rooted in infinite time Turing machines (ITTMs). It is the latter model that has sparked a renewed interest in generalized computability in the last decade. After explaining the crucial notion of "computation in the limit," various important properties of ITTMs are reviewed, in particular, their relationship to Kleene's higher type recursion. Furthermore, Welch elaborates on degree theory and the complexity of ITTM computations as well as on a close relationship between ITTMs, Burgess' quasi-inductive definitions, and the revision theory of truth. He then considers variants of the ITTM model that have longer tapes than the standard model. Afterwards Welch turns to transfinite generalizations of register machines as devised by Sheperdson and Sturgis, resulting in infinite time register machines (ITRMs) and ordinal register machines (ORMs); the latter model also has registers for ordinal values. The last models he explains are possible transfinite versions of the Blum–Shub–Smale machine, the so-called IBSSMs. Welch concludes by mentioning the extensional equivalence (on omega strings) of continuous IBSSMs, polynomial time ITTMs, and the safe recursive set functions due to Beckmann, Buss, and Friedman.

Gurevich reviews various semantics-to-syntax analyses of the so-called species of algorithms, i.e., particular classes of algorithms given by semantical constraints. In order to find a syntactic definition of such a species, e.g., a machine model, one often needs a fulcrum, i.e., a particular viewpoint to narrow down the definition of the particular species in order to make a computational analysis possible. Gurevich starts with Turing's fundamental analysis of sequential algorithms performed by idealized human computers. According to Gurevich, Turing's fulcrum was to "ignore what a human computer has in mind and concentrate on what the computer does and what the observable behavior of the computer is." Next, Gurevich turns to the analysis of digital algorithms by Kolmogorov in terms of Kolmogorov–Uspenski machines and identifies its fulcrum, namely that computation is thought of as "a physical process developing in space and time." Then Gurevich discusses Gandy's analysis of computation by discrete, deterministic mechanical devices and identifies its fulcrum in Gandy's Principle I, according to which the representation and working of mechanical devices must be expressible in the framework of hereditarily finite sets. The fourth example discussed is the author's own analysis of the species of sequential algorithms using abstract state machines. Its fulcrum is: "Every sequential algorithm has its native level of abstraction. On that level, the states can be faithfully represented by first-order structures of fixed vocabulary in such a way that the transitions can be expressed naturally in the language of that fixed vocabulary."

Turing (implicitly) introduced the notion of Turing reducibility in 1939 by making use of oracle Turing machines. This preorder induces an equivalence relation on the continuum, which identifies reals with the same information content and whose equivalence classes are the so-called Turing degrees. Barmpalias and Lewis survey order-theoretic properties of degrees of typical reals, whereby sensible notions of typicality are derived from measure and category. Barmpalias and Lewis present a detailed history of measure and category arguments in the Turing degrees as well as recent results in this area of research. Their main purpose is to provide "an explicit proposal for a systematic analysis of the order theoretically definable properties satisfied by the typical Turing degree." Barmpalias and Lewis identify three very basic questions which remain open, namely: (1) Are the random degrees dense? (2) What is the measure of minimal covers? (3) Which null classes of degrees have null upward closure?

Beklemishev's paper relates to Turing's just mentioned paper entitled "Systems of logic based on ordinals," which is very well known for its highly influential concepts of the oracle Turing machine and of relative computability. The main body of Turing's 1939 paper, however, belongs to a different area of logic, namely proof theory. It deals with transfinite recursive progressions of theories in order to overcome Gödelian incompleteness. Turing obtained a completeness result for Π_2 statements by iterating the local reflection principle, whereas Feferman in 1962 established completeness for arbitrary arithmetic statements by iteration of the uniform reflection principle. Turing's and Feferman's results have the serious drawback that the ordinal logics are not invariant under the choice of ordinal representations and their completeness results depend on artificial such representations. Beklemishev's approach is to sacrifice completeness in favor of natural choices of ordinals. He obtains a proof-theoretic analysis of the most prominent fragments of first order arithmetic by using the so-called smooth progressions of iterated reflection principles.

Juraj Hromkovic's main claim is that to properly understand Alan Turing's contribution to science, one ought to understand what mathematics is and what the role of mathematics in science is. Hromkovic answers these latter questions by comparing mathematics with a new, somewhat artificial language: "one creates a vocabulary, word by word, and uses this vocabulary to study objects, relationships" and whatever is accessible to the language of mathematics at a certain stage of its development. For Leibniz, mathematics offered an instrument for automatizing the intellectual work of humans. One expresses part of reality in the language of mathematics or in a mathematical model, and then one calculates by means of arithmetic. The result of this calculation is again a truth about the investigated part of reality. Leibniz' "dream was to achieve the same for reasoning." He was striving for a formal system of reasoning, a formal system of logic, analogous to arithmetic, the formal system for the calculation with numbers. Hilbert's dream of the perfection of mathematics involved the idea that every piece of knowledge expressible as a statement in the language of mathematics could be determined to be true or false, and that "for each problem expressible in the language of mathematics, there exists a method for solving it." In particular, there has to be a method which for every true mathematical statement yields a proof of the truth of it. Hilbert's dream was unlike Leibniz' dream restricted to mathematics. New concepts and subsequently new words are introduced into the mathematical language by a specific sort of mathematical definition, namely by axioms. "The axioms form the fundamental vocabulary of mathematics and unambiguously determine the expressive power of mathematics." Hromkovic also makes the point that inserting new axioms into mathematics increases likewise the argumentative power of mathematics: by means of these new axioms, it becomes possible to investigate and analyze things which mathematics hitherto could not study. Thus, it is possible to reformulate Hilbert's dream in the following way: that the expressive power and the argumentative power of mathematics coincide. Gödel destroyed Hilbert's dream by proving that as a matter of fact they do not coincide, that is that the expressive power of mathematics is greater than the argumentative one, and that the gap between the two is fundamental, i.e., not eliminable. Part of Hilbert's dream was also that for each problem expressible in the language of mathematics there be a method for solving it. Turing did not share this conviction and he faced the problem of how to prove the non-existence of a mathematical method of solution for a concrete problem. To solve this problem, he had to turn the concept of a mathematical method of solution (an algorithm) into a new mathematical concept and word. Doing this was, according to Hromkovic, Alan Turing's main contribution. With the new concept of algorithm added to the language of mathematics, it was now possible to study the range and limits of automation.

Starting point of Shapiro's article is the received view, initiated by Church in 1936, that Church's Thesis (CT) is not subject to rigorous mathematical proof or refutation, since CT is the identification of an informal, intuitive notion with a formal, precisely defined one. Kleene in 1952 asserts that the intuitive notion of computability is vague and he claims that it is impossible to mathematically prove things about "vague intuitive" notions. Gödel challenged the conclusion based on the supposed vagueness of computability. Gödel's own view seems to be that there is a precise property or notion resp. somehow underlying the intuitive notion of computability, which is to be captured by an analysis and progressive sharpening of the intuitive, apparently vague one. Shapiro's main question is: "How does one *prove* that a proposed sharpening of an apparently vague notion is the uniquely correct one" capturing the precise notion which was there already? The received view referred to above was not only challenged by Gödel, but eventually by several prominent philosophers, logicians, and historians such as Gandy, Mendelson, or Sieg. Since the Gödel/Mendelson/Sieg–position generated further discussions and critical reactions, the issues relating to CT "engage some of the most fundamental questions in the philosophy of mathematics" such as: "What is it to prove something? What counts as a proof? ... What is mathematics about?" Shapiro then sets out to show that the defenders of the position that CT is provable do not thereby mean provable by a ZFC-proof, nor could they possibly mean provable by a formal proof. Thus, he reaches the conclusion that even though CT is not entirely a formal or ZFC matter, this doesn't preclude it from being any less mathematical or from being provable. In order to defend this line of argument, he opposes the foundationalist model of mathematical knowledge to a holistic one. According to the latter model, it is possible to explain why Sieg's proof of CT might justifiably be called so, although it is neither a purely formal nor a ZFC proof. It can be called a proof because it is formally sound and because its premises are sufficiently evident. This according to Shapiro doesn't say that the proof may not beg any questions. But he concludes that at least "Sieg's discourse and, for that matter, Turing's are no more question-begging than any other deductive argumentation. The present theme is more to highlight the holistic elements that go into the choice of premises, both in deductions generally and in discourses that are rightly called "proofs," at least in certain intellectual contexts."

Soare sets off explaining what Turing's Thesis (TT) and what Gandy's Thesis M is. After characterizing the concept of a thesis as opposed to that of a statement of facts or a theorem, Soare considers the question whether Turing proved his assertion (that a function on the integers is effectively calculable iff it is computable by a so-called Turing machine) "beyond any reasonable doubt or whether it is merely a thesis, in need of continual verification." In his sketchy presentation of Turing's 1936 paper, Soare points out that in Turing's analysis of the notion of a mechanical procedure, Turing broke up the steps of such a procedure into the smallest steps, which could not be further subdivided. And when going through Turing's analysis, one is left with something very close to a Turing machine designed to carry out those elementary steps. Soare then relates on Gödel's, Church's, and Kleene's reaction to Turing's 1936 paper. In 1936, Post independently formulated an assertion analogous to Turing's, and he called it a working hypothesis. Somewhat in the same vein, Kleene in 1943 and especially in 1952 called Turing's assertion a thesis, which in the sequel led to the standard usage of "Turing's Thesis." Already in 1937, Church objected to Post's working hypothesis and in 1980 and 1988 Gandy challenged Kleene's claim that Turing's assertion is a thesis (i.e., could not be proved). Soare emphasizes that not only Gandy, but later on also Sieg, Dershowitz, and Gurevich as well as Kripke have presented proofs of Turing's assertion. Since Soare endorses those proofs, he gets to the conclusion that "Turing's Thesis" TT shouldn't be called that way any longer, it should rather be called Turing's Theorem.

Barry Cooper's aim is "to make clearer the relationship between the typing of information—a framework basic to all of Turing's work—and the computability theoretic character of emergent structures in the real universe." Cooper starts off with the question where incomputability comes from. He notes that there is no notion of incomputability without an underlying model of computation, which is here provided by the classical Turing machine model. He then observes that whereas from a logical point of view, a Turing machine is fairly simple, any embodiment of a particular universal Turing machine as an actual machine is highly non-trivial. "The physical complexities have been packaged in a logical structure, digitally coded," and "the logical view has reduced the type of the information embodied in the computer." Cooper first considers the relationship between incomputability and randomness. The notion of randomness used to describe the lack of predictability is sometimes taken to be more intuitive than the one of incomputability. Cooper argues that randomness turns out to be a far more complicated notion, and that all that could be substantiated "building on generally accepted assumptions about physics, was incomputability." He then proceeds from the observation that "global patterns can often be clearly observed as emergent patterns in nature and in social environments, with hard to identify global connection to the underlying computational causal context." Cooper discusses this gap by reflecting on the relationship between the halting problem (one of the classical paradigms of incomputability) and the Mandelbrot set. For him, the Mandelbrot set "provides an illuminating link between the pure abstraction of the halting problem, and the strikingly embodied examples of emergence in nature." Cooper generalizes his observations about this link by introducing the mathematics of definability. The higher properties of a structure (those important to understand in the real world) are the large-scale, emergent relations of the structure, and the connection between these and their underlying local structure is mathematically formalized in terms of definability. "Such definability can be viewed as computation over higher type data." However, as Cooper explains, "computation over higher-type information cannot be expected to have the reliability or precision of the classical model." And he uses the example of the human brain and its hosting of complex mentality to illustrate this. The mathematics to be used for this new type of computation is based on the theory of degrees of incomputability (the so-called Turing degrees) and the characterization of the Turing definable relations over the structure of the Turing degrees.

We have to close this introduction with the very sad news that Barry Cooper, the author of the last article of this volume, after a brief illness passed away on October 26, 2015. Barry has been the driving force behind the Turing Centenary celebrations in 2012 and of the Computability in Europe movement since 2005. His impact and work as a supporter of Alan Turing are ubiquitous. We dedicate our volume to Barry. His gentle and enthusiastic personality will be greatly missed.

Acknowledgements

This is our opportunity to thank those people who made a special contribution to this volume or who contributed to making this a special volume. We'd like to express our great gratitude to Martin Davis who graciously took upon himself the task of writing a wonderful preface. We'd also like to express this gratitude to Jack Copeland who contributed a couple of brilliant ideas the specific nature of which remains a secret between him and us. Moreover, we are very grateful to the referees of all the articles for their refereeing job, and especially to those who did a particularly thorough job. And we'd finally like to thank Dr. Barbara Hellriegel, Clemens Heine and Katherina Steinmetz from Birkhäuser/Springer Basel, and Venkatachalam Anand from SPi

Content Solutions/SPi Global very much for not despairing about the delay with which we delivered the goods and for making up for this delay with a very efficient and very friendly high speed production process.

Zurich
Bern
December 2015

Giovanni Sommaruga
Thomas Strahm

Contents

Part III Philosophical Reflections

Part I
Turing and the History of Computability Theory

Conceptual Confluence in 1936: Post and Turing

Martin Davis and Wilfried Sieg

Abstract In 1936, Post and Turing independently proposed two models of computation that are virtually identical. Turing refers back to these models in his (The word problem in semi-groups with cancellation. Ann. Math. **52**, 491–505) and calls them "the logical computing machines introduced by Post and the author". The virtual identity is not to be viewed as a surprising coincidence, but rather as a natural consequence of the way in which Post and Turing conceived of the steps in mechanical procedures on finite strings. To support our view of the underlying conceptual confluence, we discuss the two 1936 papers, but explore also Post's work in the 1920s and Turing's paper (Solvable and unsolvable problems. Sci. News **31**, 7–23). In addition, we consider their overlapping mathematical work on the word-problem for semigroups (with cancellation) in Post's (Recursive unsolvability of a problem of Thue. J. Symb. Log. **12**, 1–11) and Turing's (The word problem in semi-groups with cancellation. Ann. Math. **52**, 491–505). We argue that the unity of their approach is of deep significance for the theory of computability.

Keywords Ackermann • Bernays • Canonical system • Church • Computor • Correspondence decision problem • Davis • Dedekind • Dershowitz • Gandy • Gödel • Gurevich • Herbrand • Hilbert • Kleene • Kolmogorov • Mechanical procedure • Normal system • Post • Principia mathematica • Production • Recursive functions • Representation theorem • Sieg • Skolem • Tag • Thue • Turing • Turing's central thesis • Unsolvability • Uspenski • Word problem

AMS Classifications: 01A60,03-03,03D03,03D10,03D20, 20M05

M. Davis (✉)
Department of Computer Science, Courant Institute of Mathematical Sciences, New York University, 3360 Dwight Way, Berkeley, CA 94704-2523, USA
e-mail: martin@eipye.com

W. Sieg
Department of Philosophy, Carnegie Mellon University, 5000 Forbes Avenue, Pittsburgh, PA 15213, USA

G. Sommaruga, T. Strahm (eds.), *Turing's Revolution*,
DOI 10.1007/978-3-319-22156-4_1

1 Introduction

Princeton was an exciting place for logicians in the mid-1930s. Church had been appointed as assistant professor of mathematics in 1929 and, together with his students Stephen C. Kleene and J. Barkley Rosser, started to create a new subject soon to be called "recursive function theory".[1] Their work was propelled by von Neumann's presentation of Gödel's incompleteness theorems [27] in the fall of 1931 and Gödel's lectures "On undecidable propositions of formal mathematical systems" in the spring of 1934 [28]. Paul Bernays, the collaborator of David Hilbert on proof theory and the related foundational program, visited for the academic year 1935/1936 and gave the lectures [2]. Even before that visit, Bernays and Church had extensive correspondence concerning the ongoing work in recursion theory; see [59]. From September 1936 to July 1938, Alan Turing was studying in Princeton as Church's Ph.D. student; his dissertation "Systems of logic based on ordinals" was published in [73].

In 1936, Church founded a new quarterly devoted to contemporary research in logic: The Journal of Symbolic Logic. Its first volume contained a three-page paper by Emil L. Post, "Finite combinatory processes: Formulation I". The editors added a footnote to the paper that read:

> Received October 7, 1936. The reader should compare an article by A.M. Turing, On computable numbers, shortly forthcoming in Proceedings of the London Mathematical Society. The present article, however, although bearing a later date, was written entirely independently of Turing's.

Post was at the time teaching at City College in New York City and had also contact with Church, both personal and through correspondence.

As to the remark concerning Post's paper, it would indeed be readily apparent to any reader of Turing's and Post's articles that their basic idea of "computation" was the same. Turing presented a machine model; his machines were "supplied with a 'tape' ... divided into sections, called 'squares', each capable of bearing a 'symbol' " from a finite list, possibly just two different ones. In the latter case, we refer to the machine as a two-letter-machine.[2] Instead of a tape divided into squares, Post wrote of a "symbol space ... to consist of a two way infinite sequence of spaces or boxes"; each box could be either "marked" or "unmarked". In both conceptions, at a given instant, one particular square/box was in play, and the permitted basic operations were to change the symbol in the square/box and/or to move to an adjacent square/box. Turing imagined these steps carried out by a

[1]The broader history of computability has been described in a number of publications, partly by the participants in the early development in Princeton, for example, [43, 57]. Good discussions are found in, among others [9, 14, 26, 58, 62, 68].

There are many excellent books on recursion/computability theory, but not many that take Post's approach as fundamental. We just mention [13, 47, 56, 67].

[2]Machines that are deterministic are Turing's *a-machines*; if a-machines operate only on 0s and 1s they are called *computing machines*; see [71], p. 232.

simple mechanism with a finite number of states where each step included a possible change of state. In Post's formulation, the computation is carried out by a human "worker" who is following a fixed sequence of instructions including what we would call a conditional jump instruction. The jump instruction is also part of a program for Turing's machine: depending on the symbol in the square and its state, the machine may follow a different next instruction.

It seems astonishing that, on opposite sides of the Atlantic, two models of computation were proposed that are virtually identical. We suggest that this virtual identity is not to be viewed as a surprising coincidence, but rather as a natural consequence of the way in which Post and Turing conceived of steps in mechanical procedures on finite strings. To support our view of the underlying conceptual confluence, we discuss the two 1936 papers, but explore also Post's work in the 1920s and Turing's paper [75]. We consider in addition their overlapping mathematical work on the word problem for semigroups [54] and for semigroups with cancellation [74]. We argue that the unity of their approaches is of deep significance for the *theory* of computability.[3]

2 Substitution Puzzles: The Link

"Solvable and Unsolvable Problems", [75], is presumably Turing's last published paper. It appeared in the British journal *Science News*. The paper focuses on the methodological problems surrounding effective calculability, now usually discussed under the heading of Church's or Turing's Thesis. Turing argues in it for the adequacy of a rigorous notion of computability, but without ever mentioning his computing machines. That may be surprising: after all, for Church, Gödel, and many contemporary computer scientists, it is the analysis of effective, mechanical procedures in terms of "finite machines" that makes Turing's work so convincing. Instead, Turing uses here as the basic concept that of "unambiguous substitution puzzles"—a form of Post production systems. Post had used these special puzzles in [54] to describe Turing machines elegantly and to establish the unsolvability of the word problem for semigroups. Post's techniques were refined in [74] to extend the unsolvability result to semigroups with cancellation.

Underlying our presentation, as well as Turing's exposition in [75], is the recognition of the essential unity of Turing's *philosophical analysis of mechanical procedures* in [71] and Post's *mathematical work on normal forms* of production systems. Post's work was done during the 1920s, but published only in [51]. This unity goes far deeper than the confluence via extensional equivalences between different notions of computability for number theoretic functions as expounded so beautifully in [26]. It also goes deeper than just observing that the models of

[3]There are other commonalities in their approaches, e.g., in connection with relative computability, which first appeared in Turing's dissertation and which played such a key role in Post's later work. However, here we are focusing on their fundamental conceptual analysis.

computation Post and Turing introduced in 1936 are essentially the same. Here is Turing's retrospective assertion from the beginning of his [74], p. 491:

> The method [of proof] depends on reducing the unsolvability of the problem in question to a known unsolvable problem connected with the logical computing machines introduced by Post [49] and the author [71].

Turing points, in a dramatic way, to the structural identity of the computations of his two-letter machine and those of the worker in [49].

Post formulated no intrinsic reason for the model he presented in this paper, but he conjectured it to be equivalent to the "Gödel-Church development". If one considers also his work on canonical systems from the 1920s (discussed below), then there is an indication of a reason: the model may be viewed as a playful description of a simple production system to which canonical systems can be reduced. Turing's unambiguous substitution puzzles include these simple production systems. In order to specify such puzzles one is given an unlimited supply of "counters", possibly of only two distinct kinds. A finite sequence of counters is an initial configuration, and the puzzle task is to transform the given configuration into another one using substitutions from a fixed finite list of rules. Such a puzzle, though not by this name, is obtained in Sect. 9,I of Turing's [71] at the very end of his analysis of mechanical procedures; it can be carried out by a suitably generalized machine that operates on strings, a *string machine*, as presented in Sect. 5. A good example of a substitution puzzle, Turing asserts in [75], is "the task of proving a mathematical theorem within an axiomatic system". The abstract investigation of just this task for parts of Whitehead and Russell's *Principia Mathematica* was the starting point of Post's work in the 1920s, as we discuss in Sect. 4.

The "theorem-proving-puzzle" for first-order logic was not only Post's problem, but was also one of the central issues in mathematical logic during the 1920s: is it decidable by a mechanical procedure whether a particular given statement can be proved from some assumptions? Commenting on this problem, best known as Hilbert's *Entscheidungsproblem*, von Neumann conjectured in 1926 that it must have a negative solution and added, "we have no idea how to prove this". Ten years later Turing showed that the problem has indeed a negative solution, after having addressed first the key conceptual issue, namely, to give a precise explication of "mechanical procedure".[4] The explication by his machine model of computation is grounded in the analysis that leads, as mentioned, to a substitution puzzle. In parallel to what we said above about Post's 1936 model, Turing's two-letter machine may be viewed as a playful formulation of a machine to which string machines can provably be reduced and to which, in turn, the mechanical processes of a human computing agent can be reduced, if these processes (on strings or other concrete symbolic configurations) are constrained by finiteness and locality conditions; see Sect. 5.

[4]In the same year, Church established the unsolvability of the *Entscheidungsproblem*—having identified λ-definability and general recursiveness with effective calculability; [6, 7].

The conceptual confluence of Turing's work with Post's is rather stunning: Post's worker model and Turing's two-letter machine characterize exactly the same class of computations. The crucial link are the simple substitution puzzles (with just two counters) that are uniquely connected to each instance of both models. The wider confluence indicated above is discussed in detail in Sects. 4 and 5; its very special character is seen perhaps even more vividly when comparing it with the contemporaneous attempts to analyze directly the effective calculability of number theoretic functions.

3 Effectively Calculable Functions

The thoughts of mathematical logicians like Hilbert, Bernays, Herbrand, Gödel,... about mechanical procedures on syntactic configurations were not proceeding at all along the lines of Post or Turing, but were naturally informed by paradigmatic recursive procedures for calculating the values of number theoretic functions. Dedekind introduced primitive recursive functions in [19], Skolem used them systematically in [66] to develop an elementary part of number theory, and Hilbert and Bernays exploited during the 1920s the calculability of their values in proof theoretic investigations. In his [37], Herbrand expanded the class to "finitist" functions and included, in particular, the effectively calculable, but non-primitive recursive Ackermann function.

Gödel built on that work when introducing *general recursive functions* via his equation calculus in [28], the write-up of lectures given at the Institute for Advanced Study.[5] In the famous footnote 3 to this article, Gödel had suggested that a suitable generalization of the primitive recursive functions permitting all possible recursions would encompass the class of all effectively calculable number theoretic functions. His formulation via the equation calculus then came in the last section of the notes. However, despite appearances to the contrary, Gödel insisted in a letter to one of the authors (see [14]) that he was not prepared at that time to conclude that his formulation accomplished that goal, because until Kleene's elucidations [40], it was not clear that the notion was sufficiently robust.

Gödel's mathematically well-defined class of general recursive functions was identified in [5] with the informal class of effectively calculable number theoretic functions. Kleene labeled this identification as *Church's Thesis* in [41, 42]; the

[5] In the early evolution of recursion theory, Gödel's definition was viewed as being a modification of a proposal of Herbrand's—because Gödel presented it that way in his Princeton Lectures. In a letter to Jean van Heijenoort in 1964, Gödel reasserted that Herbrand had suggested, in a letter, a definition very close to the one actually presented in [28]. However, the connection of Gödel's definition to Herbrand's work is much less direct; that is clear from the two letters that were exchanged between Gödel and Herbrand in 1931. John Dawson found the letters in the Gödel Nachlass in 1986; see [17]. The letters are published in [36]; their intellectual context is discussed in [61].

Thesis is, even today, mostly supported by two kinds of reasons. There is, first of all, quasi-empirical evidence through the fact that in roughly 80 years of investigating effectively calculable functions we have not found a single one that is not recursive; that is supported further by the practical experience of hundreds of thousands of computer programmers for whom it is a mundane matter of everyday experience that even extremely complex algorithms can be carried out by an appropriate sequence of very basic operations. There is, secondly, the provable equivalence of different notions, which was already important for Church in 1935, having just proved with Kleene the equivalence of recursiveness and λ-definability. This is the "argument by confluence", highlighted in [26], and brought out clearly already by Church in footnote 3 of Church [6]:

> The fact, however, that two such widely different and (in the opinion of the author) equally natural definitions of effective calculability turn out to be equivalent adds to the strength of the reason adduced below for believing that they constitute as general a characterization of this notion as is consistent with the usual intuitive understanding of it.

The farther apart the sharply defined notions are, the more strongly does their equivalence support the claim of having obtained a most general characterization of the informal notion. In contrast, the direct conceptual confluence of Turing's and Post's work emphasizes the unity of their analyses of mechanical procedures.

The above two central reasons for supporting Church's Thesis, quasi-empirical evidence and the argument by confluence, have been complemented by sustained arguments that attempt to analyze the effective calculability of number theoretic functions. All the early attempts, from 1936 to 1946, do that in terms of a single core concept, namely, *calculation in a logic* or determination of the value of a function in a calculus or deductive system (in essence generalizing Gödel's equation calculus). Church, in Sect. 9 of his classical [6], used the "step-by-step argument" to "prove" the thesis. Calculations of function values for particular arguments are to be carried out in a logic and may use only "elementary" steps. These elementary steps are then taken to be recursive; thus, with subtle circularity, the claim has been established. Let us call the identification of "elementary" step with "recursive" step *Church's Central Thesis*. The subtle circularity is brought into the open by Hilbert and Bernays in Supplement II of the book [39]: they explicitly formulate recursiveness conditions for "deductive formalisms", define the concept of a reckonable function (regelrecht auswertbare Funktion) and then show that the functions calculable in a deductive formalism that satisfies the recursiveness conditions are exactly the general recursive functions.[6] (In fact, Hilbert and Bernays formulate *primitive* recursiveness conditions.)

[6]How far removed the considerations of Turing (and Post) were from those of Bernays, who actually wrote Supplement II of Hilbert and Bernays [39], should be clear from two facts: (i) In a letter to Church of 22 April 1937, Bernays judged Turing as "very talented" and his concept of computability as "very suggestive"; (ii) Bernays did not mention Turing's paper in Supplement II, though he knew Turing's work very well. Indeed, in his letter to Church Bernays pointed out a few errors in [71]; Church communicated the errors to Turing and, in a letter of 22 May 1937 to Bernays, Turing acknowledged them and suggested that he would write a *Correction*, [72]!

Church and Hilbert & Bernays used a form of argument that underlies the proof of Kleene's *Normal Form Theorem*. This same form of argument also easily establishes Gödel's observation in the "Remark added in proof" to his short note [29]; namely, functions that are calculable in systems of higher-order arithmetic (even of transfinite order) are already calculable in elementary arithmetic. This *absoluteness phenomenon* allowed him to think for the first time, as he mentions in a letter to Kreisel of 1 May 1968, that his concept of general recursiveness is, after all, adequate to characterize computability and thus formal theories in full generality. Ten years after the 1936 absoluteness remark Gödel asserted in his contribution to the Princeton Bicentennial [31] that kind of absoluteness for *any* formal theory extending number theory; the extending theories can now also be systems of axiomatic set theory. As in Church's step-by-step argument, the absoluteness cannot be proved, unless the formality of the extending formal theory is articulated in a rigorous way, for example by requiring that the proof predicate be (primitive) recursive.[7]

Sometime in the late 1930s, Gödel wrote the manuscript [30], apparently notes for a possible lecture. He formulated a beautifully simplified equation calculus whose postulates have two characteristics: (I) Each of them is an equation between terms (built up in the standard way from variables, numerals, and function symbols), and (II) "...the recursive postulates ... allow [one] to calculate the values of the function defined". For the calculation one needs only two rules; (R1) replaces variables by numerals, and (R2) replaces a term by a term that has been shown to be equal (to the first term). One gets to the two rules "by analyzing in which manner this calculation [according to (II)] proceeds". The two characteristics of the postulates are, Gödel claims, "exactly those that give the correct definition of a computable function". Gödel then goes on to define when a number theoretic function f is *computable*: "... if there exists a finite number of admissible postulates in f and perhaps some auxiliary functions $g_1, \ldots, g_n$ such that every true elementary equation for f can be derived from these postulates by a finite number of applications of the rules R1 and R2 and no false elementary equation for f can thus be derived." In short, Gödel views his analysis of calculating the value of a function as ultimately leading to the correct definition of a computable function.

It seems quite clear that Gödel's considerations sketched in the last paragraph are a full endorsement of Church's Thesis for the concept of general recursive functions, here called computable functions. At the beginning of Gödel [30], pp. 166–167, Gödel had emphasized the importance of characterizing "uniform mechanical procedures" on finite expressions and finite classes of expressions[8]; now

[7]In a very informative letter Church wrote on 8 June 1937 to the Polish logician Pepis, the absoluteness of general recursive functions is indirectly argued for. (Church's letter and its analysis is found in [59].) Gödel's claim, with "formality" sharpened in the way we indicated, is an almost immediate consequence of the considerations in Supplement II of Hilbert and Bernays [39].

[8]Gödel argues at the bottom of p. 166 that the expressions and finite classes of expressions can be mapped to integers ("Gödel numbering"). Thus, he asserts, a "procedure in the sense we want is nothing else but a function $f(x_1, \ldots, x_r)$ whose arguments as well as its values are integers and

he complements his analysis of effective calculability of functions by connecting it with Turing—by name, though not by substance. He claims, "That this really is the correct definition of *mechanical* [our emphasis] computability was established beyond any doubt by Turing." [30], p. 168. According to Gödel, Turing established this claim, as he had shown that [30], p. 168:

> ... the computable functions defined in this way are exactly those for which you can construct a machine with a finite number of parts which will do the following thing. If you write down any number $n_1, \ldots, n_r$ on a slip of paper and put the slip into the machine and turn the crank, then after a finite number of turns the machine will stop and the value of the function for the argument $n_1, \ldots, n_r$ will be printed on the paper.

The description of a Turing machine as a "crank machine" of finitely many parts is rather inadequate. In addition, such an equivalence result was not established in [71]; it can only be inferred from Turing's proof of the equivalence of his machine computability and λ-definability, on the one hand, and Church and Kleene's result that asserts the equivalence of λ-definability and general recursiveness, on the other hand.[9]

These are the most sophisticated arguments for Church's Thesis we know. They are important, but provide neither proofs nor convincing reasons; after all, function values are being calculated in a logic via elementary steps that are either explicitly specified (and can be shown to be primitive recursive) or they are simply asserted to be (primitive) recursive. When looking back at the original mathematical problems and the conceptual issues that had to be addressed, one notices that they are not directly concerned with such calculations. In all fairness, we have to admit that they also seem to have nothing in common with the operations of a Turing machine or the wanderings of a Post worker. Well, let's examine the latter claim by reviewing first Post's notion of a canonical system (in Sect. 4) and then Turing's 1936 analysis of calculability (in Sect. 5).

4 Canonical Systems: Post

The story has been told frequently, especially during the centenary year of 2012, of how Turing learned from Max Newman's lectures in Cambridge that although it was widely believed that there was no algorithm for provability in first-order logic, no one had proved this, and how he went on to develop his concept of

which is such that for any system of integers $n_1, \ldots, n_r$ the value can actually be calculated". So a "satisfactory definition of calculable functions" is needed, and that's what the definition of computable function yields for Gödel.

[9] In [10], p. 10 and [69], p. 214, Gödel's remark "That this really is the correct definition of *mechanical* [our emphasis] computability was established beyond any doubt by Turing." is taken as showing that computability of functions is defined here by reference to Turing machines, i.e., that Gödel at this point had taken already Turing's perspective. That view can be sustained only, if the context we sketched is left completely out of consideration.

computability as a key step in demonstrating that this is indeed the case. The story of how Post came to write [49] begins in the early 1920s and is not nearly so widely known.[10] After completing his undergraduate studies at City College in New York, he began graduate work at Columbia University, where he participated in a seminar on Whitehead and Russell's *Principia Mathematica* (PM). The seminar was directed by Cassius J. Keyser (Post's eventual thesis advisor) and was presumably devoted to studying the proofs *within* PM. Post decided for his doctoral dissertation to study PM from the outside, proving theorems as he wrote, that are "*about* the logic of propositions but are *not included* therein". This was a conception totally foreign to Whitehead and Russell, although it was close to what Hilbert was to call "metamathematics" and which he had practiced already with great success in his *Foundations of Geometry* of 1899, [38].

Post began with the first part of PM, i.e., [81], the subsystem we now call the "propositional calculus". Post proved that its derivable propositions were precisely the *tautologies*, thus showing that the propositional calculus was complete and algorithmically decidable. Independently, Bernays had already proved, in his 1918 Göttingen *Habilitationsschrift* [1], the completeness and decidability of the propositional logic of PM.[11] Next Post wanted to get similar results for first-order logic (as formulated in PM, Sects. *10 and *11). Thus he began an attack on what amounted to the same *Entscheidungsproblem* that was Turing's starting point. But working in 1921 Post could still think of finding a decision algorithm. His main idea was to hide the messy complexities of quantificational logic by seeing it as a special case of a more general type of *combinatorial structure*, and then to study the decision problem for such structures in their full generality. (Post had been influenced by C.I. Lewis in viewing the system of PM from a purely formal point of view as a "combinatorial structure"; see Post's reference in footnote 3 of Post [48] to [45], Chap. VI, Sect. III.)

Already in his dissertation, Post had introduced such a generalization that he now made the starting point of his investigation. He called this formulation *Canonical Form A* and proved that first-order logic could be expressed in a related form that he called *Canonical Form B*. He also showed that Canonical Form A was *reducible* to Canonical Form B in the sense that a decision procedure for the latter would lead to one for the former. However, it was a third formulation Post called *Canonical Form C* that has survived and plays a key role in our discussion. This formulation is defined in terms of so-called *Post canonical productions* which we now explain.

---

[10]As to Post's biography, see [15].

The mathematical and philosophical part of Post's contributions is discussed with great clarity in [26], pp. 92–98. The unity of their approaches is, however, not recognized; it is symptomatic that neither [75] nor the overlapping work in [54, 74] is even mentioned.

A very comprehensive and illuminating account of Post's work is found in [79].

[11]His work was published belatedly and only partially in [3]. In addition to the completeness question, Bernays also investigated the independence of the axioms of PM. He discovered that the associativity of disjunction is actually provable from the remaining axioms (that were then shown to be independent of each other).

Let Σ be a given finite alphabet. As usual write Σ^* for the set of finite strings on Σ. A *canonical production* on Σ has the form:

$$\begin{array}{l} g_{11}\ P_{i'_1}\ \ g_{12}\ P_{i'_2}\ \ \ldots g_{1m_1}\ P_{i'_{m_1}}\ \ g_{1(m_1+1)} \\ g_{21}\ P_{i''_1}\ \ g_{22}\ P_{i''_2}\ \ \ldots g_{2m_2}\ P_{i''_{m_2}}\ \ g_{2(m_2+1)} \\ \ldots\ \ldots\ \ \ldots\ \ldots\ \ \ldots\ldots\ \ \ldots\ \ \ldots \\ g_{k1}\ P_{i^{(k)}_1}\ g_{k2}\ P_{i^{(k)}_2}\ \ldots g_{km_k}\ P_{i^{(k)}_{m_k}}\ g_{k(m_k+1)} \\ \qquad\qquad\qquad \Downarrow \\ g_1\ \ P_{i_1}\ \ g_2\ \ P_{i_2}\ \ \ldots g_m\ \ P_{i_m}\ \ g_{m+1} \end{array}$$

Here the g's are given strings on the alphabet Σ, the P's are variables over strings, and each of the P's in the line following the $\Downarrow$ also occurs as one of the P's above the $\Downarrow$. A *system $\mathcal{S}$ in Canonical Form C* consists of a finite set of strings on Σ, *initial assertions*, together with a finite set of canonical productions. Iteratively applying the productions to the initial assertions and to the strings so obtained etc., a subset of Σ^* is obtained; Post called this subset *the set generated* by the system $\mathcal{S}$. The decision problem for a system $\mathcal{S}$ in Canonical Form C was to determine of a given $s \in \Sigma^*$ whether it is in the set generated by $\mathcal{S}$. Post was able to prove that the decision problem for Canonical Form B is reducible to that for Canonical Form C.

Thus, to solve the decision problem for first-order logic, it remained to solve the decision problem for systems in Canonical Form C. As a step in that direction, Post showed how to reduce this decision problem to that for *normal systems*, a special kind of system in Canonical Form C with only one initial assertion and productions of the simple kind

$$gP \Rightarrow P\bar{g}$$

More specifically he proved:

Post's Normal Form Theorem Let U be the set of strings on an alphabet Σ generated by some system in Canonical Form C. Then there is an alphabet $\Delta \supseteq \Sigma$ and a normal system $\mathcal{N}$ on Δ generating a set $\tilde{U} \subseteq \Delta^*$ such that

$$U = \tilde{U} \cap \Sigma^*.$$

Finally, "closing the circle", Post showed that the decision problem for normal systems is reducible to that for systems in Canonical Form A, thus showing that the decision problems for Forms A, B, and C are all equivalent.

In 1941 Post submitted an article to the *American Journal of Mathematics* in which he told the story of his work two decades earlier on unsolvability. The article was rejected as being inappropriate for a journal devoted to current research. However, the editor noted that the work on reducing canonical systems to normal form was new and suggested that Post resubmit just that part. Post followed that suggestion and the paper [51] was the result. The longer paper was eventually published in [12] when Post was no longer alive.

The story has been told elsewhere of how Post exhausted himself trying to solve the decision problem for a special kind of normal system, the so-called *tag systems*.[12] The intractability of this problem that Post had thought of as merely an initial step in his project apparently played a significant role in reversing his thinking [20]. Imagining an enumeration of all normal systems and associating with each such system the set of all strings on a distinguished symbol a, Post saw that he could diagonalize, i.e., form the set which contains the string $aa\ldots a$ with n occurrence of a just in case that string is *not* generated by the nth normal system; clearly this set can not be the set of strings on the letter a generated by any normal system. On the other hand his whole development argued for the full generating power of normal systems: beginning with the reducibility of first-order logic to Canonical Form B which Post believed could be extended to all of PM, taking into account the power of PM to encapsulate ordinary mathematical reasoning, and finally noting the reductions of Canonical Form B to Canonical Form C and then to normal systems. Since the enumeration of normal systems could easily be made quite explicit, it appeared that in apparent contradiction to the above, the set obtained by diagonalization was a set generated by a straightforward process.

Post saw that the problem lay in the tacit assumption that there was a process for determining whether a given string is generated by a given normal system. To escape this dilemma he had to either give up on the generality of normal systems or accept the conclusion that *there is no such process*. He chose the latter. Since the properties of normal systems could readily be formalized within PM, this led to the heady conclusion that there is no decision procedure for PM itself. Finally, if PM were complete with respect to propositions asserting that a given string is generated by a given normal system, then by using PM to generate all propositions that assert or deny that some particular string is generated by some particular normal system, a decision procedure for normal systems would be obtained. It followed that PM was incomplete, even for such propositions. Thus this work at least partially anticipated what Gödel, Church, and Turing were to accomplish a decade later.[13]

The very conclusion of the incompleteness of PM argued against accepting its capabilities in an argument for the generality of the generating power of normal systems. Post concluded that "...a complete analysis would have to be made of all the possible ways in which the human mind could set up finite processes for

---

[12]Tag systems may be characterized as normal systems in which for its productions $gP \Rightarrow P\bar{g}$:

1. All of the gs are of the same length;
2. the $\bar{g}$s corresponding to a given g depend only on its initial symbol;
3. if a given g occurs on the left in one of the productions, then so do all other strings of the same length having the same initial symbol as that g.

Post discusses Tag in the introductory section of Post [51] and in Sect. 3 of Post [50].

[13]Davis [15], DeMol [20], DeMol [21], and Post [50].

generating sequences."[14] Of course Post never succeeded in supplying such an analysis. Writing to Gödel he explained what he had in mind as "something of the sort of thing Turing does in his computable number paper". In a postcard sent to Gödel, shortly after their first meeting, Post wrote ruefully, " ... the best I can say is that I would have proved Gödel's Theorem in 1921—had I been Gödel".[15]

Because Post began with PM whose theorems are derived from axioms by applying rules, his canonical systems naturally took the form of strings obtained from "initial assertions" by applying rules (namely "productions"). In contrast, Church sought to identify the intuitive notion of effectively calculable *function* with lambda-definable, or equivalently general recursive, function. Post could see that the analysis in his earlier unpublished work was presented in terms of operations on strings and thus at a more fundamental level than was achieved in the newer work. In arriving in [49] at a formulation so close to Turing's, Post can be seen as bringing to bear on the concept of what can be calculated, his experience with the potency of simple manipulations of strings. Strikingly, Turing in his analysis of mechanical procedures was also to recognize, as we will see, the fundamental and powerful character of string manipulations. In describing that analysis we will throw some light on Post's allusion to "something of the sort of thing Turing does in his computable number paper".

5 Mechanical Procedures: Turing

Turing emphasizes in [71] at the very outset, in Sect. 1 and referring to Sect. 9, that he is concerned with mechanical operations on symbolic configurations—carried out by humans. Indeed, he uses *computer* to refer to human computing agents who proceed mechanically; his machines, our Turing machines, are referred to as *machines*! Gandy suggested calling a computer in Turing's sense *computor* and a computing machine, as we actually do, computer. In Sect. 9 of Turing [71], computors operate on symbolic configurations that have been written on paper; the paper is divided into squares "like a child's arithmetic book". However, the two-dimensional character of the paper is not viewed to be an "essential of computation", and the one-dimensional tape divided into squares is taken, without any argument, as the basic computing space.

Striving then to isolate computor-operations that are "so elementary that it is not easy to imagine them further divided", Turing formulates a crucial requirement: symbolic configurations relevant for a computor's actions have to be recognized immediately or at a glance. Because of the reductive step from a two-dimensional

[14]Post [50]: p. 408 in [12]; p. 387 in [16]; p. 422 in [55].

[15]Gödel [36] pp. 169, 171.

The notes that were exchanged between Gödel and Post are in this volume of Gödel's *Collected Works*.

grid to a linear tape, one has to be concerned only with immediately recognizing sequences of symbols. Turing appeals now to a crude empirical fact concerning human sensory capacities: it is impossible for a computor to determine at a glance whether 9889995496789998769 is identical with 98899954967899998769. This sensory limitation of computors leads directly to *boundedness* and *locality conditions*: (B) there is a bound on the number of symbol sequences a computor can recognize at a glance, and (L) the operations of a computor must locally modify a recognized configuration.[16]

Given that the analysis of a computor's steps leads to these restrictive conditions, it is evident that Turing machines operating on strings, *string machines*, simulate computors. Indeed, Turing having completed the analysis of the computor's calculation, asserts, "We may now construct a machine to do the work of this computer [i.e., computor]." The machine that is constructed is a string machine. Thus, the general connection of Turing with Post is clear: one just has to notice that (deterministic) string machines are (unambiguous) substitution puzzles, and that the latter are a species of Post's production systems! With respect to string machines Turing remarks,

> The machines just described [string machines] do not differ very essentially from computing machines as described in §B, and corresponding to any machine of this type a computing machine can be constructed to compute the same sequence, that is to say the sequence computed by the computer [i.e., computor].

Turing machines consequently "appear ...as a codification, of his [Turing's] analysis of calculations by humans"; that was Gandy's perspective as articulated in [26] pp. 83–84, and we share it.

Turing's analysis leads in natural steps from human mechanical procedures to those of string machines, but it is not a rigorous proof (whereas, of course, the reductive step from string to letter machines is justifiable by a proof). Turing views it as mathematically unsatisfactory because it relies on an appeal to "intuition". However, an appeal to intuition is no longer needed for what has been called Turing's Thesis, but only for a more restricted judgment: the mechanical procedures of computors can be carried out on strings, without loss of generality. Let us call this judgment *Turing's Central Thesis*. In the reflections concerning the calculability of number theoretic functions we associated with the informal notion "elementary" the sharp mathematical one "(primitive) recursive"; here we are connecting the informal notion "symbolic configuration" with the precise mathematical one of a "finite string of symbols".

---

[16]We neglect in our discussion "states of mind" of the computor. Here is the reason why. Turing argues in Sect. 9.I that the number of these states is bounded, and Post calls this Turing's "finite number of mental states" assumption. However, in Sect. 9,III Turing argues that the computor's state of mind ["mental state"] can be replaced in favor of "a more physical and definite counterpart of it". In a sense then, the essential components of a computation have been fully externalized; see [13], p. 6, how this is accomplished through the concept of an "instantaneous description".

The same argumentative move from an informal to a precise mathematical notion is made in [75]. The crucial difference lies in the greater generality of the symbolic configurations that are being considered: the puzzles are not restricted to being one- or two-dimensional, but they can even be three-dimensional. As an example of a three-dimensional puzzle Turing discusses (on p. 12 of Turing [75]) knots in some detail, in particular, their representation by sequences of letters and the operations that can be performed on those sequences. The knot problem, Turing asserts, is like a puzzle in that it asks "to undo a tangle, or more generally of trying to turn one knot into another without cutting the string".

The variant of the Central Thesis, discussed on p. 15 of Turing [75] and in [64], states that for any puzzle "we can find a corresponding substitution puzzle, which is equivalent to it in the sense that given a solution of the one we can easily find a solution of the other". The statement that any puzzle has a substitution puzzle as its normal form is puzzling in itself, and Turing admits it is "somewhat lacking in definiteness, and will remain so". He views it as being situated between a theorem and a definition:

> In so far as we know a priori what is a puzzle and what is not, the statement is a theorem. In so far as we do not know what puzzles are, the statement is a definition that tells us something about what they are.

In any event, he emphasizes, it is not a statement one should attempt to prove. We should emphasize that he considers the possibility of defining a puzzle "by some phrase beginning, for instance, 'A set of definite rules ...', but this just throws us back to the definition of 'definite rule'." He continues (pp. 15–16),

> Equally one can reduce it to the definition of 'computable function' or 'systematic procedure'. A definition of any one of these would define all the rest. Since 1935 a number of definitions have been given, explaining in detail the meaning of one or other of these terms, and these have all been proved equivalent to one another and also equivalent to the above statement. In effect there is no opposition to the view that every puzzle is equivalent to a substitution puzzle.

Turing's attitude is certainly much less definite than is Gödel's view in [32], where Turing's work is seen as giving "a precise and unquestionably adequate definition of the general concept of formal system"; such an adequate definition is provided, as Turing presents "an analysis of the concept of 'mechanical procedure .... This concept is shown to be equivalent with that of a 'Turing machine'." In the footnote attached to this remark Gödel suggests consulting not only Turing's 1936 paper but also "the almost simultaneous paper by E.L. Post (1936)".

Post's perspective on the openness of the concept "finite combinatory process", contrary to Gödel's, is strikingly indicated in the paper Gödel recommended. Post envisions there a research program that considers wider and wider formulations of such processes and has the goal of logically reducing them to formulation 1. Clearly, that is in the spirit of the investigations he had pursued in the early 1920s. (What other symbolic configurations and processes he had in mind is discussed at the beginning of our last section, entitled *Concluding Remarks*.) Post expresses in the 1936 paper his expectation that formulation 1 will "turn out to be logically equivalent to recursiveness in the sense of the Gödel-Church development". He also presents the "conclusion" that all the wider forms of finite combinatory processes are reducible to formulation 1 as a "working hypothesis". The success of the envisioned research program would "change this hypothesis not so much to a definition or to an axiom but to a natural law". We assume that Post had in mind a natural law concerning the psychology of human beings that expresses a limit on their capacity to articulate and carry out finite combinatory processes.[17] Post thought that mental states played a crucial role in that capacity, but never formulated the idea of there being only a finite number of such states—in contrast to Turing, as we just saw, cf. in particular footnote 16 above. This intellectual situation seems to have been on Post's mind when referring to "something of the sort of thing Turing does in his computable number paper"; he made the remark in the letter to Gödel we quoted in Sect. 4.[18] After all, it is only through a natural law, Post asserts in the last sentence of his paper, that

> Gödel's theorem concerning the incompleteness of symbolic logics of a certain general type and Church's results on the recursive unsolvability of certain problems [can] be transformed into conclusions concerning all symbolic logics and all methods of solvability.

It is that very conclusion Gödel saw in 1964 as being justified by Turing's work. It is remarkable how close, and yet tragically apart, Post and Turing were in these foundational, more philosophical deliberations. The sharp mathematical connections between their approaches were concretely "exploited" in [54, 74] in their achieving unsolvability results for mathematical problems that had previously arisen quite apart from mathematical logic. This will be discussed in the next section.

6 Word Problems

Alonzo Church was struck by the short paper [53] in which Post used the unsolvability of the decision problem for normal systems to prove the unsolvability of a kind of string matching problem that Post had called the *Correspondence*

[17]In footnote 8, p. 105 of Post [49], he criticizes masking the identification of recursiveness and effective calculability under a definition as Church had done. This, Post continues, "hides the fact that a fundamental discovery in the limitations of the mathematizing power of Homo Sapiens has been made and blinds us to the need of its continual verification."

[18]Post pointed this out in a number of different places: (1) Urquhart on p. 643 of Urquhart [79] quotes from Post's 1938 notebook and discusses, with great sensitivity, "an internal reason for Post's failure to stake his claim to the incompleteness and undecidability results in time" on pp. 630–633; (2) in [50], p. 377, Post refers to the last paragraph of [49] and writes: "However, should Turing's finite number of mental states hypothesis ... bear under adverse criticism, and an equally persuasive analysis be found for all humanly possible modes of symbolization, then the writer's position, while still tenable in an absolute sense, would become largely academic."

Decision Problem.[19] The paper reminded Church of a combinatorial problem involving replacements in a string of one substring by another. The problem had been formulated by the Norwegian mathematician Axel Thue and published in [70]. Church suggested to Post that methods similar to those used in [53] might yield the unsolvability of Thue's problem. Post saw Thue's problem, not as related to normal systems,[20] but rather to another special kind of canonical system characterized by a finite number of matching pairs of productions of the form:

$$PgQ \Rightarrow P\bar{g}Q \quad P\bar{g}Q \Rightarrow PgQ$$

for strings on an alphabet Σ. Post called such a set of productions a *Thue system*. Each such pair of productions enable the substitution in a given string of an occurrence of $\bar{g}$ for g or vice versa. For $u, v \in \Sigma^*$ we write $u \approx v$ to indicate that v can by obtained from u by a finite number of such substitutions; this is clearly an equivalence relation. Thue sought an algorithm that would determine for a given Thue system and a given pair of strings $u, v \in \Sigma^*$ whether $u \approx v$. In [54] it is proved that there is no such algorithm.[21]

Rather than beginning with the unsolvability of a problem concerning normal systems, as he had done in [53], Post made use of Turing machines to deal with Thue's problem. In Post's formulation, a Turing machine was defined in terms of a finite number of symbols $S_0, S_1, \ldots, S_m$ and a finite number of *states* or *internal configurations* $q_1, \ldots, q_n$. The machine was to act on a "tape" consisting of a linear two-way-infinite array of cells or squares each capable of holding a single symbol. In each of the successive operations of the machine it is in one of the given states and a single cell is distinguished as the *scanned* cell. The behavior of the machine is controlled by a finite number of *quadruples* each of which is of one of the three types:

$$q_iS_jS_kq_\ell \qquad q_iS_jRq_\ell \qquad q_iS_jLq_\ell$$

Each quadruple specifies what the machine will do when the machine is in state q_i and the scanned cell contains the symbol S_j. For a quadruple of the first type the action is to replace S_j by S_k in the scanned cell. For a quadruple of the second type it is for the machine to move to the cell immediately to the right of the current one, the new scanned square, while the tape contents remains unchanged. For a quadruple of the third type, the new scanned square, similarly, will be the one to the left. In all

[19]This problem later turned out to be a useful tool for obtaining unsolvability theorems regarding Noam Chomsky's hierarchy of formal languages, which by the way, was itself based quite explicitly on Post production systems. See also the extended discussion of the correspondence problem in [79], p. 648.

[20]It is interesting that Markov's proof [46] of the unsolvability of Thue's problem, which was quite independent of Post's, did use normal systems.

[21]Of course this is to be understood in relation to the Church-Turing Thesis.

three cases, the new state of the machine is to be q_ℓ. The deterministic behavior of the machine is enforced by requiring that no two of the quadruples are permitted to begin with the same pair q_iS_j.[22] The machine is to halt when it arrives at a state q_i scanning a symbol S_j for which no quadruple beginning q_iS_j is present.

The machine is to begin in state q_1 with all cells containing the symbol S_0, thought of as a *blank*. Hence at all stages in the machine's computation, all but a finite number of the cells will still contain S_0, as shown below:

......	S_0	S_0	S_0	S_{r_1}	S_{r_2}	...	S_j	...	S_{r_p}	S_0	S_0	S_0	......

In the diagram, the cell containing S_j is intended to represent the square currently scanned in the course of a machine computation. To represent the tape contents by a finite string, Post introduced a special symbol h to serve as beginning and end markers delimiting a region of the tape beyond which, in both directions, all cells contain S_0. To represent a situation in which the tape is as depicted and the machine is in state q_i, Post used the string:

$$hS_{r_1}S_{r_2}\ldots q_iS_j\ldots S_{r_p}h$$

The indicated initial situation can thus be represented by the string hq_1S_0h, and a machine's computation can be represented as a sequence of finite strings of this form, showing the situation at its successive stages. Post provided productions corresponding to each of the quadruples whose indicated action the machine is carrying out and that have the effect of producing the transition from one term of this sequence to the next. Thus, corresponding to each quadruple of the machine of the form $q_iS_jS_kq_\ell$, Post used the corresponding production

$$Pq_iS_jQ \Rightarrow Pq_\ell S_kQ$$

The quadruples that call for the machine to move to the left or the right require special productions to deal with the marker h and to lengthen the string to include one additional (blank) symbol from the tape. So, for quadruples of the form $q_iS_jRq_\ell$ calling for motion to the right, Post introduced the productions:

$$Pq_iS_jS_rQ \Rightarrow PS_jq_\ell S_rQ \text{ and } Pq_iS_jhQ \Rightarrow PS_jq_\ell S_0hQ$$

Likewise for quadruples $q_iS_jLq_\ell$ calling for motion to the left, Post used the productions:

$$PS_rq_iS_jQ \Rightarrow Pq_\ell S_rS_jQ \text{ and } Phq_iS_jQ \Rightarrow Phq_\ell S_0S_jQ$$

[22] In the formulation of Turing [71], the tape is infinite in only one direction and a machine's operations are specified by quintuples allowing for a change of symbol together with a motion to the left or right as a single step. Of course this difference is not significant.

Thus, Post's technique amounts to modeling the behavior of a Turing machine by a corresponding canonical production system. Since this is at the core of the striking formulations invoked by Post and Turing as already discussed, it is worthwhile to see the technique in action and therefore we proceed to outline Post's proof of the unsolvability of Thue's problem in [54]. We can begin with the well-known unsolvability of the *halting problem* which we use in the form: *to determine for a given machine whether, beginning in state q_1 with all cells containing S_0, it will ever halt.*[23]

For the present application Post introduced the special symbols $q, \bar{q}$, and the additional production $Pq_iS_jQ \Rightarrow PqQ$ for each pair q_iS_j which begins none of the quadruples. Finally, he introduced the "cleanup" productions:

$$PqS_iQ \Rightarrow PqQ \quad PqhQ \Rightarrow P\bar{q}hQ \quad PS_i\bar{q}Q \Rightarrow P\bar{q}Q$$

So beginning with the initial string hq_1S_0h and applying these productions, a string containing q will occur precisely when the given Turing machine has halted. In such a case the "cleanup" productions lead to the string $h\bar{q}h$. Thus we see that a given Turing machine beginning with a blank tape will eventually halt if and only if, beginning with the initial string hq_1S_0h and applying this system of productions, the string $h\bar{q}h$ is eventually reached. But in fact, and crucially, we can claim more: namely, a Turing machine beginning with a blank tape will eventually halt if and only if $hq_1S_0h \approx h\bar{q}h$ in the Thue system obtained by adding the productions obtained by interchanging the left and right side of each of the above productions. Thus an algorithm to solve Thue's problem could be used to determine whether the given Turing machine will ever halt, and hence, *there can be no such algorithm.*

It will be helpful in discussing the above claim to refer to the original productions as the *forward* productions and the new productions we have added as the *reverse* productions. So, suppose that $hq_1S_0h \approx h\bar{q}h$, and let the sequence of strings

$$hq_1S_0h = u_1, u_2, \ldots, u_n = h\bar{q}h$$

be the successive steps applying a mix of forward and reverse productions that demonstrates that this is the case. Post showed how to eliminate the use of reverse productions; namely, let the transition from u_{s-1} to u_s be the last occurrence of a use of a reverse production. Then $s < n$ because no reverse production can lead to $h\bar{q}h$. So the transition from u_s to u_{s+1} is via a forward production. But u_{s-1} can also

[23]Post works with the closely related unsolvability of the problem of whether a particular distinguished symbol will ever appear on the tape, because unlike the halting problem, it appears explicitly in [71]. But dissatisfied with the rigor of Turing's treatment, Post outlines his own proof of that fact. He also includes a careful critique pointing out how a convention that Turing had adopted for convenience in constructing examples had been permitted to undermine some of the key proofs. Anyhow, for the present application to Thue's problem, Post begins by deleting all quadruples for which the distinguished symbol is the third symbol of the four, thus changing the unsolvability to one of halting.

be obtained from u_s via the forward production from which the reverse production was formed that had enabled the transition from u_{s-1} to u_s. And by the construction, corresponding to the deterministic character of Turing machines, only one forward production is applicable to u_s. Hence, $u_{s-1} = u_{s+1}$ and nothing is lost if u_s and u_{s+1} are simply deleted from the sequence. Repeating this process all uses of reverse productions are successively eliminated.

Thue's problem is also known as the *word problem for semigroups*. This is because the concatenation of strings is an associative operation. The word problem for *cancellation semigroups* is obtained if one adds the conditions

$$uv \approx uw \text{ implies } v \approx w, \quad vu \approx wu \text{ implies } v \approx w,$$

where u, v, w are any strings. Adding this condition makes the word problem much more complicated and therefore it is far more difficult to prove its unsolvability. The proof of this unsolvability given in [74] is too intricate to say much about the detailed constructions.[24] But what is very relevant to the theme of this paper is the extent to which Turing placed himself in the same tradition as Post. To begin with he acknowledged the formulation in [49] as making Post an equal co-originator of what we (and Post) call the Turing machine. He adopted much of Post's methodology including a two-way-infinite tape and the representation of a stage in a computation by a finite string that begins and ends with the same special symbol and includes a representation of the scanned square, the machine's state, and the tape content. In fact he makes a point of acknowledging Post's invention of this device as a significant contribution. One complication in Turing's proof that might be mentioned is that strings representing a particular stage of a computation now come in two flavors: in addition to Post's strings in which the symbol representing a machine state is placed to the left of the scanned symbol, Turing also uses such a string in which the state symbol is to the right of the scanned symbol. Turing did acknowledge the validity of the critique in [54] referred to above, but clearly did not attach the importance to it that Post apparently did.

We have seen earlier that in [75] we find a remarkable confluence in the conceptual apparatus employed by Post and Turing. The technical mathematics discussed in this section indicates how this apparatus allowed the intelligible and vivid framing of specific mathematical problems and helped to make fruitful this seeming synchronicity. In [75] (discussed extensively above) Turing had argued that if there were a general method for determining whether the goal of any given substitution puzzle is reachable, that method would itself take the form of a substitution puzzle. Then using the Cantor diagonal method he concluded that there is no such method, that the problem of finding such a method is unsolvable. Combining all this with Turing's aside to the effect that the task of finding a proof

[24]Turing's proof was published in the prestigious *Annals of Mathematics*. Eight years later an analysis and critique of Turing's paper [4] was published in the same journal. Boone found the proof to be essentially correct, but needing corrections and expansions in a number of the details.

(from appropriate axioms) of a desired mathematical proposition is itself a puzzle, we find ourselves in an eerie situation; we are back with Post in 1921: Formal logics for mathematics equivalent to production systems; arbitrary production systems reducible to a simple normal form; diagonalization applied to the normal form yielding undecidability and unsolvability theorems. How wonderful to see this convergence of the ideas of these two remarkable thinkers.

7 Concluding Remarks

In the previous section, we saw that important unsolvability results were obtained by reducing Turing machine computations to moves in suitable "unambiguous substitution puzzles" and by exploiting the unsolvability of the halting or printing problem. The joining of complementary techniques, perspectives and results is truly amazing. Let us briefly return to the foundational problem that was addressed in Sect. 5. The issue of Turing's Central Thesis, associating with the informal notion "finite symbolic configuration" the precise mathematical of "finite string of symbols", was not resolved. In their 1936 papers, both Turing and Post consider or suggest considering larger classes of mathematical configurations; that move is to make Turing's Central Thesis inductively more convincing and turn Post's working hypothesis into a natural law. In [79], p. 643, it is asserted that "Post evidently had plans to continue in a series of articles on the topic ... showing how 'Formulation 1' could be extended to broader computational models." Urquhart reports that some work on a "Formulation 2" is contained in Post's notebooks in Philadelphia. Post considers there in particular "rules operating in a two-dimensional symbol space, somewhat reminiscent of later work on cellular automata, such as Conway's Game of Life ...".

Let us mention some later concrete work that involves such more general classes of configurations. Kolmogorov and Uspenski considered in their [44] particular kinds of graphs. Sieg and Byrnes [65] generalized the K&U-graphs to K-graphs and conceived operations on them as being given by generalized Post production rules. Finally, Gandy in [25] introduced discrete dynamical systems that also permitted, ironically, modeling the parallel computations of Conway's Game of Life and other cellular automata. However, in these examples an appeal to the appropriate Central Thesis can't be avoided, if one wants to argue for the full adequacy of the mathematically rigorous notion. The open-endedness of considering ever more encompassing classes of configurations may have been the reason, why Turing in [75] thought that the variant of his Thesis must remain indefinite and that this very statement is one "which one does not attempt to prove".[25] This may also have

[25]Gandy, in [25], uses particular discrete dynamical systems to analyze "machines", i.e., discrete mechanical devices, and "proves" a mechanical thesis (M) corresponding to Turing's thesis. Dershowitz and Gurevich in [22] give an axiomatization and claim to prove Church's Thesis. (For

been the reason for Post's call for the "continual verification" of the natural law that has been discovered. For Post this "continual verification" took the form of an implied duty to explicitly prove that any process claimed on intuitive grounds to be mechanically calculable be accompanied by a rigorous proof that the process falls under one of the various equivalent explications that have been set forth. In [52] this duty is not fulfilled in the printed account based on an invited address to the American Mathematical Society. However, Post assures readers that "... with a few exceptions ... we have obtained formal proofs of all the consequently mathematical theorems here developed informally." Post goes on to say, "Yet the real mathematics involved must lie in the informal development. For ... transforming [the informal proof] into the formal proof turned out to be a routine task." Researchers have long since given up any pretense that they subject their complex arguments to this "continual verification" and no one suggests that this raises any doubt about the validity of the results obtained.

Can one avoid the appeal to a "Central Thesis"? Sieg suggested in [60] a positive answer, namely, by introducing a more abstract concept of computable processes; that concept is rooted in Post and Turing's way of thinking about mechanical, local manipulations of finite symbolic configurations. Analogous abstraction steps were taken in nineteenth century mathematics; a pertinent example is found already in [18]. Call a set O an "ordered system" if and only if there is a relation R on O such that (i) R is transitive, (ii) if for two elements x and y of O the relation R holds, then there are infinitely many elements between x and y, and (iii) every element x of O determines a cut.[26] Consider the rational numbers with their ordinary "$x < y$" relation and the geometric line with the relation "p is to the left of q". Both the specific sets with their relations fall under this abstract concept. There is an abundance of such structural definitions throughout modern mathematics; for example, groups, fields, topological spaces.

For some of the abstract notions representation theorems can be established stating that every model of the abstract notion is isomorphic to a "concrete" model. Here are two well known examples: every group is isomorphic to a permutation group; every Boolean algebra is isomorphic to a Boolean algebra of sets. A suitable representation theorem can also be proved for *computable discrete dynamical systems*. In [60] the abstract notion of a Turing computor was introduced as a computable discrete dynamical system (over hereditarily finite sets with a countably infinite set of urelements), and it was established that the computations of any model of the abstract notion can be reduced to computations of a Turing machine. What has been achieved? Hilbert called the characteristic defining conditions for structures

a discussion of this claim, see [63].) In the context of our discussion here, one can say that Gandy and Dershowitz & Gurevich introduce very general models of computations—"Gandy Machines" in Gandy's case, "Abstract State Machines" in Dershowitz and Gurevich's case—and reduce them to Turing machine computations.

[26]A cut in O (determined by x) is a partition of O into two non-empty parts O_1 and O_2, such that all the elements of O_1 stand in the relation R to all the elements of O_2 (and x is taken to be in O_1).

"axioms", and so do we when talking about the axioms for groups or rings. In that sense, an axiomatic analysis of "mechanical procedures that can be carried out by computors" can be given—building on natural boundedness and locality conditions. The methodological problems have not been removed, but they have been deeply transformed: they concern axioms, no longer statements whose status is somewhere between a definition and a theorem; they are no longer unusual, but rather common and difficult, as they ask us to assess the correctness and appropriateness of axioms for an intended, albeit abstract concept. The central role Turing machines and Post systems play for the theory of computability is secured by the representation theorem.

References

1. P. Bernays, Beiträge zur axiomatischen Behandlung des Logik-Kalküls; Habilitationsschrift. Presented to the Georg-August-Universität zu Göttingen on 9 July 1918 [Reprinted [24], pp. 222–268]
2. P. Bernays, *Logical Calculus*. Lecture Notes at the Institute for Advanced Study (1935/1936) [Notes by P. Bernays, with the assistance of F.A. Ficken; Princeton 1936. The notes can be found in the Bernays Nachlass of the ETH Zürich, HS973:6]
3. P. Bernays, Axiomatische Untersuchung des Aussagen-Kalküls der Principia Mathematica. Math. Z. **25**, 305–320
4. W.W. Boone, An analysis of Turing's 'The word problem in semi-groups with cancellation'. Ann. Math. **67**, 195–202 [Reprinted [76], pp. 175–182]
5. A. Church, An unsolvable problem of elementary number theory (abstract). Bull. Am. Math Soc. **41**, 333
6. A. Church, An unsolvable problem of elementary number theory. Am. J. Math. **58**, 345–363 [Reprinted [12, 16], pp. 89–107]
7. A. Church, A note on the Entscheidungsproblem. J. Symb. Log. **1**, 40–41 [Correction, vol. 1, pp. 101–102. Reprinted with the correction incorporated in the text, [12, 16], pp. 110–115]
8. B. Cooper, J. van Leeuwen (eds.), *Alan Turing: His Work and Impact* (Elsevier, Amsterdam, 2013)
9. J. Copeland (ed.), *The Essential Turing* (Clarendon Press, Oxford, 2004)
10. J. Copeland, O. Shagrir, Turing versus Gödel on computability and the mind, in [11], pp. 1–33
11. J. Copeland, C. Posy, O. Shagrir (eds.), *Computability: Turing, Gödel, Church and Beyond* (MIT Press, Cambridge)
12. M.D. Davis (ed.), *The Undecidable* (Raven Press, Hewlett, 1965)
13. M.D. Davis, *Computability and Unsolvability* (McGraw-Hill) [Reprinted with an additional appendix, Dover 1983]
14. M.D. Davis, Why Gödel didn't have Church's thesis. Inf. Control **54**, 3–24
15. M.D. Davis, *Emil L. Post: His Life and Work*, [55], pp. xi-xxviii
16. Improved reprint of [12] (Dover, 2004)
17. J. Dawson, Prelude to recursion theory: the Gödel-Herbrand correspondence, in *First International Symposium on Gödel's Theorems*, ed. by Z.W. Wolkowski (World Scientific Publishing Co.) pp. 1–13
18. R. Dedekind, *Stetigkeit und irrationale Zahlen* (Vieweg) [Translated in [23] pp. 765–779]
19. R. Dedekind, *Was sind und was sollen die Zahlen?* (Vieweg) [Translated in [23] pp. 787–833]
20. L. DeMol, Closing the circle. An analysis of Emil Post's early work. Bull. Symb. Log. **12**, 267–289

21. L. DeMol, Generating, solving and the mathematics of Homo Sapiens. Emil Post's views on computation, in *A Computable Universe: Understanding and Exploring Nature as Computation*, ed. by H. Zenil, World Scientific Publishing, Singapore, 2013, pp. 45–62
22. N. Dershowitz, Y. Gurevich, A natural axiomatization of computability and proof of Church's thesis. Bull. Symb. Log. **14**, 299–350
23. W.B. Ewald (ed.), *From Kant to Hilbert: A Source Book in the Foundations of Mathematics*, vol. II (Oxford University Press, Oxford)
24. W.B. Ewald, W. Sieg (eds.), *David Hilbert's Lectures on the Foundations of Arithmetic and Logic, 1917–1933* (Springer, Berlin)
25. R. Gandy, Church's thesis and principles for mechanisms, in *The Kleene Symposium*, ed. by J. Barwise, H.J. Keisler, K. Kunen (North-Holland Publishing Company, Amsterdam), pp. 123–148
26. R. Gandy, The confluence of ideas in 1936, in *The Universal Turing Machine: A Half-Century Survey*, ed. by R. Herken (Oxford University Press, Oxford), pp. 55–111
27. K. Gödel, Über formal unentscheidbare Sätze der Principia Mathematica und verwandter Systeme I. Monatshefte für Mathematik und Physik **38**, 173–198. [Reprinted [33] pp. 144–194 (even numbered pages). Translated in: [80] pp. 596–628; [12, 16] pp. 5–38; [33] pp. 145–195 (odd numbered pages)]
28. K. Gödel, *On Undecidable Propositions of Formal Mathematical Systems*. Notes on Lectures at the Institute for Advanced Study, Princeton, ed. by S.C. Kleene, J.B. Rosser [Reprinted [12, 16] pp. 41–74; [33] pp. 346–369]
29. K. Gödel, Über die Länge von Beweisen, *Ergebnisse eines mathematischen Kolloquiums* Heft, vol. 7, pp. 23–24 [Reprinted [33] pp. 396–398 (even numbered pages) Translated in: [12, 16] pp. 82–83; [33] pp. 397–399 (odd numbered pages)]
30. K. Gödel, Undecidable diophantine propositions, in [35], pp. 164–175
31. K. Gödel, Remarks before the Princeton bicentennial conference on problems in mathematics, in [34], pp. 150–153
32. K. Gödel, Postscriptum to [28], in [12, 16], pp. 71–73; [33] pp. 369–371
33. K. Gödel, *Collected Works*, vol. I, ed. by S. Feferman et al. (Oxford University Press, Oxford)
34. K. Gödel, *Collected Works*, vol. II, ed. by S. Feferman et al. (Oxford University Press, Oxford)
35. K. Gödel, *Collected Works*, vol. III, ed. by S. Feferman et al. (Oxford University Press, Oxford)
36. K. Gödel, *Collected Works*, vol. V, ed. by S. Feferman et al. (Oxford University Press, Oxford)
37. J. Herbrand, Sur la non-contradiction de l'arithmétique. Crelles Journal für die reine und angewandte Mathematik **166**, 1–8 [Translated in [80], pp. 618–628]
38. D. Hilbert, Grundlagen der Geometrie, in *Festschrift zur Feier der Enthüllung des Gauss-Weber-Denkmals in Göttingen* (Teubner, Leipzig), pp. 1–92
39. D. Hilbert, P. Bernays, *Grundlagen der Mathematik*, vol. II (Springer, Berlin)
40. S.C. Kleene, General recursive functions of natural numbers. Mathematische Annalen **112**, 727–742 [Reprinted [12, 16], pp. 236–253]
41. S.C. Kleene, Recursive predicates and quantifiers. Trans. Am. Math. Soc. **53**, 41–73
42. S.C. Kleene, *Introduction to Metamathematics* (Wolters-Noordhoff Publishing, Amsterdam)
43. S.C. Kleene, Origins of recursive function theory. Ann. Hist. Comput. **3**, 52–67
44. A. Kolmogorov, V. Uspensky, On the definition of an algorithm. AMS Transl. **21**(2), 217–245
45. C.I. Lewis, *A Survey of Symbolic Logic* (University of California Press, Berkeley)
46. A.A. Markov, On the impossibility of certain algorithms in the theory of associative systems. Doklady Akademii Nauk S.S.S.R., n.s., 1951, **77**, 19–20 (Russian); C. R.. Acad. Sci. de l'U.R.S.S. , n.s., **55**, 583–586 (English translation)
47. P. Martin-Löf, *Notes on Constructive Mathematics* (Almqvist & Wiksell, Stockholm)
48. E.L. Post, Introduction to a general theory of elementary propositions. Am. J. Math. **43**, 163–165 [Reprinted [80], pp. 265–283. Reprinted [55], pp. 21–43]
49. E.L. Post, Finite combinatory processes. Formulation I. J. Symb. Log. **1**, 103–105 [Reprinted [12, 16], pp. 289–291. Reprinted [55], pp.103–105]
50. E.L. Post, Absolutely unsolvable problems and relatively undecidable propositions: account of an anticipation, in [12], pp. 340–433, [16], pp. 340–406, [55], pp. 375–441

51. E.L. Post, Formal reductions of the general combinatorial decision problem. Am. J. Math. **65**, 197–215 [Reprinted [55], pp. 442–460]
52. E.L. Post, Recursively enumerable sets of positive integers and their decision problems. Bull. Am. Math Soc. **50**, 284–316 [Reprinted [12, 16], pp.305–337; [55], pp.461–494]
53. E.L. Post, A variant of a recursively unsolvable problem. Bull. Am. Math. Soc. **52**, 264–268 [Reprinted [55], pp. 495–500]
54. E.L. Post, Recursive unsolvability of a problem of thue. J. Symb. Log. **12**, 1–11 [Reprinted [12, 16], pp. 293–303; [55], pp. 503–513]
55. E.L. Post, *Solvability, Provability, Definability: The Collected Works of Emil L. Post*, ed. by D. Martin (Birkhäuser, Basel)
56. P. Rosenbloom, *The Elements of Mathematical Logic* (Dover Publications, New York)
57. J.B. Rosser, Highlights of the history of the Lambda-Calculus. Ann. Hist. Comput. **6**(4), 337–349
58. W. Sieg, Mechanical procedures and mathematical experience, in *Mathematics and Mind*, ed. by A. George (Oxford University Press, Oxford), pp. 71–117
59. W. Sieg, Step by recursive step: Church's analysis of effective calculability. Bull. Symb. Log. **3**, 154–180 [Reprinted (with a long Postscriptum) in *Turing's Legacy: Developments from Turing's Ideas in Logic*, ed. by R. Downey. Lecture Notes in Logic (Cambridge University Press), to appear in 2013]
60. W. Sieg, Calculations by man and machine: mathematical presentation, in *In the Scope of Logic, Methodology and Philosophy of Science*. Volume one of the 11th International Congress of Logic, Methodology and Philosophy of Science, Cracow, August 1999; P. Gärdenfors, J. Wolenski, K. Kijania-Placek (eds.), *Synthese Library*, vol. 315 (Kluwer), pp. 247–262
61. W. Sieg, Only two letters: the correspondence between Herbrand and Gödel. Bull. Symb. Log. **11**, 172–184 [Reprinted in *K. Gödel - Essays for His Centennial*, ed. by S. Feferman, C. Parsons, S.G. Simpson. Lecture Notes in Logic (Cambridge University Press, 2010) pp. 61–73]
62. W. Sieg, On computability, in *Philosophy of Mathematics*, ed. by A. Irvine (Elsevier, Amsterdam), pp. 535–630
63. W. Sieg, Axioms for computability: do they allow a proof of Church's thesis? in *A Computable Universe: Understanding and Exploring Nature as Computation*, ed. by H. Zenil, World Scientific Publishing, Singapore, 2013, pp. 99–123
64. W. Sieg, Normal forms for puzzles: a variant of Turing's thesis", in [8], pp. 332–339
65. W. Sieg, J. Byrnes, K-graph machines: generalizing Turing's machines and arguments, in *Gödel '96*, ed. by P. Hajek. Lecture Notes in Logic, vol. 6 (A.K. Peters, Natick), pp. 98–119
66. T. Skolem, Begründung der elementaren Arithmetik durch die rekurrierende Denkweise ohne Anwendung scheinbarer Veränderlichen mit unendlichem Ausdehnungsbereich. Videnskapsselkapets sktifter, I. Matematisk-naturvidenskabelig klass, no. 6 [Translated in [80] pp. 302–333]
67. R. Smullyan, *Theory of Formal Systems*. Annals of Mathematics Studies, vol. 47 (Princeton University Press, Princeton)
68. R. Soare, Computability and recursion. Bull. Symb. Log. **2**(3), 284–321
69. R. Soare, Interactive computing and relativized computability, in [11], pp. 203–260
70. A. Thue, Probleme über Veränderungen von Zeichenreihen nach gegebenen Regeln. Skrifter utgit av Videnskapsselskapet i Kristiania, I. Matematisk-naturvidenskabelig klasse, no. 10
71. A.M. Turing, On computable numbers with an application to the Entscheidungsproblem. Proc. Lond. Math. Soc. **42**(2), 230–267 [Reprinted [12, 16] pp. 116–154. Reprinted [78] pp. 18–56. Reprinted [9] pp. 58–90; 94–96. Reprinted [8] pp. 16–43]
72. A.M. Turing, Correction to [71]. Proc. Lond. Math. Soc. **43**(2), 544–546
73. A.M. Turing, Systems of logic based on ordinals. Proc. Lond. Math. Soc. **45**(2), 161–228 [Reprinted [12, 16], pp.154–222]
74. A.M. Turing, The word problem in semi-groups with cancellation. Ann. Math. **52**, 491–505 [Reprinted [77], pp. 63–78. Reprinted [8], pp. 345–357]
75. A.M. Turing, Solvable and unsolvable problems. Sci. News **31**, 7–23 [Reprinted [76], pp. 187–203. Reprinted [9], pp. 582–595. Reprinted [8], pp. 322–331]

76. A.M. Turing, *Mechanical Intelligence: Collected Works of A.M. Turing*, ed. by D.C. Ince (North-Holland, Amsterdam)
77. A.M. Turing, *Pure Mathematics: Collected Works of A.M. Turing*, ed. by J.L. Britton (North-Holland, Amsterdam)
78. A.M. Turing, *Mathematical Logic: Collected Works of A.M. Turing*, ed. by R.O. Gandy, C.E.M. Yates (North-Holland, Amsterdam)
79. A. Urquhart, Emil Post, in *Handbook of the History of Logic*, ed. by D.M. Gabbay, J. Woods. Logic from Russell to Church, vol. 5 (Elsevier, Amsterdam), pp. 617–666
80. J. van Heijenoort (ed.), *From Frege to Gödel: A Sourcebook in Mathematical Logic, 1879–1931* (Harvard, Cambridge, 1967)
81. A.N. Whitehead, B. Russell, *Principia Mathematica*, vol. 1 (Cambridge University Press, Cambridge)

Algorithms: From Al-Khwarizmi to Turing and Beyond

Wolfgang Thomas

Abstract The foundational work of Alan Turing and contemporaries on computability marked a turning point in the development of mathematical sciences: It clarified in a rather absolute sense what is computable in the setting of symbolic computation, and it also opened the way to computer science where the use of algorithms and the discussion on their nature was enriched by many new facets. The present essay is an attempt to address both aspects: We review the historical development of the concept of algorithm up to Turing, emphasizing the essential role that logic played in this context, and we discuss the subsequent widening of understanding of "algorithm" and related "machines", much in the spirit of Turing whose visions we see realized today.

Keywords Algorithm • Al-Khwarizmi • Computability • Logic • Leibniz • Turing

1 Prologue

Alan Turing's contributions to science and engineering are extraordinary, both in depth and in breadth. His work spans scientific fields as represented in the four volumes of his Collected Works (pure mathematics, mathematical logic, machine intelligence, and morphogenesis), and it also covers less well documented work in "engineering" disciplines, such as computer architecture.[1] The present paper focuses on a very specific but most prominent aspect of Turing's heritage, namely his analysis of symbolic computation and the perspectives he connected with the idea of "machines" for intellectual tasks. The pivot element in this discussion

[1]On the occasion of the Alan Turing Year 2012, a new presentation of Turing's work, including hitherto unpublished papers, is available in the volume [5].

W. Thomas (✉)
RWTH Aachen, Lehrstuhl Informatik 7, 52056 Aachen, Germany
e-mail: thomas@informatik.rwth-aachen.de

G. Sommaruga, T. Strahm (eds.), *Turing's Revolution*,
DOI 10.1007/978-3-319-22156-4_2

is his pioneering paper "On computable numbers, with an application to the Entscheidungsproblem" of 1936 [24].

In a first part we describe the central role that logic played in the long process of clarification of the concept of "algorithm" that culminated in the work of Turing and his contemporaries Church, Kleene, and Post. This role of logic is worth being emphasized since "algorithms" were originally understood as procedures for numeric calculation. In the second part we discuss the development towards today's systems of information processing that has led to a much more comprehensive view on "algorithms", not just in scientific discourse but much more so in cultural and political discussions in the wider public.

This paper offers observations that cannot be called original; and also the historical sketch we provide is rough and does not touch a number of stages that may also be considered relevant.[2] Nevertheless, we found it worth trying to develop the larger picture around the idea of algorithm and—in terms of time—to address an extended historical axis, centered at 1936, the year when Turing's landmark paper appeared.

2 Some Prehistory: Al-Khwarizmi and Leibniz

In the work of Turing and his contemporaries, the terms "procedure", "finite process", and (as mostly used by Turing) "machine" occur more often than "algorithm". All these terms, however, point to the same idea: a process of symbolic computation fixed by an unambiguous and finite description.

The word "algorithm" originates in the medieval "algorism" as a recipe to perform calculations with numbers, originally just natural numbers. "Algorism" goes back to one of the most brilliant scientists of the islamic culture, Al-Khwarizmi (around 780–850), who worked in the "House of Wisdom" of the Chalif of Bagdad.[3] In this academy, founded by the Chalif Harun Al-Rashid and brought to culmination by his son Al-Mamun, scientists were employed for a wide spectrum of activities, among them translations (e.g. from Greek and Persian to Arabic), construction of scientific instruments, expeditions, and—quite important—advice to the Chalif. Al-Khwarizmi must have been a leading member. His full name (adding together all name ingredients we know of) was Muhammad Abu-Abdullah Abu-Jafar ibn Musa Al-Khwarizmi Al-Majusi Al-Qutrubbulli. The attribute "Al-Khwarizmi" points to the province of "Choresmia", located in today's Usbekistan, where he probably was born and grew up. He was sent by the Caliph to Egypt for an exploration the giza pyramids, he undertook measurements (e.g., executing the experiment of Eratosthenes to determine the diameter of the earth), and he wrote treatises.[4]

[2]Among the more comprehensive sources we mention [6].

[3]For an interesting account on the "House of Wisdom", we recommend [11].

[4]For a more detailed summary of Al Khwarizmi's life see, e.g., [27].

The most influential ones were his book on algebra ("Kitāb al-mukhtasar fi hisab al-jabr wa'l-muqabala") and his text "Computing with the Indian Numbers" ("Kitāb al-Jamʿ wa-l-tafrīq bi-ḥisāb al-Hind"). We concentrate here on the latter, in which he describes the execution of the basic operations of arithmetic (addition, multiplication, and others) in the decimal number system. The Indian sources he used are not known. Also the original text of Al-Khwarizmi seems to be lost. We have translations into Latin, for example the famous manuscript of the thirteenth century kept at the library of the University of Cambridge (England).[5] This text, however, is a bad example of scientific literature: Citations and comments are mixed into a conglomerate, and also many places where the decimal ciphers should appear remain empty. Probably the monk who wrote this text was eager to put everything into a solid theological context, and he was not comfortable with writing down these strange decimal symbols. Thus one has to guess at several places how the missing example computations would look like that should clarify the textual descriptions. It is amusing to read the phrase "but now let us return to the book", indicating that the author comes back to Al-Khwarizmi. And thus, many paragraphs start with the repetitive phrase "Dixit Algorizmi"—which motivated the term "algorism" for the procedures described in this work.

It is noteworthy that this concept of "algorithm" clearly refers to a process of symbol manipulation, in contrast to calculations performed on the abacus. The arrangement of pieces on the abacus also reflects the decimal system, but the computation process there is not symbolic in the proper sense of the word.

A new dimension to symbolic computation was added by Gottfried Wilhelm Leibniz (1646–1716). Extending ideas of precursors (among them Ramon Llull and Anastasius Kircher), he developed the vision of calculating truths (true statements) and not just numerical values. This vision was partly motivated by the fierce theological disputes of his time, a phenomenon which was not just academic but penetrated politics. Leibniz was born at the very end of the 30 years' war that had devastated Germany and that was partly rooted in theological conflicts between the catholic and the protestant. Leibniz dreamed of a universal calculus that would help philosophers in their disputes by just following the call "Calculemus!" He hints at his concept of a "characteristica universalis" in a letter to Duke Johann Friedrich of Braunschweig-Lüneburg[6]:

> In philosophy I found some means to do, what Descartes und others did via Algebra and Analysis in Arithmetic and Geometry, in all sciences by a combinatorial calculus ["per

---

[5] A full presentation in facsimile with transcription to Latin is given in [26]; a translation to English in [2].

[6] Leibniz wrote this letter [13] of 1671 to a duke and not to a colleague; hence he used German rather than Latin, with some Latin words inserted: "In Philosophia habe ich ein Mittel funden, dasjenige was Cartesius und andere per Algebram et Analysin in Arithmetica et Geometria gethan, in allen scientien zuwege zu bringen per Artem Combinatoriam [...]. Dadurch alle Notiones compositae der ganzen welt in wenig simplices als deren Alphabet reduciret, und aus solches alphabets combination wiederumb alle dinge, samt ihren theorematibus, und was nur von ihnen zu inventiren müglich, ordinata methodo, mit der zeit zu finden, ein weg gebahnet wird."

> Artem Combinatoriam"] [...]. By this, all composed notions of the world are reduced to few simple parts as their Alphabet, and from the combination of such alphabet [letters] a way is opened to find again all things, including their truths ["theorematibus"], and whatever can be found about them, with a systematic method in due time.

Leibniz undertook only small steps in this huge project, but in a methodological sense he was very clear about the task. As he suggests, logic should be applied by procedures of "alphabet's combination", i.e., symbolic computation. And he was very definite about his proposal to join the algorithmic procedures known from arithmetic with logic. This idea of "arithmetization of logic" (which later Hilbert pursued in his program to show the consistency of mathematics) is raised in two ways:

In his paper "Non inelegans specimen demonstrandi in abstractis" of 1685 [15] ("A not inelegant example of abstract proof method"), he develops the rudiments of Boolean algebra, using equations such as "$A + A = A$" with "+" as a sign for union. As an example, let us state the last theorem (XIII) of his note:

> Si coincidentibus addendo alia fiant coincidentia, addita sunt inter se communicantia

i.e.,

> If from two equal entities we get, by adjoining something, other but again equal entities, then among the added parts there must be something in common

or in a more formal set theoretic terminology, using the symbols of Leibniz:

> If to A we add B, respectively N, and we obtain again "coincidentia", i.e. $A + B = A + N$, and these are "alia", i.e. proper supersets of A, then B and N must have a nonempty intersection.

Clearly this approach to reasoning prepares Boolean algebra as a calculus, using notation of arithmetic.

The second idea on the arithmetization of logic appears in his note "Elementa calculi" (Elements of a calculus) of 1679 [16], where we find the following passage[7]:

> For example, since man is a rational animal (and since gold is the heaviest metal), if hence the number for animal (for metal) is a (such as 2) (m such as 3) and of rational (heaviest) r such as 3 (p such as 5), then the number for man or h will be the same as ar, which in our example is 2×3, i.e. 6 (and the number for gold, s, will be the same as mp, which in this example is 3×5, i.e. 15).

We see very clearly the idea to represent elementary concepts by prime numbers and their conjunction by products of prime numbers, which allows to reestablish the factors. This prepares the idea of Gödel numbering that entered the stage again 250

[7]"Verbi gratia quia Homo est Animal rationale (et quia Aurum est metallum ponderosissimum) hinc si sit Animalis (metalii) numerus a ut 2 (m ut 3) Rationalis (ponderosissimi) vero numerus r ut 3 (p ut 5) erit numerus hominis seu h idem quot ar id est in hoc exemplo 2,3 seu 6 (et numerus auri solis s idem quot mp id est in hoc exemplo 3,5 seu 15."

years later—using number theoretic facts to code complex objects (like statements or proofs) by numbers—in a way that allows unique decomposition.

It is somewhat breathtaking to see how optimistic Leibniz was about the realization of his ideas. In a note[8] of 1677 he writes

> When this language is introduced sometime by the missionaries, then the true religion which is unified to the best with rationality, will be founded firmly, and one does not need to fear a renunciation of man from it in the future, just as one does not need to fear a renunciation from algebra and geometry.

This idea of a rational theory of ethics was shared by many of Leibniz's contemporaries. As examples we just mention Spinoza's treatise "Ethica. Ordine geometrio demonstrata" (1677) and the dissertation "Philosophia practica universalis, methodo mathematica conscripta" (1703) of Leibniz's student Christian Wolff.

But more than his colleagues, Leibniz formulated rather bold promises—in a very similar way as we do today when we apply for project money[9]:

> I think that some selected people can do the job in five years, and that already after two years they will reach a stage where the theories needed most urgently for life, i.e., moral and metaphysics, are manageable by an unfallible calculus.

3 Towards Hilbert's Entscheidungsproblem

Leibniz's dream in its full generality remained (and remains) unrealized. Surprisingly, however, it was materialized in the domain of mathematics. This process started with George Boole who developed the vague sketch of Leibniz into a proper theory: "Boolean algebra". The breakthrough in devising a universal scientific calculus was then achieved by Gottlob Frege. His "Begriffsschrift" (1879) introduces a formal language in which mathematical statements can be expressed, the essential innovation being a clarification of the role of quantifiers and quantification. His own work on the foundations of arithmetic, and in particular the subsequent enormous effort undertaken by Russell and Whitehead in their "Principia Mathematica", opened a way to capture mathematics in a formal system.

But in this development the objectives connected with formalization shifted dramatically. The objective was no longer the Leibnizian approach to compute truths needed in life (or just in mathematics) but of a more methodological nature. The shift occurred with the discovery of contradictions ("paradoxes") in Frege's system. The most prominent problem was the contradiction found independently by Russell and

[8]From [14]: "Nam ubi semel a Missionariis haec lingua introduce poterit, religio vera quae maxime rationi consentanea est, stabilia erit et non magis in posterum metuanda erit Apostasia, quam ne hominess Arithmeticam et Geometriam, quam semel dedicere, mox damnent."

[9]From [14]: "Aliquot selectos homines rem intra quinquennium absolvere posse puto; intra biennium autem doctrinas, magis in vita frequentalas, id est Moralem et Metaphysicam, irrefragabile calculo exhibebunt."

Zermelo inherent in the concept of a "set of those sets that do not contain themselves as elements". The formalization of mathematics was now pursued as a way to show its consistency. As Hilbert formulated in his program on the foundations of mathematics, the task was to analyze the combinatorial processes in formal proofs and by such an analysis arrive at the result that the arithmetical equation $0 = 1$, for example, cannot be derived. In pursuing this program, the key issues were axiomatizations of theories, the consistency of theories, and the soundness and completeness of proof calculi.

The fundamental results of Gödel (completeness of the first-order proof calculus and incompleteness of any axiomatic system of arithmetic) made it clear that only in a fragmentary way there was hope to fulfill Hilbert's program. An essential ingredient in Gödel's approach was the arithmetization of logic (today called "Gödelization"), transforming Leibniz's hint mentioned above into a powerful method.

However, a positive result of the foundational research of the early twentieth century was that the "atomic ingredients" of mathematical proofs, as condensed in the rules of the proof calculus of first-order logic, were established. Together with the axiomatization of set theory, a framework emerged in which most of mathematics could be formally simulated. This framework clarified to a large extent which kind of symbolic manipulations are necessary to do logic algorithmically—as Al-Khwarizmi had explained this centuries before for numeric calculations. Most remarkably, it was an algorithmic problem in the domain of logic (and not in the domain of arithmetic) which motivated a general analysis of computability and hence of "algorithm".

This was "Hilbert's Entscheidungsproblem", as formulated in the monograph "Einführung in die theoretische Logik" by Hilbert and Ackermann (1928).

> Das Entscheidungsproblem ist gelöst, wenn man ein Verfahren kennt, das bei einem vorgelegten logischen Ausdruck durch endlich viele Operationen die Entscheidung über die Allgemeingültigkeit bzw. Erfüllbarkeit erlaubt.

Turing's paper "On computable numbers, with an application to the Entscheidungsproblem" [24] solves this problem in the negative, and it does so by a radical reduction of "algorithm" to very elementary steps of symbol manipulation.

Before discussing this work in more detail, let us emphasize again that the objectives of Hilbert and his colleagues were quite distinct from the visions that Leibniz had in mind, although one might say that axiomatic set theory is the fulfillment of Leibniz's project of devising a "characteristica universalis", restricted to the field of mathematics. Rather than studying global properties of formal systems, such as consistency and completeness, Leibniz wanted to use formal rules as a "knowledge engineer": to find and verify interesting statements by calculation—primarily in areas far beyond mathematics. Hilbert did—as far as we know—not adopt the view that formalization would help in any way to solve concrete mathematical problems, by performing the algorithmic execution of proofs. For him and most logicians in his tradition, the formalization of mathematics is an approach to understand its methodological foundations.

Only today, in computer science, both traditions of “formal logic” are merged again: In computer science, formal systems are set up in many ways, for example as programming languages or as query languages for data bases, and in this design questions of soundness, completeness, and complexity have to be addressed. But we see at the same time the application of these formal systems of data processing to solve concrete problems in virtually all sciences and domains of human life, very much in the spirit of Leibniz.

4 Turing’s Breakthrough

Turing learned about Hilbert’s Entscheidungsproblem in lectures of Max Newman in Cambridge, after 4 years of (very successful) studies of mathematics. It was the fresh look of a young genius that helped to settle the problem. The paper he wrote is one of the most remarkable documents of mathematical literature of the twentieth century. In fact, the solution of the Entscheidungsproblem (which was solved independently by Alonzo Church [3]) is only one of at least seven innovations which Turing offered in his paper:

1. A machine model capturing computability
2. Its justification
3. Conception and implementation of a universal program
4. Establishment of a non-solvable problem
5. Proof that Hilbert’s Entscheidungsproblem is undecidable
6. Equivalence between Turing machines and λ-calculus
7. Initial steps to computable analysis

We do not repeat here the precise formulation of the model of Turing machine. It should be noted that a variant of this model (“finite combinatory processes”) was presented in the same year 1936 by Emil Post [17]. What makes Turing’s paper so brilliant is the mature and conceptually tight justification of his model. This justification starts with phrases which remind us precisely of the algorithms that were the subject of Al-Khwarizmi:

> Computing is normally done by writing certain symbols on paper. We may suppose this paper is divided into squares like a child’s arithmetic book. . . .

A second remarkable aspect in Turing’s paper is the fact that after presenting his model of Turing machine, he immediately exhibits a problem that is not solvable with this model. For this, he develops the idea of a universal machine, enters the technicalities of actually constructing one (and, as an aside, introduces the programming technique today called “macros” for this purpose), and then applies a diagonalization argument. This appearance of a powerful model connected immediately with a corresponding unsolvability result should be compared with the centuries that elapsed between the clear understanding of algebraic expressions (in

Vieta's time) and the proof of Abel that for polynomials of degree 5 one cannot in general find solutions in this format.

In fact, the mere possibility to envisage algorithmically unsolvable problems emerged only at a rather late stage. In 1900, in the formulation of Hilbert's 10th problem

> Eine diophantische Gleichung mit irgendwelchen Unbekannten und mit ganzen rationalen Zahlenkoeffizienten sei vorgelegt: Man soll ein Verfahren angeben, nach welchem sich mittels einer endlichen Anzahl von Operationen entscheiden lässt, ob die Gleichung in ganzen rationalen Zahlen lösbar ist.

One just finds the task to develop a "procedure" ("Verfahren"). The earliest place in mathematical literature where the certainty about algorithmic solutions is put into doubt seems to be a most remarkable paper by Axel Thue of 1910 ("Die Lösung eines Spezialfalles eines allgemeinen logischen Problems" [23]). He formulates the fundamental problem of term rewriting: Given two terms s, t and a set of axioms as equations between terms, decide whether from s one can obtain t by a finite number of applications of the axioms. He resorts to a special case in order to provide a partial solution. About the general case one finds the following prophetic remark:

> A solution of this problem in the most general case might perhaps be connected with unsurmountable difficulties.[10]

It is a pity that this brilliant paper remained unnoticed for decades; one reason for this is perhaps its completely uninformative title. (A detailed discussion is given in [21].)

The work of Turing and his contemporaries Church, Kleene, Post finished a struggle of many centuries for an understanding of "algorithm" and its horizon of applicability, termed "computability". This success was possible by a merge of two traditions in symbolic computation: arithmetic and logic. The impression that an unquestionable final point was reached with this work was underlined by Gödel, who stated in 1946, 10 years after Turing's breakthrough [7]:

> Tarski has stressed [. . .] (and I think justly) the great importance of the concept of general recursiveness (or Turing's computability). It seems to me that this importance is largely due to the fact that with this concept one has for the first time succeeded in giving an absolute definition of an interesting epistemological notion, i.e., one not depending on the formalism chosen. [. . .] By a kind of miracle it is not necessary to distinguish orders.

5 Moves Towards Computer Science

The year 1936 not only marks a point of final achievement but is at the same time the initialization of a new and rapidly developing discipline: computer science (or

[10] "Eine Lösung dieser Aufgabe im allgemeinsten Falle dürfte vielleicht mit unüberwindlichen Schwierigkeiten verbunden sein."

"informatics"). Each of the pioneers mentioned above in connection with Turing, namely Church, Kleene, and Post, as well es Turing himself, were active in this launch of a new scientific subject.

Turing himself turned, for example, to questions that are hosted today in the field of "computer architecture", but he also addressed many further issues, such as program verification. In 1957, Church formulated a fundamental problem beyond program verification—"Application of recursive arithmetic to the problem of circuit synthesis" [4]—thus opening a fascinating branch of computer science, in which today game theoretic methods are used for the automatic construction of interactive programs (see, e.g. [22]). Kleene should be noted for his path-breaking work [12] on regular events and finite automata, establishing the basic equivalence result in automata theory. Finally, Post was the first to exhibit purely combinatorial (and thus purely "mathematical", as opposed to logical) undecidable problems (among them Post's Correspondence Problem [19]), and he developed in [18] a theory of problem reduction that today underlies much of complexity theory.

The subsequent rise of computer science changed our views on algorithms in two ways: First, algorithmic methods in computer science transcend the framework of symbol manipulation inherent in Turing's analysis. Secondly, "algorithms" are understood today in the context of the highly complex software systems that govern our life (e.g., in enterprise software, internet applications, etc.) Let us address them both.

6 New Facets of "Algorithm"

Turing's analysis (as well as the parallel work of Post) refers to procedures that work on finite words composed from symbols taken from a finite alphabet. As noted by Turing, the algorithms of basic arithmetic can be treated in this framework. However, a standard first-year course in computer science, usually titled "Algorithms and data structures", already shows several chapters that go beyond this domain. In particular, we deal there with algorithms over trees and over graphs rather than words. In this generalized setting, some features of algorithms arise that are hidden when we work over words. For example, over graphs we observe the lack of a natural ordering (e.g. for the set of vertices). This lack of order allows to say "Pick an element in the set V of vertices ..." without the requirement (and indeed, without the possibility) of fixing a particular element. Over words, the situation is different: Picking a letter in a word always implies the possibility to pick a particular (e.g., the first) letter. As Gurevich and others remarked, the Turing model working with the substrate of words on a tape does not allow us to deal with algorithms on the adequate level of abstraction. The machinery of coding (by words) that enables us to make a bridge to Turing's model spoils this adequacy. To settle this problem, a generalized view on algorithms was developed in the model of "abstract state machine" [8]. It has the flexibility that is needed to answer the challenge of very diverse kinds of algorithms as they are specified, for example, in programming languages.

In some domains of algorithmic mathematics, a more abstract view (than that of computations over words) is unavoidable even when allowing "codings by words". An example is the domain of classical Euclidean geometry, where algorithms in the form of "geometric constructions" are familiar since antiquity. The data handled by these constructions (points, lines, etc.), are infinite objects, as are the real numbers. Points of Euclidean space and real numbers cannot serve as "inputs" to an algorithm in Turing's sense, since in their generality they are not codable by finite words. Abstract state machines over the space of (tuples of) reals can be invoked to handle geometric algorithms and algorithms over the reals; for the latter, also the Blum-Shub-Smale model [1] provides an adequate framework.

Apart from this, a host of new algorithmic concepts was developed in computer science, often termed as "schemes", "processes", etc., which are not immediately covered by the model of Turing machine but constitute clearly mechanisms to transform data according to finite recipes. One can give numerous examples: Classification procedures for image and speech processing (as needed in visual computing, data mining, and automatic speech translation), distributed algorithms that guarantee a convergence of network states to prescribed equilibria, or algorithms for planning and search as used in robotics. A field of growing importance is the use of algorithms in non-terminating reactive systems (such as communication protocols or control systems); here the idea of Turing of a nonterminating process (e.g., to generate the decimal expansion of a real number) is enhanced by the feature that an infinite computation may be subject to a non-termining sequence of receipts of inputs while it works.

These procedures are far away from the simple set-up of symbolic computation considered by Turing in his paper of 1936. But it was exactly Turing who was aware of the perspective of a widening of the horizon of algorithmic methods—at a very early stage. His term was "machine", allowing him to cover the software and hardware aspects simultaneously. Let us document this by two citations:

> We may hope that machines will eventually compete with men in all purely intellectual fields. [25]
>
> One way to setting about our task of building a "thinking machine" would be to take a man as a whole and try to replace all parts of him by machinery. This would include television cameras, microphones, loudspeakers, wheels, and "handling servo-mechanisms", as well as some sort of "electronic brain".[11]

[11] Cited from [10, p. 117].

7 Algorithms as Molecules in Large Organisms

The vision of "thinking machine" as sketched by Turing is a reality today. Such systems are developed in the field of robotics; they are serious approximations of living organisms.

Indeed, computer science has created hierarchies and networks of algorithms of such a huge complexity that their description necessarily makes use of terms familiar from biology. "Life cycle", "software evolution", "virus", "worm", "infection" are common terms in the discussion of large software systems or networks. One may say that the instructions for Turing machines represent the atomic level of data processing, and algorithms (in the classical sense) the molecular level. Based on this, much larger systems are designed in which the global structure and sensory elements for the interaction with the environment (e.g., humans) are often more essential features than the individual components.

A methodological problem of computer science is the span of orders of magnitude realized in the huge data conglomerates and software systems as we see them today. If we consider the physical world, start from the atoms as basic units, and proceed in stages to larger physical objects, where one step involves an enlargement by a factor 1000, we reach, successively in four steps, objects of the following kind: a molecule, a living cell, a small animal, a small village. These kinds of objects are studied in separate scientific disciplines, ranging from physics via chemistry and biology to sociology. Each of these disciplines has specific—and rather different—models that are suitable for the respective objects under consideration. In computer science we may start with a byte corresponding to the atomic level (describing, for example, an alphanumeric symbol as handled in a Turing machine instruction), and proceed in four steps to a Kilobyte (corresponding to a small program of a dozen lines), a Megabyte (corresponding to a book, or to code produced by a team of students during a software project), a Gigabyte (a size of data comparable to the Windows operating system), and a Terabyte (comparable to the stock of the Library of Congress). Much larger are the "big data" handled in the world wide web. Although this huge span of orders of magnitude is studied in one science, computer science, it is obvious that rather diverse methodologies (as in the natural sciences) are needed for an adequate study. A common misunderstanding of the role of Turing machines is the remark that this model "does not match the reality of computer science systems today". This remark is as misplaced as is the rather empty statement that atomic physics does not capture the reality of living organisms.

Despite this richness of the landscape of computer systems and computer science, the public view on the intelligent machinery designed by computer scientists puts the term "algorithm" at its center, most often in the plural form. "Algorithms" is what make devices and processes of information technology run. Any conglomerate of units of information processes is covered by this term. This is different from the situation in biology where one does not say that biological systems are "molecules".

Let us illustrate this observation on the colloquial use of "algorithms" by three headlines the author noted in 2012–2013, taken from the press, respectively the web.

The deeply troubling perspective of programmed robots designed for military combat (on earth and on air) was discussed with the subtitle "The moral of algorithms" in a leading German newspaper.[12] In connection with the comprehensive analysis of data on the web (covering millions of persons) by government agencies, the term "the tyranny of algorithms" was used in an article of the same newspaper,[13] and, finally, the controversy in this discussion was condensed by "Spiegel Online" into the remarkable headline "Freedom against algorithms".[14]

While it is clear to the experts that current implementations of computer systems ultimately rely on small computation steps in microprocessors and are thus in principle reducible to Turing machine computations, we see that in the public discussion the actual understanding of "algorithms" drastically exceeds the content of the Turing model—it is today located on a much more general level.

8 Returning to Leibnizian Visions?

The current colloquial usage of "algorithms" seems approaching the visions of Leibniz on treating central questions of human behavior and social disputes by calculation. Indeed, in an unexpected sense, our contemporary information processing systems are coming close to Leibniz's idea—envisaging the solution of moral, social, and political questions by algorithms. We observe today that software systems used in the financial markets and in military and government institutions are installed precisely for the purpose of decision finding in questions of economics, politics, and even war.[15]

Thus, "algorithms" in this general view, i.e., the algorithmic abilities of the most sophisticated software and hardware systems that computer scientists develop today, silently put into reality a fair amount of Leibnizian utopia. But in contrast to Leibniz's hopes, it seems that the "unfallible calculus" underlying these systems can no more be seen as a secure means to get nearer to the "best of all worlds"; rather we are confronted with new troubling questions on estimating the power and justifying the use of algorithms.

[12] F. Rieger, Das Gesicht unserer Gegner von morgen, Frankfurter Allgemeine Zeitung, 20th Sept. 2012.

[13] G. Baum, Wacht auf, es geht um die Menschenwürde, Frankfurter Allgemeine Zeitung, 16th June 2013.

[14] "Freiheit gegen Algorithmen", Spiegel Online, 21st June 2013.

[15] This aspect, with a focus on the role of algorithmic game theory, is developed at length by F. Schirrmacher, a leading German journalist, in [20], a bestseller on the German book market.

References

1. L. Blum, M. Shub, S. Smale, On a theory of computation and complexity over the real numbers: NP-completeness, recursive functions and universal machines. Bull. Am. Math. Soc. **21**, 1–46 (1989)
2. J.N. Crossley, A.S. Henry, Thus spake al-Khwārizmī: a translation of the text of Cambridge University Library Ms. Ii.vi.5. Hist. Math. **17**, 103–131 (1990)
3. A. Church, A note on the Entscheidungsproblem. J. Symb. Log. **1**, 40–41 (1936)
4. A. Church, in: *Summaries of the Summer Institute of Symbolic Logic*. Application of recursive arithmetic to the problem of circuit synthesis, vol. I (Cornell University, Ithaca, 1957), pp. 3–50
5. B. Cooper, J.V. Leeuwen (eds.), *Alan Turing: His Work and Impact* (Elsevier, Amsterdam, 2013)
6. M. Davis, *The Universal Computer – The Road from Leibniz to Turing*. Turing Centennial Edition (CRC Press, Boca Raton, 2012)
7. K. Gödel, Remarks before the Princeton bicentennial conference on problems in mathematics, in *Kurt Gödel, Collected Works*, ed. by S. Feferman et al., vol. II (Oxford University Press, Oxford, 1990), pp. 150–153
8. Y. Gurevich, Sequential abstract-state machines capture sequential algorithms. ACM Trans. Comput. Log. **1**, 77–111 (2000)
9. H. Herring (ed.), *G.W. Leibniz Schriften zur Logik und zur philosophischen Grundlegung von Mathematik und Naturwissenschaft* (lat. u. deutsch) (Suhrkamp, Frankfurt, 1996)
10. A. Hodges, *Alan Turing: The Enigma* (Vintage, London, 1992)
11. J. Al-Khalili, *The House of Wisdom: How Arabic Science Saved Ancient Knowledge and Gave Us the Renaissance* (Penguin Press, New York, 2011)
12. S.C. Kleene, Representation of events in nerve nets and finite automata, in *Automata Studies*, ed. by C.E. Shannon, J. McCarthy (Princeton University Press, Princeton, 1956), pp. 3–41
13. G.W. Leibniz, Brief an Herzog Johann Friedrich von Braunschweig-Lüneburg (Okt. 1671), in *Philosophische Schriften von Gottfried Wilhelm Leibniz*, ed. by C.I. Gerhardt, vol. 1 (Weidmannsche Buchhandlung, Berlin, 1875), pp. 57–58
14. G.W. Leibniz, Anfangsgründe einer allgemeinen Charakteristik, in [9], pp. 39–57
15. G.W. Leibniz, Ein nicht unelegantes Beispiel abstrakter Beweisführung, in [9], pp. 153–177
16. G.W. Leibniz, Elemente eines Kalküls, in [9], pp. 67–91
17. E.L. Post, Finite combinatory processes – formulation 1. J. Symb. Log. **1**, 103–105 (1936)
18. E.L. Post, Recursively enumerable sets of positive integers and their decision problems. Bull. Am.. Math. Soc. **50**, 284–316 (1944)
19. E.L. Post, A variant of a recursively unsolvable problem. Bull. Am. Math. Soc. **52**, 264–268 (1946)
20. F. Schirrmacher, *EGO: Das Spiel des Lebens* (Karl Blessing-Verlag, München, 2013)
21. M. Steinby, W. Thomas, Trees and term rewriting in 1910: on a paper by Axel Thue. Bull. Eur. Assoc. Theor. Comput. Sci. **72**, 256–269 (2000)
22. W. Thomas. Infinite games and verification, in *Proceedings of International Conference on Computer Aided Verification CAV'02*. Lecture Notes in Computer Science, vol. 2404 (Springer, Berlin, Heidelberg, New York, 2002), pp. 58–64
23. A. Thue, Über die Lösung eines Spezialfalls eines allgemeinen logischen problems. *Kristiania Videnskabs-Selskabets Skrifter*. I. Mat. Nat. Kl. 1910, No. 8
24. A.M. Turing, On computable numbers, with an application to the Entscheidungsproblem. Proc. Lond. Math. Soc. **42**, 230–265 (1936)
25. A.M. Turing, Computing machinery and intelligence. Mind **59**, 433–460 (1950)

26. K. Vogel, *Mohammed ibn Musa Alchwarizmi's Algorismus. Das früheste Lehrbuch zum Rechnen mit indischen Ziffern* (Zeller, Aalen, 1963)
27. H. Zemanek, Dixit algorizmi: his background, his personality, his work, and his influence, in *Algorithms in Modern Mathematics and Computer Science*, ed by A Ershov, D Knuth. Proceedings, Urgench, Uzbek SSR, 16–22 September 1979. Springer Lecture Notes in Computer Science, vol. 122 (Springer, Berlin, 1981), pp. 1–81

The Stored-Program Universal Computer: Did Zuse Anticipate Turing and von Neumann?

B. Jack Copeland and Giovanni Sommaruga

Abstract This chapter sets out the early history of the stored-program concept. The several distinct 'onion skins' making up the concept emerged slowly over a ten-year period, giving rise to a number of different programming paradigms. A notation is developed for describing different aspects of the stored-program concept. Theoretical contributions by Turing, Zuse, Eckert, Mauchly, and von Neumann are analysed, followed by a comparative study of the first practical implementations of stored-programming, at the Aberdeen Ballistic Research Laboratory in the US and the University Manchester in the UK. Turing's concept of universality is also examined, and an assessment is provided of claims that various historic computers—including Babbage's Analytical Engine, Flowers' Colossus and Zuse's Z3—were universal. The chapter begins with a discussion of the work of the great German pioneer of computing, Konrad Zuse.

Keywords ACE • Alan turing • Analytical engine • Automatic sequence controlled calculator • BINAC • Charles Babbage • Colossus • EDVAC • ENIAC • F.C. Williams • Henschel AG • History of computing • History of hardware • History of software • Howard Aiken • J. Presper Eckert • John Mauchly • John V. Atanasoff • John von Neumann • Konrad Zuse • M.H.A. Newman • Manchester Baby • Plankalkül • Richard Clippinger • Stored–program concept • Thomas H. Flowers • Thomas Haigh • Tom Kilburn • Universal Turing machine • Zuse KG • Z1 • Z2 • Z3 • Z4 • Z5 • Z11 • Z22 • Z23

B.J. Copeland (✉)
Department of Philosophy, University of Canterbury, New Zealand

The Turing Centre, ETH Zurich, Zurich, Switzerland

School of Historical and Philosophical Inquiry, University of Queensland, Australia
e-mail: jack.copeland@canterbury.ac.nz

G. Sommaruga
The Turing Centre, ETH Zurich, Zurich, Switzerland

Department of Humanities, Social and Political Sciences, ETH Zurich, Zurich, Switzerland
e-mail: sommarug@ethz.ch

G. Sommaruga, T. Strahm (eds.), *Turing's Revolution*,
DOI 10.1007/978-3-319-22156-4_3

1 Introduction

To Konrad Zuse belongs the honour of having built the first working program-controlled general-purpose digital computer. This machine, later called Z3, was functioning in 1941.[1] Zuse was also the first to hire out a computer on a commercial basis: as Sect. 2 explains, Zuse's Z4 was rented by the Swiss Federal Institute of Technology (ETH Zurich) for five years, and provided the first scientific computing service in Continental Europe.

Neither Z3 nor Z4 were electronic computers. These machines were splendid examples of pre-electronic relay-based computing hardware. Electromechanical relays were used by a number of other early pioneers of computing, for example Howard Aiken and George Stibitz in the United States, and Alan Turing at Bletchley Park in the United Kingdom. Bletchley's relay-based 'Bombe' was a parallel, special-purpose electromechanical computing machine for codebreaking (though some later-generation Bombes were electronic).[2] Aiken's giant relay-based Automatic Sequence Controlled Calculator, built by IBM in New York and subsequently installed at Harvard University (known variously as the IBM ASCC and the Harvard Mark I) had much in common with the earlier Z3.

Alan Turing, 1912–1954. *Credit: King's College Library, Cambridge*

From an engineering point of view, the chief differences between the electromagnetic relay and electronic components such as vacuum tubes stem from the fact that, while the vacuum tube contains no moving parts save a beam of electrons, the relay contains mechanical components that move under the control of an

[1]Zuse, K. 'Some Remarks on the History of Computing in Germany', in Metropolis, N., Howlett, J., Rota, G. C. (eds) *A History of Computing in the Twentieth Century* (New York: Academic Press, 1980).

[2]For additional information about the Bombe, including Gordon Welchman's contributions, and the earlier Polish Bomba, see Copeland, B. J., Valentine, J., Caughey, C. 'Bombes', in Copeland, B. J., Bowen, J., Sprevak, M., Wilson, R., et al., *The Turing Guide* (Oxford University Press), forthcoming in 2016.

electromagnet and a spring, in order to make and break an electrical circuit. Vacuum tubes achieve very much faster digital switching rates than relays can manage. Tubes are also inherently more reliable, since relays are prone to mechanical wear (although tubes are more fragile). A small-scale electronic digital computer, containing approximately 300 tubes, was constructed in Iowa during 1939–42 by John V. Atanasoff, though Atanasoff's machine never functioned satisfactorily. The first large-scale electronic digital computer, Colossus, containing about 1600 tubes, was designed and built by British engineer Thomas H. Flowers during 1943, and was installed at Bletchley Park in January 1944, where it operated 24/7 from February of that year.[3]

Zuse's Z3 and Z4, like Aiken's ASCC, and other relay-based computers built just prior to, or just after, the revolutionary developments in digital electronics that made the first electronic computers possible, were a final luxuriant flowering of this soon-to-be-outdated computing technology (though for purposes other than computing, relays remained in widespread use for several more decades, e.g. in telephone exchanges and totalisators). Outmatched by the first-generation electronic machines, Zuse's computer in Zurich and Aiken's at Harvard nevertheless provided sterling service until well into the 1950s. While relay-based computers were slower than their electronic rivals, the technology still offered superhuman speed. Electromechanical computers carried out in minutes or hours calculations that would take human clerks weeks or months.

It was not just the absence of digital electronics that made Z3, Z4 and ASCC pre-modern rather than modern computers. None incorporated the *stored-program* concept, widely regarded as the sine qua non of the modern computer. Instructions were fed into the ASCC on punched tape. This programming method echoed Charles Babbage's nineteenth-century scheme for programming his Analytical Engine, where instructions were to be fed into the Engine on punched cards connected together with ribbon so as to form a continuous strip—a system that Babbage had based on the punched-card control of the Jacquard weaving loom. If the calculations that the ASCC was carrying out required the repetition of a block of instructions, this was clumsily achieved by feeding the same instructions repeatedly through the ASCC's tape-reader, either by punching multiple copies of the relevant block of instructions onto the tape or, if the calculation permitted it, by gluing the tape ends together to form a loop.[4] Zuse's Z3 and Z4 also had punched tape programming (Zuse preferred cine film to paper).[5] In a stored-program computer, on the other hand, the same instructions can be selected repeatedly and fed from

[3]Copeland, B. J. et al. *Colossus: The Secrets of Bletchley Park's Codebreaking Computers* (Oxford: Oxford University Press, 2006, 2010).

[4]Campbell, R. V. D. 'Aiken's First Machine: The IBM ASCC/Harvard Mark I', in Cohen, I. B., Welch, G. W. (eds) *Makin' Numbers: Howard Aiken and the Computer* (Cambridge, Mass.: MIT Press, 1999), pp. 50–51; Bloch, R. 'Programming Mark I', in Cohen and Welch, *Makin' Numbers*, p. 89.

[5]Zuse, 'Some Remarks on the History of Computing in Germany', p. 615; see also the photograph of a segment of Zuse's cine film tape in the *Konrad Zuse Internet Archive*, http://zuse.zib.de/

memory, providing an elegant solution to the problem of how to loop through a subroutine a number of times.

Konrad Zuse, 1910–1995.
Credit: ETH Zurich

Although Z3 and Z4 (like their predecessors Z1 and Z2) used punched-tape program control, there have always been rumours in the secondary literature that Zuse independently invented the stored-program concept, perhaps even prior to Turing's classic 1936 exposition of the concept and the extensive further development of it by both Turing and John von Neumann in 1945. Nicolas Jequier, for example, described Z3 as the first computer to have an '[i]nternally stored program'.[6] Jürgen Schmidhuber recently wrote in *Science*:

> By 1941, Zuse had physically built the first working universal digital machine, years ahead of anybody else. Thus, unlike Turing, he not only had a theoretical model but actual working hardware.[7]

In *Nature* Schmidhuber wrote:

> Zuse's 1936 patent application (Z23139/GMD Nr. 005/021) also described what is commonly called a 'von Neumann architecture' (re-invented in 1945), with program and data in modifiable storage.[8]

---

item/8OeSo6XPtIV2X44R (thanks to Matthias Röschner, vice-director of the Deutsches Museum Archiv, for information).

[6]Jequier, N. 'Computer Industry Gaps', *Science and Technology*, vol. 93 (Sept. 1969), pp. 30–35 (p. 34).

[7]Schmidhuber, J. 'Turing in Context', *Science*, vol. 336 (29 June 2012), pp. 1638–1639. sciencescape.org/paper/22745399.

[8]Schmidhuber, J. Comments on a review by John Gilbey (*Nature*, vol. 468, 9 December 2010, pp. 760–761), at www.nature.com/nature/journal/v468/n7325/abs/468760a.html.

Computer historians Brian Carpenter and Robert Doran are more cautious, saying only that

> The stored program concept—that a computer could contain its program in its own memory—derived ultimately from Turing's paper *On Computable Numbers*, and Konrad Zuse also developed it in Germany, in the form of his *Plankalkül* language, without having read *On Computable Numbers*.[9]

John von Neumann, 1903–1957. *Credit: Photographer unknown. From the Shelby White and Leon Levy Archives Center, Institute for Advanced Study, Princeton, NJ, USA*

Zuse described his sophisticated *Plankalkül* programming language (discussed in Sects. 2 and 7) as embodying 'the notation and results of the propositional and the predicate calculus' and he called it 'the first programming language'.[10]

Fascinated by these persistent rumours about Zuse, we investigated his unpublished writings from the period 1936–1945, in order to discover what he had actually said about universality and the stored-program concept. We studied especially his April 1936 patent application Z23139, his December 1936 patent application Z23624, entries from June 1938 in his workbook, his 1941 patent application Z391 (setting out the design of Z3[11]), and his 1945 manuscript 'Der Plankalkül'.[12]

---

[9]Carpenter, B. E., Doran, R. W. 'Turing's Zeitgeist', in Copeland, Bowen, Sprevak, Wilson et al., *The Turing Guide*.

[10]Zuse, 'Some Remarks on the History of Computing in Germany', p. 621; Zuse interviewed by Christopher Evans in 1975 ('The Pioneers of Computing: An Oral History of Computing', London: Science Museum; © Board of Trustees of the Science Museum).

[11]Konrad Zuse interviewed by Uta Merzbach in 1968 (Computer Oral History Collection, Archives Centre, National Museum of American History, Washington D.C.).

[12]Zuse, K. Patent Application Z23139, 'Verfahren zur selbsttätigen Durchführung von Rechnungen mit Hilfe von Rechenmaschinen' [Procedure for the automatic execution of calculations with the aid of calculating machines], 9 April 1936, Deutsches Museum Archiv, document reference NL 207/00659; Zuse, K. Patent Application Z23624, 'Rechenmaschine' [Calculating machine], 21 December 1936, Deutsches Museum Archiv, NL 207/0991; Zuse, K. Patent Application Z391, 'Rechenvorrichtung' [Calculating device], 1941, in the *Konrad Zuse Internet Archive*, http://zuse.

The fact that Zuse wrote in German has always presented an obstacle to the dissemination of his achievements among Anglophone historians. Indeed, much of Zuse's unpublished work is hand written, in a form of old-fashioned German shorthand. We present English translations of key passages from Zuse's documents (so far as we know, for the first time).

Our conclusion will be that the truth lies somewhere between Schmidhuber's statements and the more cautious statement by Carpenter and Doran. Their cautious statement is true, but there is more to be said. We cannot, however, endorse Schmidhuber's or Jequier's claims.

The structure of this chapter is as follows. After a short overview of Zuse's life and work in Sect. 2, based largely on Zuse's own accounts of events in tape-recorded interviews given in 1968 and 1975, Sect. 3 goes on to provide a comparative account of Turing's and von Neumann's contributions to the stored-program concept. Both men made fundamental and far-reaching contributions to the development of this keystone concept. In the voluminous secondary literature, however, von Neumann's contributions are generally exaggerated relative to Turing's, even to the extent that many accounts describe von Neumann as the inventor of the stored-program concept, failing altogether to mention Turing. Section 3 explains why this von Neumann-centric view is at odds with the historical record, and describes in detail the respective contributions made by Turing and von Neumann to the development of the concept during the key decade 1936–1946. Section 3 also discusses aspects of the work of the many others who contributed, in one way or another, to the development of this concept, including Eckert, Mauchly, Clippinger, Williams, and Kilburn.

Section 4 offers a fresh look at the stored-program concept itself. Six programming paradigms, that existed side by side during the decade 1936–1946, are distinguished: these are termed **P1**–**P6**. **P3**–**P6** form four 'onion skins' of the stored-program concept and are of special interest. Equipped with this logical analysis, and also with the historical contextualization provided in Sect. 3, Sects. 5 and 6 turn to a detailed examination of unpublished work by Zuse, from the period 1936–1941. Finally, Sect. 7 summarizes our conclusions concerning the multifaceted origins of the stored-program concept.

zib.de/item/axy7eq6AntFRwuJv; Zuse, K. 'Der Plankalkül', manuscript, no date, in the *Konrad Zuse Internet Archive*, http://zuse.zib.de/file/1rUAfKDkirW803gT/a3/1c/07/c3-af26-4522-a2e5-6cdd96f40568/0/original/aa32656396be2df124647a815ee85a61.pdf; Zuse, K. 'Der Plankalkül', typescript, no date, Deutsches Museum Archiv, NL 207/0235.

2 Zuse: A Brief Biography

'It was a foregone conclusion for me, even in childhood, that I was to become an engineer', Zuse said.[13] Born in Berlin on 22 June 1910, he grew up in East Prussia and then Silesia (now lying mostly within the borders of Poland).[14] His father was a civil servant in the German Post Office. Young Konrad initially studied mechanical engineering at the Technical University in Berlin-Charlottenburg, but switched to architecture and then again to civil engineering.[15] Graduating from the Technical University in 1935, with a diploma in civil engineering, he obtained a job as a structural engineer at *Henschel-Flugzeugwerke AG* (Henschel Aircraft Company) in Schonefeld, near Berlin.

A determined young man with a clear vision of his future, Zuse left Henschel after about a year, in order to pursue his ambition of building an automatic digital binary calculating machine.[16] As a student, Zuse had become painfully aware that engineers must perform what he called 'big and awful calculations'.[17] 'That is really not right for a man', he said.[18] 'It's beneath a man. That should be accomplished with machines.' He started to rough out designs for a calculating machine in 1934, while still a student, and with his departure from Henschel set up a workshop in the living room of his parents' Berlin apartment.[19] There Zuse began constructing his first calculator, in 1936.[20] His 'parents at first were not very delighted', Zuse said drily.[21] Nevertheless they and his sister helped finance the project, and Kurt Pannke, the proprietor of a business manufacturing analogue calculating machines, helped out as well with small amounts of money.[22] Some of Zuse's student friends chipped in, too, and they also contributed manpower—half a dozen or more pairs of hands assisted with the construction of Zuse's first machine.[23]

---

[13]Zuse interviewed by Merzbach.

[14]Zuse interviewed by Merzbach.

[15]Zuse, K. *Der Computer – Mein Lebenswerk* [The computer—my life's work] (Berlin: Springer, 4th edn, 2007), p. 13.

[16]Zuse interviewed by Merzbach; Zuse, 'Some Remarks on the History of Computing in Germany', p. 612.

[17]Zuse, 'Some Remarks on the History of Computing in Germany', p. 611.

[18]Zuse interviewed by Merzbach.

[19]Zuse interviewed by Merzbach; Zuse interviewed by Evans; Zuse, 'Some Remarks on the History of Computing in Germany', p. 612.

[20]Zuse, 'Some Remarks on the History of Computing in Germany', pp. 612–613.

[21]Zuse interviewed by Evans.

[22]Zuse interviewed by Merzbach; Zuse interviewed by Evans.

[23]Zuse interviewed by Merzbach.

Zuse's Z1 computer, in a Berlin apartment belonging to Zuse's parents. *Credit: ETH Zurich*

Later named Z1, his first calculator was completed in 1938, but never worked properly.[24] Z1 was purely mechanical.[25] Zuse said that its storage unit functioned successfully, and although the calculating unit could, Zuse recollected, multiply binary numbers and do floating-point arithmetic, it was prone to errors.[26] A significant problem was that the punched-tape program control was defective and Z1's various units never functioned together as a whole.[27]

Zuse believed initially that a mechanical calculator would be more compact than a relay-based machine.[28] Nevertheless, he set out detailed ideas concerning an electromechanical computer as early as 1936, as Sect. 6 describes. By 1938, the difficulties with Z1's calculating unit had convinced him of the need to follow an electromechanical path, and he built Z2, a transitional machine.[29] While the storage unit remained mechanical, the calculating unit was constructed from relays.[30] According to his son, Horst, Zuse used 800 telephone relays in the calculating unit.[31]

---

[24]Zuse interviewed by Merzbach. See also Rojas, R. 'Konrad Zuse's Legacy: The Architecture of the Z1 and Z3', *IEEE Annals of the History of Computing*, vol. 19 (1997), pp. 5–16.

[25]Zuse, 'Some Remarks on the History of Computing in Germany', pp. 613, 615.

[26]Zuse interviewed by Merzbach; Zuse interviewed by Evans.

[27]Zuse interviewed by Merzbach; Zuse interviewed by Evans.

[28]Zuse interviewed by Evans.

[29]Zuse interviewed by Merzbach; Zuse, 'Some Remarks on the History of Computing in Germany', p. 613.

[30]Zuse, 'Some Remarks on the History of Computing in Germany', p. 615.

[31]Zuse, H. 'Konrad Zuse Biographie', www.horst-zuse.homepage.t-pnline.de/kz-bio.html, p. 1.

Z2 was completed in 1939, the same year that Aiken and IBM produced the first detailed circuit drawings for the ASCC.[32] Z2 was binary, controlled by punched tape, and offered fixed-point arithmetic. Little more than an experiment in relay technology, the tiny computer had only 16 binary digits of storage.[33] It 'didn't work properly', Zuse said.[34] The problem was the relays. For economy's sake, he had bought used ones which he attempted to refurbish, but he set the contact pressure too low.[35]

When war came, in 1939, Zuse was drafted into the army.[36] But he saw no fighting, and in fact spent less than a year as a soldier.[37] Herbert Wagner, head of a department at Henschel that was developing flying bombs, urgently needed a statistician, and managed to arrange for Zuse to be released from the military.[38] Back in Berlin, Zuse was able to start work again on his calculating machine. At first this was only 'at night, Saturday afternoons and Sundays', he said, but then Henschel's aviation research and development group became interested.[39] Suddenly Zuse was given additional resources.

By 1941 he was able to set up a business of his own, while continuing to work part-time as a statistician. *K. Zuse Ingenieurbüro und Apparatebau* (K. Zuse Engineering and Machine-Making Firm) had a workshop in Berlin and ultimately a staff of about twenty.[40] The workshop had to be moved three or four times, as buildings succumbed to the bombing.[41] According to Zuse, his *Ingenieurbüro* was the only company in wartime Germany licensed to develop calculators.[42] Various armament factories financed his work, as well as the *Deutsche Versuchsanstalt für Luftfahrt* (German Institute for Aeronautical Research, DVL). According to a 2010 article in *Der Spiegel*, the DVL provided over 250,000 Reichsmarks for Zuse's calculator research (approximately 2 million US dollars in today's terms).[43] *Der Spiegel* wrote: 'The civil engineer was more deeply involved in the NS [National Socialist] arms industry than was believed hitherto. His calculating machines were

---

[32]Campbell, 'Aiken's First Machine', p. 34. Zuse wrote that Z2 was completed in 1939 (in 'Some Remarks on the History of Computing in Germany', p. 615); Rojas, however, gave 1940 as the completion date (Rojas, R. 'Zuse, Konrad', p. 3, zuse.zib.de/item/RuvnRJScXfvdt7BA).

[33]Zuse interviewed by Merzbach; Zuse interviewed by Evans.

[34]Zuse interviewed by Merzbach.

[35]Zuse interviewed by Merzbach; Zuse interviewed by Evans.

[36]Zuse interviewed by Merzbach.

[37]Zuse, *Der Computer – Mein Lebenswerk*, pp. 50, 57.

[38]Zuse interviewed by Merzbach; Zuse, 'Some Remarks on the History of Computing in Germany', p. 612.

[39]Zuse interviewed by Merzbach.

[40]Zuse interviewed by Merzbach.

[41]Zuse interviewed by Merzbach.

[42]Zuse, *Der Computer – Mein Lebenswerk*, p. 68.

[43]Schmundt, H. 'Rassenforschung am Rechner', *Der Spiegel*, Nr. 24 (2010) (14 June 2010), pp. 118–119.

considered important for the "final victory"'.[44] German historian Hartmut Petzold previously gave a lower figure, saying the DVL provided 50,000 Reichsmarks for Zuse's work.[45]

At Henschel, Zuse was involved with the Hs 293 rocket-propelled missile. He designed two special-purpose calculating machines, named S1 and S2, to assist with the manufacture of these weapons.[46] Air-launched and radio-controlled, the missiles were built on an assembly line (at a rate of one every ten minutes, Zuse estimated), and then during a final stage of production, a complicated system of sensors monitored the wings, while their geometry was fine-tuned.[47] Several hundred sensors were used to achieve the aerodynamic accuracy required for guiding the missile by radio. Initially, calculations based on the sensor data were done by hand, using a dozen Mercedes calculating machines and working day and night.[48] Each individual calculation 'took hours and hours', Zuse remembered.[49] He built S1 to automate these calculations. S1 was a relay-based binary calculator with a wired program, set by means of rotary switches.[50] He recollected completing the prototype, containing about 800 relays, in 1942.[51] Eventually there were three S1s, 'running day and night for several years', Zuse said.[52] Operators still had to enter the sensor data by hand, using a keyboard. The later S2 was designed to eliminate this data-entry stage, by connecting the sensors' outputs directly to the calculator, via a form of analog-to-digital converter.[53] Zuse explained that S2 was completed in 1944, but never became operational, because the factory (in Sudetenland) was dismantled just as the computer became ready.[54]

Zuse began building his fully electromechanical Z3 in 1940, again in his parents' living room, and completed it in 1941.[55] Z3's speed was on average one operation per second, Zuse recollected, and the memory unit had 64 storage cells.[56] The

---

[44]Schmundt, 'Rassenforschung am Rechner', p. 119.

[45]Petzold, H. *Moderne Rechenkünstler: Die Industrialisierung der Rechentechnik in Deutschland* (Munich: C. H. Beck, 1992), pp. 193–4, 201 ff.

[46]Zuse, *Der Computer – Mein Lebenswerk*, p. 54.

[47]Zuse interviewed by Merzbach; Zuse interviewed by Evans; Zuse, 'Some Remarks on the History of Computing in Germany', p. 619.

[48]Zuse interviewed by Merzbach; Zuse interviewed by Evans.

[49]Zuse interviewed by Merzbach.

[50]Zuse interviewed by Merzbach; Zuse, 'Some Remarks on the History of Computing in Germany', p. 615.

[51]Zuse interviewed by Evans.

[52]Zuse, 'Some Remarks on the History of Computing in Germany', p. 615; Zuse interviewed by Merzbach.

[53]Zuse interviewed by Merzbach.

[54]Zuse interviewed by Evans.

[55]Zuse, 'Some Remarks on the History of Computing in Germany', pp. 613, 615; Zuse interviewed by Merzbach; Zuse interviewed by Evans.

[56]Zuse interviewed by Evans.

time of large-scale electromechanical computing machines had come. By 1941 Turing's Bombes were turning Bletchley Park into a codebreaking factory, while at IBM progress was running slightly slower: Aiken's ASCC was partially working in 1942, and the computer solved its first practical problem on 1 January 1943.[57] Z3 contained some 2000 relays, 1400 in the storage unit and 600 in the calculating unit (which was capable of floating-point operations).[58] By comparison, Turing's Bombe contained 111 electromechanical 26-point rotary switches and approximately 150 relays; and the ASCC contained 3300 relays, as well as 2200 electromechanical 10-point rotary switches for storing decimal numbers.[59]

Zuse recollected that Z3 cost about 20,000 Reichsmarks to build, in an era when top-of-the-range calculating machines of the type used by engineers cost at most around 3000 Reichsmarks.[60] The DVL financed Z3 (and according to Zuse also provided about 10,000 Reichsmarks for his wing-geometry calculators).[61] Z3 was never used on a day-to-day basis, Zuse said, but did carry out a number of 'small test calculations': 'things that were interesting for aerodynamics, for airplane construction'.[62] In 1943, bombs destroyed Z3 as it stood in Zuse's workshop.[63] 'Z3 was total loss', he said.[64] All documentation was also destroyed in the bombing; only his 1941 patent application remained as a record of the machine.[65]

Z3, used only for 'program testing', was a step on the way to a larger and better computer, Z4, which Zuse had begun working on in 1942.[66] He returned to the idea of a mechanical store, but retained relays for the calculating unit.[67] Curiously, while building Z4 he heard from the German Secret Service about Aiken's ASCC; after the war, the two met when Aiken visited Zurich, and then again when Zuse visited New York in 1950 (at the invitation of Remington Rand) and made a detour to Boston to pay a call on Aiken.[68]

---

[57]Campbell, 'Aiken's First Machine', p. 55; Bashe, C. 'Constructing the IBM ASCC (Harvard Mark I)', in Cohen and Welch, *Makin' Numbers*, p. 74.

[58]Zuse, *Der Computer – Mein Lebenswerk*, p. 55; Zuse, 'Some Remarks on the History of Computing in Germany', p. 615.

[59]'Operations of the 6812th Signal Security Detachment, ETOUSA', 1 October 1944 (US National Archives and Records Administration, College Park, Maryland, RG 457, Entry 9032, Historic Cryptographic Collection, Pre–World War I Through World War II, Box 970, Nr. 2943), pp. 82–84; Campbell, R. V. D., Strong, P. 'Specifications of Aiken's Four Machines', in Cohen and Welch, *Makin' Numbers*, p. 258.

[60]Zuse interviewed by Merzbach.

[61]Zuse interviewed by Merzbach.

[62]Zuse interviewed by Merzbach.

[63]Zuse interviewed by Merzbach.

[64]Zuse interviewed by Merzbach.

[65]Zuse interviewed by Merzbach.

[66]Zuse interviewed by Evans.

[67]Zuse interviewed by Merzbach.

[68]Zuse interviewed by Merzbach.

Z4 was almost complete when, as Germany teetered on the brink of collapse, Berlin became too perilous for Zuse to remain.[69] With Z4 strapped to an army truck he fled to Gottingen. Zuse put the machine back together again in the DVL's Gottingen laboratory, and according to Fritz Bauer it was in Gottingen that Z4 was put into operation for the first time.[70]

As Soviet troops drew ever closer to Gottingen, the Air Ministry ordered Zuse to move his computer into the vast subterranean tunnels housing the factories for V1 and V2 flying bombs. He visited the tunnels and found 'horrible conditions', where 'twenty thousand people who had been inmates of concentration camps' worked as slaves.[71] 'Anywhere at all, only not here', he said.[72] Then he was offered a place on a special convoy taking rocket scientist Wernher von Braun to Bavaria.[73] After only six weeks in Gottingen, Zuse loaded his computer onto a truck again, and travelling with about a dozen of his staff was transported south through Munich and Ettal, heading for the relative safety of the mountains by 'a very adventurous route', he said.[74]

Zuse's first hiding place was the tiny mountain village of Hinterstein, lying at the head of a remote valley in the Bavarian Alps, and only 5 km from the alpine border with Austria. He concealed the computer in a barn covered with hay.[75] There was 'no possibility to continue the work with hardware', he said, and with time on his hands Zuse decided to turn to developing his *Plankalkül*. He had begun thinking about a logical programming calculus during the war, producing sheaves of rough handwritten notes, partly in shorthand and partly in programming notation of his own devising.[76] It was in Hinterstein, during 1945, that he 'put together' his 'theoretical ideas ... and made a real calculus of it'.[77] Zuse produced a manuscript, titled 'Der Plankalkül', of some 250 handwritten folios divided into five chapters. This was subsequently typed, on paper headed 'K. Zuse Ingenieurbüro und Apparatebau, Berlin'. Some extracts from his 1945 manuscript are translated in Sect. 7.

In 1946, once American troops occupied the mountainous area, it seemed safe to shift the computer to the larger village of Hopferau, some 25 km away, where Zuse

[69]Zuse interviewed by Merzbach.

[70]Zuse interviewed by Merzbach; Bauer, F. L. 'Between Zuse and Rutishauser—The Early Development of Digital Computing in Central Europe', in Metropolis, Howlett and Rota, *A History of Computing in the Twentieth Century*, p. 505.

[71]Zuse interviewed by Merzbach; Zuse interviewed by Evans.

[72]Zuse interviewed by Merzbach.

[73]Zuse interviewed by Merzbach; Zuse interviewed by Evans.

[74]Zuse interviewed by Merzbach; Zuse interviewed by Evans.

[75]Horst Zuse in conversation with Copeland.

[76]Zuse, K. Rough notes on the Plankalkül, no date, probably 1944 or the early months of 1945, Deutsches Museum Archiv, NL 207/0783 and NL 207/0887.

[77]Zuse interviewed by Evans.

remained until 1949.[78] An option with the German branch of Hollerith helped tide him over, he said—a couple in Hopferau had told an acquaintance at Hollerith of the 'strange inventor' in the village.[79] 1946 saw the start of his *Zuse-Ingenieurbüro* in Hopferau.[80] Another early contract was with Remington Rand Switzerland. He explained that the money enabled him to enlarge his company and to employ two or three men. His small factory produced a program-controlled relay calculator for Remington Rand. Named the M9, this was attached to punched card equipment. According to Zuse about thirty were delivered and Remington's clients used them in Switzerland, Germany and Italy.[81]

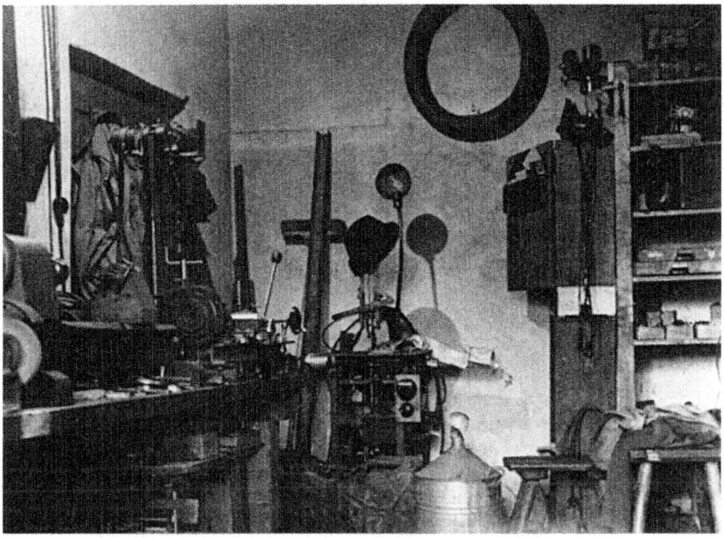

Zuse's computer workshop in the alpine village of Hopferau. *Credit: ETH Zurich*

It was while Zuse was in Hopferau that a 'gentleman from Zurich' visited him, starting the chain of events that led to Z4's delivery to ETH.[82] The visitor was Eduard Stiefel, founder of ETH's Institute for Applied Mathematics. Stiefel wanted a computer for his Institute and heard about Z4.[83] Both Stiefel and his assistant Heinz Rutishauser had recently visited the US and were familiar with Aiken's

[78]Zuse interviewed by Merzbach; Zuse, *Der Computer – Mein Lebenswerk*, p. 96.

[79]Zuse interviewed by Merzbach.

[80]Zuse, 'Konrad Zuse Biographie', p. 2.

[81]Zuse interviewed by Merzbach.

[82]Zuse interviewed by Merzbach.

[83]Bruderer, H. *Konrad Zuse und die Schweiz. Wer hat den Computer erfunden?* [Konrad Zuse and Switzerland. Who invented the computer?] (Munich: Oldenbourg, 2012), p. 5.

work.[84] They knew the worth of the large German electromechanical computer which the vagaries of war had delivered almost to their doorstep. ETH offered Zuse a rental agreement.

Shortly after Stiefel's visit, in 1949, Zuse moved to the small town of Neukirchen, about 50 km north of Dusseldorf, and there founded *Zuse Kommanditgesellschaft* (Zuse KG), increasing his staff to five.[85] Zuse KG would supply Europe with small, relatively cheap computers. Zuse's first task at Neukirchen was to restore and enlarge Z4 for ETH. He related that a second tape reader (or 'scanner') was attached, enabling numbers as well as programs to be fed in on punched tape, and circuitry was added for conditional branching (none of Z1–Z3 had been equipped with conditional branching).[86] He said the storage unit was enlarged from 16 cells to 64.[87] Rented by ETH from July 1950 until April 1955, Z4 was the first large-scale computer to go into regular operation in Continental Europe; and Stiefel's Institute for Applied Mathematics became a leading centre for scientific and industrial calculation. Despite assorted problems with the relays, Z4 was reliable enough to 'let it work through the night unattended', Zuse remembered.[88]

Now that Z4 had a home, Zuse moved on to Z5.[89] The German company Leitz, manufacturer of Leica cameras, needed a computer for optical calculations, and commissioned Z5. According to Petzold, the computer cost 200,000 Deutschmarks (about US$650,000 in today's terms) and was six times faster than Z4.[90] Next came Z11, a small relay-based wired-program computer that Zuse developed for the Géodésie company, again used mainly for optical calculations.[91] About 50 Z11s were built.[92] Applications included surveying and pension calculations.[93]

In 1957, Zuse moved his growing company to Bad Hersfeld, 50 km south of Kassel, and the following year embarked on Z22, his first vacuum tube computer.[94] Zuse had come round to tubes just as they were becoming outmoded for computer use—MIT's TX-0 transistorized computer first worked in 1956. Nevertheless,

---

[84]Zuse interviewed by Merzbach.

[85]Zuse interviewed by Merzbach.

[86]Zuse interviewed by Merzbach; Zuse, 'Some Remarks on the History of Computing in Germany', p. 616.

[87]Zuse interviewed by Merzbach.

[88]Zuse, 'Some Remarks on the History of Computing in Germany', p. 619. Urs Hochstrasser, one of the leading users of Z4 at ETH, gave an account of the problems associated with Z4's relays; see Bruderer, *Konrad Zuse und die Schweiz. Wer hat den Computer erfunden?*, pp. 19–27.

[89]Zuse interviewed by Merzbach.

[90]Petzold, *Moderne Rechenkünstler*, p. 216.

[91]Zuse interviewed by Merzbach.

[92]Zuse interviewed by Merzbach.

[93]Petzold, *Moderne Rechenkünstler*, pp. 216–217.

[94]Zuse, 'Konrad Zuse Biographie', p. 2.

Zuse's Z4 computer at ETH in Zurich. *Credit: ETH Zurich*

Zuse KG's Bad Hersfeld factory turned out more than 50 of the low-priced Z22 computers.[95] A transistorized version, Z23, went on the market in 1961.[96] Other electronic computers followed, the Z25 and Z64.[97] Oddly, Petzold says that with 'the step to electronic technology, Zuse KG also made the step to modifiable stored programs and thus to the von Neumann concept'.[98] As we explain, this concept is hardly von Neumann's, and in any case Zuse himself wrote of storing programs as early as 1936.

According to Horst Zuse, Zuse KG produced a total of 250 computers, with a value of more than 100 million Deutschmarks.[99] In 1964, however, Zuse and his wife relinquished ownership of the company.[100] By that time, despite Zuse KG's rapid growth, the company was overburdened by debt, and the Zuses put their shares on the market. German engineering company Brown Boveri purchased

[95]'Zuse Computers', Computer History Museum, www.computerhistory.org/revolution/early-computer-companies/5/108.

[96]Zuse, *Der Computer – Mein Lebenswerk*, p. 125; Bruderer, *Konrad Zuse und die Schweiz. Wer hat den Computer erfunden?*, p. 63.

[97]Zuse, *Der Computer – Mein Lebenswerk*, pp. 126, 131–132.

[98]Petzold, *Moderne Rechenkünstler*, p. 217.

[99]Zuse, 'Konrad Zuse Biographie', p. 2.

[100]Zuse, *Der Computer – Mein Lebenswerk*, p. 137.

Zuse KG, with Zuse staying on as a consultant. Another sale in 1967 saw Siemens AG owning Brown Boveri.[101] Following further sales, a distant successor of Zuse's former company still exists today on Bad Hersfeld's Konrad-Zuse-Strasse, ElectronicNetwork GmbH, a contract electronics manufacturer.

As Sect. 1 mentioned, Zuse applied for a number of patents on his early computer designs (his most important patent applications, discussed in detail in Sects. 5 and 6, were in 1936 and 1941). However, the German authorities never granted Zuse a patent. During the war, he said, 'nothing much' happened regarding his patent, and then in the postwar years 'nothing whatever happened': his application 'lay around, gathering dust in a drawer of the patent office for years'.[102] When things finally did get moving, his efforts to patent his inventions came to the attention of IBM. Zuse explained that IBM worked through another company, Triumph Corporation, 'who lodged the protest'.[103] A 'serious legal battle' followed, Zuse said, and things dragged on until 1967, when the German federal patent court finally and irrevocably declined a patent.[104] The problem, according to the judge, was the patent's lack of *Erfindungshöhe*, literally 'invention height'. As Zuse explained matters, the judge stated that 'the requisite invention value has not been attained'.[105]

Konrad Zuse died in Huhnfeld on 18 December 1995.

3 Turing, von Neumann, and the Universal Electronic Computer

This section outlines the early history of the stored-program concept in the UK and the US, and compares and contrasts Turing's and John von Neumann's contributions to the development of the concept.[106] Although von Neumann is routinely said to be the originator of the stored-program concept, we find no evidence in favour of this common view. Turing described fundamental aspects of the concept in his 1936 article 'On Computable Numbers', which von Neumann had read before the war. When von Neumann arrived at the University of Pennsylvania's Moore

[101] Zuse, H. 'Historical Zuse-Computer Z23', 1999, www.computerhistory.org/projects/zuse_z23/index.shtml.

[102] Zuse interviewed by Merzbach.

[103] Zuse interviewed by Merzbach.

[104] Zuse interviewed by Merzbach; Zuse, *Der Computer – Mein Lebenswerk*, pp. 97–100. See also Petzold, H. *Die Ermittlung des 'Standes der Technik' und der 'Erfindungshöhe' beim Patentverfahren Z391. Dokumentation nach den Zuse-Papieren* [Establishing the 'state of the technological art' and 'inventiveness' in patent application Z391. Documentation from the Zuse papers] (Bonn: Selbstverlag, 1981).

[105] Zuse interviewed by Merzbach.

[106] Von Neumann was an alumnus of ETH Zurich, graduating as a chemical engineer in October 1926.

School of Electrical Engineering, in 1944, he recognized the potential of applying Turing's concept to practical computing. His own principal original contribution was devising practical coding schemes for stored programming. We also discuss the view, surprisingly popular, that what Turing termed a 'machine' in 1936 was a mathematical abstraction—essentially a set of quintuples—and that he made no connection between his abstract 'machine' and real computers.

At Cambridge, during the first three months of 1935, the young Alan Turing attended a course of advanced lectures on the Foundations of Mathematics, given by Max Newman, a Fellow of St John's College.[107] It was in these lectures that Turing heard of David Hilbert's *Entscheidungsproblem*, or *decision* problem. Yorick Smythies attended the lectures in 1934 and took detailed notes. Newman covered the Hilbert programme, propositional and predicate calculus, cardinals, theory of types and the axiom of reducibility, Peano arithmetic, Hilbert on proving consistency, and Gödel's first and second incompleteness theorems; and he mentioned that the *Entscheidungsproblem* had been settled only in the special case of monadic expressions.[108]

As stated by Turing, Hilbert's *Entscheidungsproblem* for the functional calculus (first-order predicate calculus) is this: *Is there a general (mechanical) process for determining whether a given formula A of the functional calculus **K** is provable.*[109] As everyone knows, Turing took on the *Entscheidungsproblem* and showed it to be unsolvable, along the way inventing the universal Turing machine. In a single, brilliant paper Turing ushered in both the modern computer and the mathematical study of unsolvability.

Newman's own contribution was not limited to bringing the *Entscheidungsproblem* to Turing's notice. In his lectures, Newman defined a *constructive* process as one that a *machine* can carry out. He explained in an interview:

> And this of course led [Turing] to the next challenge, what sort of machine, and this inspired him to try and say what one would mean by a perfectly general computing machine.[110]

Turing's 1936 paper 'On Computable Numbers' is the birthplace of the fundamental logical principles of the modern computer, and in particular the two closely

---

[107]Cambridge University Reporter, 18 April 1935, p. 826.

[108]Notes taken by Yorick Smythies during Newman's Foundations of Mathematics lectures in 1934 (St John's College Library, Cambridge).

[109]Turing, A. M. 'On Computable Numbers, with an Application to the Entscheidungsproblem', in Copeland, B. J. (ed.) *The Essential Turing: Seminal Writings in Computing, Logic, Philosophy, Artificial Intelligence, and Artificial Life* (Oxford: Oxford University Press, 2004), p. 84. 'On Computable Numbers' was published in 1936 but in the secondary literature the date of publication is often given as 1937, e.g. by Andrew Hodges in his biography *Alan Turing: The Enigma* (London: Vintage, 1992). The circumstances of publication of 'On Computable Numbers' are described on p. 5 of *The Essential Turing*.

[110]Newman interviewed by Christopher Evans ('The Pioneers of Computing: An Oral History of Computing', London: Science Museum); quoted in *The Essential Turing*, p. 206 (transcription by Copeland).

related logical ideas on which modern computing is based. We call these the 'twin pillars'. They are the concepts of (1): a *universal* computing machine, that operates by means of (2): a program of instructions *stored in the computer's memory* in the same form as data.[111] If different programs are placed on the memory-tape of the universal Turing machine, the machine will carry out different computations. Turing proved that the universal machine could obey any and every 'table of instructions'—any and every program expressed in the programming code introduced in his 1936 paper. His machine was universal in the sense that it could carry out *every* mechanical (or 'effective') procedure, if appropriately programmed.

The stored-program universal Turing machine led ultimately to today's archetypical electronic digital computer: the single slab of hardware, of fixed structure, that makes use of internally stored instructions in order to become a word-processor, or desk calculator, or chess opponent, or photo editor—or any other machine that we have the skill to create in the form of a program. Since these electronic machines necessarily have limited memories (unlike the universal Turing machine, with its indefinitely extendible tape), each is what Turing called 'a universal machine with a given storage capacity'.[112]

Turing's universal machine has changed the world. Yet nowadays, when nearly everyone owns a physical realization of one, his idea of a universal computer is apt to seem as obvious as the wheel and the arch. Nevertheless, in 1936, when engineers thought in terms of building different machines for different purposes, Turing's vision of a universal machine was revolutionary.

Zuse also briefly outlined a computing machine that would make use of programs stored in memory, in a few handwritten pages and a sequence of diagrams contained in a 1938 notebook, two years after Turing gave his extensive and detailed treatment of the idea. There is, however, no evidence that Zuse also formulated the concept of

[111]The first historians to insist that the stored-program concept originated in Turing's 1936 paper were (so far as is known) Brian Carpenter and Bob Doran, in a classic article that is one of New Zealand's earliest and greatest contributions to the history of computing: Carpenter, B. E., Doran, R. W. 'The Other Turing Machine', *The Computer Journal*, vol. 20 (1977), pp. 269–279. They said: 'It is reasonable to view the universal Turing machine as being programmed by the description of the machine it simulates; since this description is written on the memory tape of the universal machine, the latter is an abstract stored program computer' (p. 270). In the United States, Martin Davis has been advocating powerfully for the same claim since 1987; see Davis, M. D. 'Mathematical Logic and the Origin of Modern Computers', in Herken, R. (ed.) *The Universal Turing Machine: A Half-Century Survey* (Oxford: Oxford University Press, 1988); and Davis, M. D. *Engines of Logic: Mathematicians and the Origin of the Computer* (New York: Norton, 2000). However, the proposition that the stored-program concept originated in 'On Computable Numbers' is far from being a historians' reconstruction: as the present chapter explains, this was common knowledge among Turing's post-war colleagues at the National Physical Laboratory, and it was obvious to members of Max Newman's wartime group at Bletchley Park that digital electronics could be used to implement practical forms of Turing's universal machine of 1936.

[112]Turing, A. M. 'Intelligent Machinery', in *The Essential Turing*, p. 422.

a universal machine, as distinct from a general-purpose computer, as we explain in Sect. 5.

As early as the 1830s Charles Babbage—one of the first to appreciate the vast potential of computing machinery—had also envisioned a general-purpose computer, his Analytical Engine. Babbage said that the 'conditions which enable a finite machine to make calculations of unlimited extent are fulfilled in the Analytical Engine'.[113] Zuse first learned of Babbage's work (probably in 1938) from the US Patent Office which, on the basis of a comparison with Babbage's plans, declined Zuse's application for a patent.[114] In 1950, Turing stated that Babbage's Analytical Engine was universal, and the same has recently been proved of a modified form of Zuse's Z3 by Raul Rojas (see Sect. 5).[115] Such judgments are possible only from the vantage point of 'On Computable Numbers'—Babbage himself did not have the concept of a universal machine, a machine that is able to carry out all effective procedures.[116]

Nor did Babbage have the stored-program concept. As Sect. 1 mentioned, the Analytical Engine's program resided on punched cards and, as each card entered the Engine, the instruction marked on that card would be obeyed. True, the cards, strung together with ribbon, bore some resemblance to the universal Turing machine's tape; but in the Analytical Engine there was a distinction of kind between program and data, and this is dispensed with in the universal Turing machine. As Turing's friend and colleague the mathematician Robin Gandy put the point, in the universal Turing machine 'the mechanisms used in reading a program are of the same kind as those used in executing it'.[117]

Newman reported that Turing was interested 'right from the start' in building a universal computing machine.[118] However, Turing knew of no suitable technology.

[113] Babbage, C. *Passages from the Life of a Philosopher*, vol. 11 of Campbell-Kelly, M. (ed.) *The Works of Charles Babbage* (London: William Pickering, 1989), p. 97.

[114] Zuse interviewed by Merzbach.

[115] Rojas, R. 'How to Make Zuse's Z3 a Universal Computer', *IEEE Annals of the History of Computing*, vol. 20 (1998), pp. 51–54.

[116] Historian Thomas Haigh, in his impassioned outburst 'Actually, Turing Did Not Invent the Computer' (*Communications of the ACM*, vol. 57 (2014), pp. 36–41), confuses the logical distinction between, on the one hand, the universal machine concept and, on the other, the concept 'of a single machine that could do different jobs when fed different instructions'. Talking about this second concept, Haigh objects that it was not Turing but Babbage who 'had that idea long before' (pp. 40–41). Babbage did indeed have that idea; the point, however, is that although Babbage had the concept of a general-purpose computing machine, the universal machine concept originated with Turing. (All this is explained in Copeland's 'Turing and Babbage' in *The Essential Turing*, pp. 27–30.)

[117] Gandy, R. 'The Confluence of Ideas in 1936', in Herken, R. (ed.) *The Universal Turing Machine: A Half-Century Survey* (Oxford: Oxford University Press, 1998), p. 90. Emphasis added.

[118] Newman interviewed by Evans; Newman, M. H. A. 'Dr. A. M. Turing', *The Times*, 16 June 1954, p. 10.

Relays, he thought, would not be adequate.[119] So, for the next few years, Turing's revolutionary ideas existed only on paper. A crucial moment came in 1944, when he set eyes on Flowers' racks of high-speed electronic code-cracking equipment, at Bletchley Park. Colossus was neither stored-program nor general-purpose, but it was clear to Turing (and to Newman) that the technology Flowers was pioneering, large-scale digital electronics, was the way to build a miraculously fast universal computer, the task to which Turing turned in 1945. Meanwhile, Zuse had pressed ahead with relays and had built a general-purpose computer, but neither knew of the work of the other at that time.

Did Zuse and Turing meet postwar? Probably not. Zuse said (in 1992) that he had no knowledge of Turing's 'On Computable Numbers' until 1948.[120] This recollection of Zuse's, if correct, makes it unlikely that he and Turing met the previous year at a colloquium in Gottingen, as German computer pioneer Heinz Billing reported in his memoirs.[121] A more likely connection is Turing's colleague from the National Physical Laboratory (NPL) Donald Davies, who interrogated Zuse in England.[122] Zuse was invited to London in 1948 and placed in a large house in Hampstead, where a number of British computer experts arrived to question him.[123] Zuse remembered it as a 'very nice trip'.[124] Quite likely Davies—who, along with Turing's other colleagues in NPL's ACE section, saw 'On Computable Numbers' as containing the 'key idea on which every stored-program machine was based'—would have mentioned Turing's paper to Zuse.[125] Davies recollected that the interview did not go particularly well: Zuse eventually 'got pretty cross', and things 'degenerated into a glowering match'. Zuse was 'quite convinced', Davies said, that he could make a smallish relay machine 'which would be the equal of any of the electronic calculators we were developing'.

Although Turing completed his design for an electronic stored-program computer in 1945, another four years elapsed before the first universal Turing machine in electronic hardware ran the first stored program, on Monday 21 June 1948. It was the first day of the modern computer age. Based on Turing's ideas, and almost

---

[119] Robin Gandy interviewed by Copeland, October 1995.

[120] Zuse in conversation with Brian Carpenter at CERN on 17 June 1992; Copeland is grateful to Carpenter for sending him some brief notes on the conversation that Carpenter made at the time. See also Carpenter, B. E. *Network Geeks: How They Built the Internet* (London: Springer, 2013), p. 22.

[121] Jänike, J., Genser, F. (eds) *Ein Leben zwischen Forschung und Praxis—Heinz Billing* [A Life Between Research and Practice—Heinz Billing] (Dusseldorf: Selbstverlag Friedrich Genser, 1997), p. 84; Bruderer, *Konrad Zuse und die Schweiz. Wer hat den Computer erfunden?*, pp. 64–66.

[122] Davies interviewed by Christopher Evans in 1975 ('The Pioneers of Computing: An Oral History of Computing', London: Science Museum; © Board of Trustees of the Science Museum).

[123] Zuse, *Der Computer – Mein Lebenswerk*, p. 101; Zuse interviewed by Evans; Davies interviewed by Evans.

[124] Zuse interviewed by Evans.

[125] Davies interviewed by Evans.

big enough to fill a room, this distant ancestor of our mainframes, laptops, tablets and phones was called 'Baby'.[126] Baby was built by radar engineers F. C. Williams and Tom Kilburn, in Newman's Computing Machine Laboratory at the University of Manchester, in the north of England.[127]

However, historians of the computer have often found Turing's contributions hard to place, and many histories of computing written during the six decades since his death sadly do not so much as mention him. Even today there is still no real consensus on Turing's place in computing history. In 2013, an opinion piece by the editor of the Association for Computing Machinery's flagship journal objected to the claim that Turing invented the stored-program concept. The article's author, Moshe Vardi, dismissed the claim as 'simply ahistorical'.[128] Vardi emphasized that it was not Turing but the Hungarian-American mathematician John von Neumann who, in 1945, 'offered the first explicit exposition of the stored-program computer'. This is true, but the point does not support Vardi's charge of historical inaccuracy. Although von Neumann did write the first paper explaining how to convert Turing's ideas into electronic form, the fundamental conception of the stored-program universal computer was nevertheless Turing's.

Von Neumann was close to the centre of the American effort to build an electronic stored-program universal computer. He had read Turing's 'On Computable Numbers' before the war,[129] and when he became acquainted with the U.S. Army's ENIAC project in 1944, he discovered that the stored-program concept could be applied to electronic computation.[130] ENIAC was designed by Presper Eckert and John Mauchly at the Moore School of Electrical Engineering (part of the University of Pennsylvania), in order to calculate the complicated tables needed by gunners to aim artillery, and the computer first ran in 1945. Like Colossus before it, ENIAC was programmed by means of re-routing cables and setting switches, a process that could take as long as three weeks.[131] Viewed from the modern stored-program world, this conception of programming seems unbearably primitive. In the taxonomy of programming paradigms developed in Sect. 4, this method of programming is **P1**, the most rudimentary level in the taxonomy.

---

[126]Tootill, G. C. 'Digital Computer—Notes on Design & Operation', 1948–9 (National Archive for the History of Computing, University of Manchester).

[127]For more about Baby, see Copeland, B. J. 'The Manchester Computer: A Revised History', *IEEE Annals of the History of Computing*, vol. 33 (2011), pp. 4–37; Copeland, B. J. *Turing, Pioneer of the Information Age* (Oxford: Oxford University Press, 2012, 2015), ch. 9.

[128]Vardi, M. Y. 'Who Begat Computing?', *Communications of the ACM*, vol. 56 (Jan. 2013), p. 5.

[129]Stanislaw Ulam interviewed by Christopher Evans in 1976 ('The Pioneers of Computing: an Oral History of Computing', Science Museum: London).

[130]Goldstine, H. *The Computer from Pascal to von Neumann* (Princeton: Princeton University Press, 1972), p. 182.

[131]Campbell-Kelly, M. 'The ACE and the Shaping of British Computing', in Copeland, B. J. et al. *Alan Turing's Electronic Brain: The Struggle to Build the ACE, the World's Fastest Computer* (Oxford: Oxford University Press, 2012; a revised and retitled paperback edition of the 2005 hardback *Alan Turing's Automatic Computing Engine*), p. 151.

Conscious of the need for a better method of programming, the brilliant engineer Eckert had the idea of storing instructions in the form of numbers as early as 1944, inventing a high-speed recirculating memory.[132] This was based on apparatus he had previously used for echo cancellation in radar, the mercury delay line. Instructions and data could be stored uniformly in the mercury-filled tube, in the form of pulses—binary digits—that were 'remembered' for as long as was necessary. This provided the means to engineer the stored-program concept, and mercury delay lines were widely employed as the memory medium of early computers—although in fact the first functioning electronic stored-program computer used not delay lines but an alternative form of memory, the Williams tube. Based on the cathode ray tube, the Williams tube was invented by Williams and further developed by Kilburn.

Along with Zuse, Eckert has a strong claim to be regarded as a co-originator of the stored-program paradigm that in Sect. 4 is denoted '**P3**'. Eckert said that the stored-program concept was his 'best computer idea'—although, of course, he arrived at the idea approximately eight years after the publication of Turing's 'On Computable Numbers'.[133] Endorsing Eckert's claim, Mauchly commented that they were discussing 'storing programs in the same storage used for computer data' several months before von Neumann first visited their ENIAC group.[134] Art Burks, a leading member of the ENIAC group and later one of von Neumann's principal collaborators at the Princeton computer project, also explained that—long before von Neumann first visited them—Eckert and Mauchly were 'saying that they would build a mercury memory large enough to store the program for a problem as well as the arithmetic data'.[135]

Maurice Wilkes, who visited the Moore School group in 1946 and who went on to build the Cambridge EDSAC delay-line computer (see the timeline in Fig. 3), gave this first hand account of the roles of Eckert, Mauchly and von Neumann:

> Eckert and Mauchly appreciated that the main problem was one of storage, and they proposed ... ultrasonic delay lines. Instructions and numbers would be mixed in the same memory. ... Von Neumann ... appreciated at once ... the potentialities implicit in the stored program principle. That von Neumann should bring his great prestige and influence to bear was important, since the new ideas were too revolutionary for some, and powerful voices were being raised to say that ... to mix instructions and numbers in the same memory was going against nature.[136]

Burks presented a similar picture:

> The second revolution [at the Moore School] was the stored-program computer. ... There were two main steps. Pres [Eckert] and John [Mauchly] invented the circulating mercury

[132]Eckert, J. P. 'The ENIAC', in Metropolis, Howlett and Rota, *A History of Computing in the Twentieth Century*, p. 531.

[133]Eckert, 'The ENIAC', p. 531.

[134]Mauchly, J., commenting in Eckert, 'The ENIAC', pp. 531–532.

[135]Letter from Burks to Copeland, 16 August 2003.

[136]Wilkes, M. V. *1967 ACM Turing Lecture*: 'Computers Then and Now', *Journal of the Association for Computing Machinery*, vol. 15 (1968), pp. 1–7 (p. 2).

> delay line store, with enough capacity to store program information as well as data. Von Neumann created the first modern order code and worked out the logical design of an electronic computer to execute it.[137]

Von Neumann, then, did not originate the stored-program concept, but contributed significantly to its development, both by championing it in the face of conservative criticism, and, even more importantly, by designing an appropriate instruction code for stored programming. As Tom Kilburn said, 'You can't start building until you have got an instruction code'.[138]

During the winter of 1944 and spring of 1945, von Neumann, Eckert and Mauchly held a series of weekly meetings, working out the details of how to design a stored-program electronic computer.[139] Their proposed computer was called the EDVAC. In effect they designed a universal Turing machine in hardware, with instructions stored in the form of numbers, and common processes reading the data and reading and executing the instructions. While there are no diary entries to prove the point beyond any shadow of doubt, nor statements in surviving letters written by von Neumann at this precise time, his appreciation of the great potentialities inherent in the stored-program concept could hardly fail to have been influenced by his knowledge of Turing's 'On Computable Numbers', nor by his intimate knowledge of Kurt Gödel's 1931 demonstration that logical and arithmetical sentences can be expressed as numbers.[140] Von Neumann went on to inform electronic engineers at large about the stored-program concept.

In his 1945 document titled 'First Draft of a Report on the EDVAC', von Neumann set out, in rather general terms, the design of an electronic stored-program computer.[141] However, shortly after this appeared in mid 1945, his collaboration with Eckert and Mauchly came to an abrupt end, with the result that the ill-fated EDVAC was not completed until 1952.[142] Trouble arose because von Neumann's colleague Herman Goldstine had circulated a draft of the report before Eckert's and

---

[137]Burks, A. W. 'From ENIAC to the Stored-Program Computer: Two Revolutions in Computers', in Metropolis, Howlett, and Rota, *A History of Computing in the Twentieth Century*, p. 312.

[138]Kilburn interviewed by Copeland, July 1997.

[139]Von Neumann, J., Deposition before a public notary, New Jersey, 8 May 1947; Warren, S. R. 'Notes on the Preparation of "First Draft of a Report on the EDVAC" by John von Neumann', 2 April 1947. Copeland is grateful to Harry Huskey for supplying him with copies of these documents.

[140]Gödel, K, 'Uber formal unentscheidbare Sätze der Principia Mathematica und verwandter Systeme I.' [On formally undecidable propositions of Principia Mathematica and related systems I], *Monatshefte für Mathematik und Physik*, vol. 38 (1931), pp. 173–198.

[141]Von Neumann, J. 'First Draft of a Report on the EDVAC', Moore School of Electrical Engineering, University of Pennsylvania, 1945; reprinted in full in Stern, N. *From ENIAC to UNIVAC: An Appraisal of the Eckert-Mauchly Computers* (Bedford, Mass.: Digital Press, 1981).

[142]Huskey, H. D. 'The Development of Automatic Computing', in *Proceedings of the First USA-JAPAN Computer Conference*, Tokyo, 1972, p. 702.

Mauchly's names were added to the title page.[143] Bearing von Neumann's name alone, the report was soon widely read. Eckert and Mauchly were furious but von Neumann was unrepentant.

'My personal opinion', von Neumann said defiantly in 1947, 'was at all times, and is now, that this [the distribution of the report] was perfectly proper and in the best interests of the United States'.[144] Widespread dissemination of the report had, he said, furthered 'the development of the art of building high speed computers'. Perhaps he was hinting that Eckert and Mauchly would have opposed widespread distribution of the report. It would be perfectly understandable if they had, since the report's entering the public domain effectively prevented them from patenting their ideas. Eckert later wrote bitterly of 'von Neumann's way of taking credit for the work of others'.[145] Jean Jennings, one of ENIAC's programmers and a member of the ENIAC group from early 1945, noted that von Neumann 'ever afterward accepted credit—falsely—for the work of the Moore School group. ... [He] never made an effort to dispel the general acclaim in the years that followed'.[146]

After a dispute with the University of Pennsylvania about intellectual property rights, Eckert and Mauchly formed their own Electronic Control Company, and began work on their EDVAC-like BINAC. Meanwhile, von Neumann drew together a group of engineers at the Institute for Advanced Study in Princeton. He primed them by giving them Turing's 'On Computable Numbers' to read.[147] Julian Bigelow, von Neumann's chief engineer, was well aware of the influence that 'On Computable Numbers' had had on von Neumann. The reason that von Neumann was the 'person who really ... pushed the whole field ahead', Bigelow explained, was because 'he understood a good deal of the mathematical logic which was implied by the [stored program] idea, due to the work of A. M. Turing ... in 1936'.[148] 'Turing's machine does not sound much like a modern computer today, but nevertheless it was', Bigelow said—'It was the germinal idea'. The physical embodiment of Turing's universal computing machine that von Neumann's engineers built at Princeton began working in 1951. Known simply as the 'Princeton

[143]See e.g. Stern, N. 'John von Neumann's Influence on Electronic Digital Computing, 1944–1946', *Annals of the History of Computing*, vol. 2 (1980), pp. 349–362.

[144]Von Neumann, Deposition, 8 May 1947.

[145]Eckert, 'The ENIAC', p. 534.

[146]Jennings Bartik, J. *Pioneer Programmer: Jean Jennings Bartik and the Computer that Changed the World* (Kirksville, Missouri: Truman State University Press: 2013), pp. 16, 18.

[147]Letter from Julian Bigelow to Copeland, 12 April 2002; see also Aspray, W. *John von Neumann and the Origins of Modern Computing* (Cambridge, Mass.: MIT Press, 1990), p. 178.

[148]Bigelow in a tape-recorded interview made in 1971 by the Smithsonian Institution and released in 2002; Copeland is grateful to Bigelow for previously sending him a transcript of excerpts from the interview.

computer', it was not the first of the new stored-program electronic computers, but it was the most influential.[149]

Although Turing is not mentioned explicitly in von Neumann's papers developing the design for the Princeton computer, von Neumann's collaborator Burks told Copeland that his, von Neumann's, and Goldstine's key 1946 design paper did contain a reference to Turing's 1936 work.[150] Von Neumann and his co-authors emphasized that 'formal-logical' work—by which they meant in particular Turing's 1936 investigation—had shown 'in abstracto' that stored programs can 'control and cause the execution' of any sequence (no matter how complex) of mechanical operations that is 'conceivable by the problem planner'.[151]

Meanwhile, in 1945, Turing joined London's National Physical Laboratory, to design an electronic universal stored-program computer. John Womersley, head of NPL's newly formed Mathematics Division, was responsible for recruiting him. Womersley had read 'On Computable Numbers' shortly after it was published, and at the time had considered building a relay-based version of Turing's universal computing machine. As early as 1944 Womersley was advocating the potential of electronic computing.[152] He named NPL's projected electronic computer the Automatic Computing Engine, or ACE—a deliberate echo of Babbage.

Turing studied 'First Draft of a Report on the EDVAC', but favoured a radically different type of design. He sacrificed everything to speed, launching a 1940s version of what is today called RISC (Reduced Instruction Set Computing).[153] In order to maximise the speed of the machine, Turing opted for a decentralised architecture, whereas von Neumann described a centralised design that foreshadowed the modern central processing unit (cpu).[154] Turing associated different arithmetical and logical functions with different delay lines in the ACE's Eckert-type mercury memory, rather than following von Neumann's model of a single central unit in

[149]The Princeton computer is described in Bigelow, J. 'Computer Development at the Institute for Advanced Study', in Metropolis, Howlett, Rota, *A History of Computing in the Twentieth Century*.

[150]Letter from Arthur Burks to Copeland, 22 April 1998.

[151]Burks, A. W., Goldstine, H. H., von Neumann, J. 'Preliminary Discussion of the Logical Design of an Electronic Computing Instrument', Institute for Advanced Study, 28 June 1946, in vol. 5 of Taub, A. H. ed. *Collected Works of John von Neumann* (Oxford: Pergamon Press, 1961), section 3.1 (p. 37).

[152]See Copeland, B. J. 'The Origins and Development of the ACE Project', in Copeland et al., *Alan Turing's Electronic Brain*.

[153]Doran, R. W. 'Computer Architecture and the ACE Computers', in Copeland et al., *Alan Turing's Electronic Brain*.

[154]The terms 'decentralised' and its opposite 'centralised' are due to Jack Good, who used them in a letter to Newman about computer architecture on 8 August 1948; the letter is in Good, I. J. 'Early Notes on Electronic Computers' (unpublished, compiled in 1972 and 1976; a copy is in the National Archive for the History of Computing, University of Manchester, MUC/Series 2/a4), pp. 63–4.

which all the arithmetical and logical operations take place.[155] Turing was (as his colleague James Wilkinson observed[156]) 'obsessed' with making the computations run as fast as possible, and once a pilot version of the ACE was operational, it could multiply at roughly 20 times the speed of its closest competitor.[157]

The pilot model of Turing's Automatic Computing Engine, in the Mathematics Division of London's National Physical Laboratory. *Credit: National Physical Laboratory © Crown copyright*

Turing described his design in a report titled 'Proposed Electronic Calculator', completed by the end of 1945.[158] The proposals in the report were in fact much more concrete than those contained in von Neumann's rather abstract treatment in the 'First Draft'. The 'First Draft' hardly mentioned electronics, and Harry Huskey, the engineer whose job it was to draw up the first hardware designs for the EDVAC, said he found the 'First Draft' to be of 'no help'.[159] Turing, on the other hand,

[155]For additional detail concerning the differences between Turing's decentralized architecture and the centralized architecture favoured by von Neumann, see Copeland et al., *Alan Turing's Electronic Brain*; and Copeland, 'The Manchester Computer: A Revised History'.

[156]Wilkinson interviewed by Christopher Evans in 1976 ('The Pioneers of Computing: An Oral History of Computing', London: Science Museum).

[157]See the table by Martin Campbell-Kelly on p. 161 of Copeland et al., *Alan Turing's Electronic Brain*.

[158]Turing, A. M. 'Proposed Electronic Calculator', ch. 20 of Copeland et al., *Alan Turing's Electronic Brain*.

[159]Letter from Huskey to Copeland, 4 February 2002.

gave detailed specifications of the various hardware units, and even included sample programs in machine code.

Turing was content to borrow some of the elementary design ideas in von Neumann's report (and also the notation, due originally to McCulloch and Pitts, that von Neumann used to represent logic gates—a notation that Turing considerably extended in 'Proposed Electronic Calculator'). One example of a borrowing is Turing's diagram of an adder, essentially the same as von Neumann's diagram.[160] This borrowing of relatively pedestrian details is probably what Turing was referring to when he told a newspaper reporter in 1946 that he gave 'credit for the donkey work on the A.C.E. to Americans'.[161] Yet, the similarities between Turing's design and the von Neumann-Eckert-Mauchly proposals are relatively minor in comparison to the striking differences.

In their 1945 documents 'Proposed Electronic Calculator' and 'First Draft of a Report on the EDVAC', Turing and von Neumann both considerably fleshed out the stored-program concept, turning it from the bare-bones logical idea of Turing's 1936 paper into a fully articulated, electronically implementable design concept. The 1945 stored-program concept included:

- dividing stored information into 'words' (the term is used by Eckert and Mauchly in a September 1945 progress report on the EDVAC[162])
- using binary numbers as addresses of sources and destinations in memory
- building arbitrarily complex stored programs from a small stock of primitive expressions (as in 'On Computable Numbers').

Each document set out the basis for a very different practical version of the universal Turing machine (and the von Neumann design was indebted to extensive input from Eckert and Mauchly). Each document also replaced Turing's pioneering programming code of 1936 with a form of code more appropriate for high-speed computing. Again, each presented a very different species of code, von Neumann favouring instructions composed of operation codes followed by addresses, while Turing did not use operation codes: the operations to be performed were implied by the source and destination addresses.

Turing pursued the implications of the stored-program idea much further than von Neumann did at that time. As has often been remarked, in the 'First Draft' von Neumann blocked the wholesale modification of instructions by prefixing them

[160]Compare Fig. 10 of Turing's 'Proposed Electronic Calculator' (on p. 431 of Copeland et al., *Alan Turing's Electronic Brain*) with Fig. 3 of von Neumann's report (on p. 198 of Stern, *From ENIAC to UNIVAC*).

[161]*Evening News*, 23 December 1946. The cutting is among a number kept by Sara Turing and now in the Modern Archive Centre, King's College, Cambridge (catalogue reference K 5).

[162]Eckert, J. P., Mauchly, J. W. 'Automatic High Speed Computing: A Progress Report on the EDVAC', Moore School of Electrical Engineering, Sept. 1945. http://archive.computerhistory.org/resources/text/Knuth_Don_X4100/PDF_index/k-8-pdf/k-8-u2736-Report-EDVAC.pdf. Copeland is grateful to Bob Doran for pointing out this early occurrence of the term 'word' (in correspondence).

with a special tag. Only the address bits could be modified. Carpenter and Doran pointed out in their classic 1977 paper that, because von Neumann 'gave each word a nonoverrideable tag, he could not manipulate instructions', and they emphasized that it was Turing, and not von Neumann, who introduced 'what we now regard as one of the fundamental characteristics of the von Neumann machine'.[163]

The manipulation of instructions as if they were numbers was fundamental to the computer design that Turing put forward in 'Proposed Electronic Calculator'. He described program storage in editable memory as giving 'the machine the possibility of constructing its own orders'.[164] His treatment of conditional branching involved performing arithmetical operations on instructions considered as numbers (e.g. multiplying an instruction by a given digit).[165] Carpenter and Doran emphasized, 'Von Neumann does not take this step' (the step of manipulating instructions as if they were numbers) in the 'First Draft'.[166] The idea that instructions and data are common coin was taken for granted by ACE's programmers at the NPL. Sometimes instructions were even used as numerical constants, if an instruction considered as a number happened to equate to the value required.[167]

Furthermore, Turing recognized in 'Proposed Electronic Calculator' that a program could manipulate other programs.[168] As Carpenter and Doran again say, 'The notion of a program that manipulates another program was truly spectacular in 1945'.[169] Zuse had similar ideas in 1945, envisaging what he called a '*Planfertigungsgerät*' [plan producing machine], a 'special computer to make the program for a numerical sequence controlled computer'.[170] He added: 'This device was intended to do about the same as sophisticated compilers do today'.[171] Zuse discussed automated programming in his 1945 manuscript 'Der Plankalkül', describing this process as 'calculating calculating plans'.[172] Turing even envisaged programs that are able to rewrite their own instructions in response to experience. 'One can imagine', he said in a lecture on the ACE, 'that after the machine had been operating for some time, the instructions would have altered out of all recognition'.[173]

---

[163]Carpenter and Doran, 'The Other Turing Machine', p. 270; see also Carpenter and Doran, 'Turing's Zeitgeist'.

[164]Turing, 'Proposed Electronic Calculator', p. 382.

[165]Turing, 'Proposed Electronic Calculator', pp. 382–383.

[166]Carpenter and Doran, 'The Other Turing Machine', p. 270.

[167]Vickers, T. 'Applications of the Pilot ACE and the DEUCE', in Copeland et al., *Alan Turing's Electronic Brain*, p. 277.

[168]Turing, 'Proposed Electronic Calculator', p. 386.

[169]Carpenter and Doran, 'Turing's Zeitgeist'.

[170]Zuse, 'Some Remarks on the History of Computing in Germany', pp. 616–617.

[171]Zuse, 'Some Remarks on the History of Computing in Germany', p. 617.

[172]Zuse, 'Der Plankalkül' (manuscript), pp. 30–31.

[173]Turing, A. M. 'Lecture on the Automatic Computing Engine', in *The Essential Turing*, p. 393.

The two 1945 documents by Turing and von Neumann each had a very different philosophy. Essentially, von Neumann's 'First Draft' presented a novel form of numerical *calculator*. Right at the beginning of 'First Draft', in the course of what he called 'some general explanatory remarks', he offered this 'definition' of his subject matter: 'An *automatic computing system* is a (usually highly composite) device, which can carry out instructions to perform calculations of a considerable order of complexity'.[174] The EDVAC, he explained, would be a 'very high speed' automatic digital calculator. Turing, on the other hand, was envisaging a different kind of beast. For instance, he listed in 'Proposed Electronic Calculator' an assortment of *non*-numerical problems suitable for the ACE. These included solving a jig-saw, a problem that he described as 'typical of a very large class of non-numerical problems that can be treated', adding: 'Some of these have great military importance, and others are of immense interest to mathematicians.'[175] By this time Turing already had significant experience with non-numerical computation: his Bombe was designed to solve a specific type of non-numerical problem.[176] Turing also mentioned chess in 'Proposed Electronic Calculator', making his famous remark that 'There are indications ... that it is possible to make the machine display intelligence at the risk of its making occasional serious mistakes'.[177] Despite its modest title, 'Proposed Electronic Calculator' offered far more than a numerical calculator.

In January 1947 Turing travelled to the United States, to attend the Harvard Symposium on Large-Scale Digital Calculating Machinery. Organized by Aiken at his Harvard Computation Laboratory, this was the world's second sizable computing conference, with more than 300 delegates attending (a smaller conference with around 85 delegates was held at MIT in fall 1945).[178] With the birth of the stored-program electronic computer expected imminently, the time was ripe for a collection of visionary lectures; but Aiken, the leading champion in the US of electromechanical program-controlled computation, did not seize the moment. With the exception of Mauchly, Goldstine, and Jay Forrester—who at MIT was planning the Whirlwind I computer, one of the first stored-program machines to run (see the timeline in Fig. 3)—none of the leading pioneers of the new stored-program technology lectured at the symposium, not even the physically present Turing. His contributions were confined to astute comments during the brief post-lecture

[174]Von Neumann, 'First Draft of a Report on the EDVAC', p. 181 in Stern, *From ENIAC to UNIVAC*.

[175]Turing, 'Proposed Electronic Calculator', pp. 388–9.

[176]Turing, A. M. 'Bombe and Spider', in *The Essential Turing*.

[177]Turing, 'Proposed Electronic Calculator', p. 389.

[178]'Members of the Symposium', *Proceedings of a Symposium on Large-Scale Digital Calculating Machinery. Jointly Sponsored by The Navy Department Bureau of Ordnance and Harvard University at The Computation Laboratory 7–10 January 1947*. Vol. 16 of *The Annals of the Computation Laboratory of Harvard University* (Cambridge, MA: Harvard University Press, 1948), pp. xvii–xxix.

discussions, including the following succinct expression of what we now see as the central feature of universal computers:

> We [at the National Physical Laboratory] are trying to make greater use of the facilities available in the machine to do all kinds of different things simply by programming ... This is an application of the general principle that any particular operation of physical apparatus can be reproduced ... simply by putting in more programming.[179]

To sum up, Turing's vision transcended the numerical calculator of von Neumann's 'First Draft'. Turing was planning an entirely new kind of machine, one capable of rewriting its own programs, of reproducing the behaviour of a wide range of different forms of physical apparatus, and even of exhibiting intelligence. As philosopher Teresa Numerico put it, 'In Turing's project, but not von Neumann's, we are confronted by a machine different from all previous machines.'[180]

In the end, it was von Neumann's simple, centralized design rather than Turing's quirky decentralized design that went on to become the industry standard, the ubiquitous 'von Neumann machine'. Von Neumann, though, was actually very clear—both in private and in public—in attributing the twin logical pillars to Turing. It is unfortunate that his statements are not more widely known. He explained in a letter in November 1946 that Turing's 'great positive contribution' was to show that 'one, definite mechanism can be "universal"'[181]; and in a 1949 lecture he emphasized the crucial importance of Turing's research, which lay, he said, in Turing's 1936 demonstration that a single appropriately designed machine 'can, when given suitable instructions, do anything that can be done by automata at all'.[182] Von Neumann's friend and scientific colleague Stanley Frankel recollected that von Neumann 'firmly emphasized to me, and to others I am sure, that the fundamental conception is owing to Turing'.[183] Frankel added, 'In my view von Neumann's essential role was in making the world aware of these fundamental concepts introduced by Turing ...'. IBM's Cuthbert Hurd, who also worked closely with von Neumann, emphasized 'I never heard him make the claim that he invented stored programming'.[184]

---

179 'Sheppard Discussion', *Proceedings of a Symposium on Large-Scale Digital Calculating Machinery*, p. 273.

180 Numerico, T. 'From Turing Machine to "Electronic Brain"', in Copeland et al., *Alan Turing's Electronic Brain*, p. 182.

181 Letter from von Neumann to Norbert Wiener, 29 November 1946; in the von Neumann Archive at the Library of Congress, Washington, D.C. (quoted on p. 209 of *The Essential Turing*).

182 'Rigorous Theories of Control and Information', in von Neumann, J. *Theory of Self-Reproducing Automata* (Urbana: University of Illinois Press, 1966; ed. Burks A. W.), p. 50.

183 Letter from Frankel to Brian Randell, 1972 (first published in Randell, B. 'On Alan Turing and the Origins of Digital Computers', in Meltzer, B., Michie, D. (eds) *Machine Intelligence 7* (Edinburgh: Edinburgh University Press, 1972)). Copeland is grateful to Randell for giving him a copy of this letter.

184 Hurd, C., Comments on Eckert, 'The ENIAC', in Metropolis, Howlett, and Rota, *A History of Computing in the Twentieth Century*, p. 536.

Returning to Moshe Vardi's efforts to refute the claim that Turing originated the stored-program concept, Vardi states—defending von Neumann's corner—that 'we should not confuse a mathematical idea with an engineering design'. So at best Turing deserves the credit for an abstract mathematical idea? Not so fast. Vardi is ignoring the fact that some inventions do belong equally to the realms of mathematics and engineering. The universal Turing machine of 1936 was one such, and this is part of its brilliance.

What Turing described in 1936 was not an abstract mathematical notion but a solid three-dimensional machine (containing, as he said, wheels, levers, and paper tape[185]); and the cardinal problem in electronic computing's pioneering years, taken on by both 'Proposed Electronic Calculator' and the 'First Draft', was just this: How best to build a practical electronic form of the universal Turing machine?

The claim that in 1936 Turing came up merely with an abstract mathematical idea, and moreover without perceiving any connection between it and potential real computing machinery, is a persistent one. Notoriously, 'Proposed Electronic Calculator' did not so much as mention the universal machine of 1936, leading some commentators to wonder whether even in Turing's mind there was any connection between the ACE and his earlier abstract machine (a doubt forcefully expressed by George Davis, a pioneer of computing from the pilot ACE era).[186] Computer historian Martin Campbell-Kelly also doubted that the universal Turing machine was a 'direct ancestor of the ACE', pointing out that the memory arrangements of the 1936 machine and of the ACE were very different, with the ACE's 'addressable memory of fixed-length binary numbers' having 'no equivalent in the Turing Machine'.[187]

However, some fragments of an early draft of 'Proposed Electronic Calculator' cast much new light on this issue.[188] The fragments survive only because Turing used the typed sheets as scrap paper, covering the reverse sides with rough notes on circuit design; his assistant, Mike Woodger, happened to keep the rough notes. In these fragments, Turing explicitly related the ACE to the universal Turing machine, explaining why the memory arrangement described in his 1936 paper required modification when creating a practical design for a computer. He wrote:

> In 'Computable numbers' it was assumed that all the stored material was arranged linearly, so that in effect the accessibility time was directly proportional to the amount of material stored, being essentially the digit time multiplied by the number of digits stored. This was

---

[185]Turing, A. M., draft précis (in French) of 'On Computable Numbers' (undated, 2 pp.; in the Turing Papers, Modern Archive Centre, King's College Library, Cambridge, catalogue reference K 4).

[186]George Davis, verbal comments at the ACE 2000 Conference, National Physical Laboratory, Teddington, 2000; and also at a seminar on Turing organised by the British Computer Conservation Society, Science Museum, London, 2005.

[187]Campbell-Kelly, 'The ACE and the Shaping of British Computing', pp. 156–157.

[188]These fragments were published for the first time as Turing, A. M. 'Notes on Memory', in Copeland et al., *Alan Turing's Automatic Computing Engine*, Oxford: Oxford University Press, 2005).

> the essential reason why the arrangement in 'Computable numbers' could not be taken over as it stood to give a practical form of machine. Actually we can make the digits much more accessible than this, but there are two limiting considerations to the accessibility which is possible, of which we describe one in this paragraph. If we have N digits stored then we shall need about log2N digits to describe the place in which a particular digit is stored. This will mean to say that the time required to put in a request for a particular digit will be essentially log2N × digit time. This may be reduced by using several wires for the transmission of a request, but this might perhaps be considered as effectively decreasing the digit time.[189]

Arguments that the ACE cannot have been inspired by the universal machine of 1936, since Turing did not mention his 1936 machine in 'Proposed Electronic Calculator', are plainly non-starters. It must also be remembered that the NPL hired Turing for the ACE project precisely because Womersley was familiar with, and had been inspired by, 'On Computable Numbers'.

Historian Thomas Haigh, who, like Vardi, is fighting in von Neumann's corner, weighs in on the side of the sceptics, attempting to raise doubt about whether 'Turing was interested in building an actual computer in 1936'.[190] He tries to undermine Max Newman's testimony on this point (almost as though he were von Neumann's lawyer), writing that the information 'is sourced not to any diary entry or letter from the 1930s but to the recollections of one of Turing's former lecturers made long after real computers had been built'.[191] Haigh fails to inform his readers that this former lecturer was none other than Newman, who (as explained above) played a key role in the genesis of the universal Turing machine, and for that matter also in the development of the first universal Turing machine in electronic hardware (to the point of securing the transfer, from Bletchley Park to the Manchester Computing Laboratory, of a truckload of electronic and mechanical components from dismantled Colossi).[192] Even in the midst of the attack on the German codes, Newman was thinking about the universal Turing machine: when Flowers was designing Colossus in 1943, Newman showed him Turing's 1936 paper, with its key idea of storing symbolically-encoded instructions in memory.[193] Donald Michie, a member of Newman's wartime section at Bletchley Park, the 'Newmanry'—home to nine Colossi by 1945—recollected that, in 1944–45, the Newmanry's mathematicians were 'fully aware of the prospects for implementing physical embodiments of the UTM [universal Turing machine] using vacuum-tube technology'.[194]

In fact, Newman's testimony about the foregoing point is rather detailed. He explained in a tape-recorded interview that when he learned of Turing's universal

[189]Turing, 'Notes on Memory', p. 456.

[190]Haigh, 'Actually, Turing Did Not Invent the Computer', p. 39.

[191]Haigh, 'Actually, Turing Did Not Invent the Computer', p. 39.

[192]Copeland et al., *Colossus*, p. 172; for a detailed account of Newman's role in the Manchester computer project, see Copeland, 'The Manchester Computer: A Revised History'.

[193]Flowers in interview with Copeland, July 1996.

[194]Letter from Michie to Copeland, 14 July 1995.

computing machine, early in 1936, he developed an interest in computing machinery that he described as being, at this time, 'rather theoretical'. Whereas, Newman continued, 'Turing himself, right from the start, said it would be interesting to try to *make* such a machine'.[195] Newman emphasized this same point in his obituary of Turing in *The Times*:

> The description that he then [1936] gave of a 'universal' computing machine was entirely theoretical in purpose, but Turing's strong interest in all kinds of practical experiment made him even then interested in the possibility of actually constructing a machine on these lines.[196]

Donald Davies, in 1947 a young member of Turing's ACE group at NPL, emphasized in a 1975 interview that the stored-program concept originated in 'On Computable Numbers'.[197] It seems to have been common knowledge among those involved with Turing at the NPL that the fundamental idea of the ACE derived from 'On Computable Numbers'. Sir Charles Darwin, Director of the NPL, wrote in a 1946 note titled 'Automatic Computing Engine (ACE)': 'The possibility of the new machine started from a paper by Dr. A. M. Turing some years ago'.[198] In 1947 Turing himself gave a clear statement of the connection, as he saw it, between the universal computing machine of 1936 and the electronic stored-program universal digital computer:

> Some years ago I was researching on what might now be described as an investigation of the theoretical possibilities and limitations of digital computing machines. I considered a type of machine which had a central mechanism, and an infinite memory which was contained on an infinite tape. ... [D]igital computing machines ... are in fact practical versions of the universal machine. There is a certain central pool of electronic equipment, and a large memory, [and] the appropriate instructions for the computing process involved are stored in the memory.[199]

4 A Hierarchy of Programming Paradigms

In a recent critique of the stored-program concept ('"Stored Program Concept" Considered Harmful'), Thomas Haigh maintains that 'discussion of the "stored program concept" has outlived the purpose for which it was created and provides

[195]Newman interviewed by Evans.

[196]Newman, M. H. A. 'Dr. A. M. Turing', *The Times*, 16 June 1954, p. 10.

[197]Davies interviewed by Evans.

[198]Darwin, C. 'Automatic Computing Engine (ACE)', National Physical Laboratory, 17 April 1946 (National Archives, document reference DSIR 10/385); published in Copeland et al., *Alan Turing's Automatic Computing Engine*, pp. 53–57.

[199]Turing, 'Lecture on the Automatic Computing Engine', pp. 378, 383.

a shortcut to confusion'.[200] He observes that the phrase 'stored program' rarely appeared in early documents, saying that this 'fairly obscure term' originated in an IBM internal memo in 1949.[201] Haigh continues:

> In the late 1970s and 1980s ... the idea of the 'stored program computer' was borrowed from technical discourse, developing from a fairly obscure term into a central concept in heated debates over what should be considered the first true computer and why.[202]

This 'resurgence of the stored program concept ... as a concept for historical discussion' was 'harmful', Haigh says, and '[t]he time has come to replace it'.[203] His readers are told that there is 'endemic confusion surrounding the stored-program concept'; and that the term 'stored program' is 'hopelessly overloaded with contradictory meanings' and is 'unhelpfully imprecise'.[204] However, Haigh offers no substantial arguments to support these views. Indeed, the main purpose of his diatribe against the stored-program concept appears to be to pave the way for his proposal to replace traditional terminology by a suite of neologisms that are designed to position von Neumann centre stage. If Haigh got his way we would all be speaking of 'the von Neumann architectural paradigm' and the 'modern code paradigm'—the latter a term explicitly introduced in order to 'describe the program-related elements of the 1945 First Draft'.[205]

Haigh's claim that the stored-program concept was by and large a construction of 1970s and 1980s historians does not withstand scrutiny. Far from the concept's being 'fairly obscure' before historians latched onto it, the concept in fact played a central role in numerous key documents from the early years of electronic computing. The reason that the *phrase* 'stored program' generally did not appear in these documents is simply that the founding fathers tended not to use the word 'program'. Von Neumann preferred 'orders', as did Zuse (*Befehle*).[206] Zuse also used the term 'calculating plan' (*Rechenplan*). Turing, in 1936, introduced the term 'instruction table' for what we would now call a program, and in 'Proposed Electronic Calculator' he continued to use this natural term, speaking of storing an instruction table or simply of storing instructions. In the introduction to the first of his Adelphi Lectures (the British counterpart of the Moore School Lectures,

---

[200]Haigh, T. '"Stored Program Concept" Considered Harmful: History and Historiography', in Bonizzoni, P., Brattka, V., Löwe, B. (eds) *The Nature of Computation. Logic, Algorithms, Applications* (Berlin: Springer, 2013), p. 247. See also Haigh, T., Priestley, M., Rope, C. 'Reconsidering the Stored-Program Concept', *IEEE Annals of the History of Computing*, vol. 36 (2014), pp. 4–17.

[201]Haigh, '"Stored Program Concept" Considered Harmful', pp. 243–244.

[202]Haigh, '"Stored Program Concept" Considered Harmful', p. 244.

[203]Haigh, '"Stored Program Concept" Considered Harmful', pp. 245, 247; Haigh, Priestley and Rope, 'Reconsidering the Stored-Program Concept', p. 12.

[204]Haigh, Priestley and Rope, 'Reconsidering the Stored-Program Concept', pp. 4, 14, 15.

[205]Haigh, '"Stored Program Concept" Considered Harmful', pp. 247, 249; Haigh, Priestley and Rope, 'Reconsidering the Stored-Program Concept', p. 12.

[206]Von Neumann, 'First Draft of a Report on the EDVAC', Sections 14–15 (pp. 236 ff in Stern, *From ENIAC to UNIVAC*).

although on a much smaller scale), Turing explained that 'the machine will incorporate a large "Memory" for the storage of both data and instructions'.[207] In an early homage to the joys of stored programming, he remarked enthusiastically that the 'process of constructing instruction tables should be very fascinating', continuing: 'There need be no real danger of it ever becoming a drudge, for any processes that are quite mechanical may be turned over to the machine itself'.[208]

Others followed Turing's usage. In their famous 1948 letter to *Nature*, announcing the birth of the Manchester Baby, Williams and Kilburn explained that the 'instruction table' was held in the computer's 'store'.[209] In the same letter they also used the term 'programme of instructions', and emphasized that 'the programme can be changed without any mechanical or electro-mechanical circuit changes'. Their selection of the terms 'store' and 'programme' proved to be a way of speaking that many others would also find natural, and by 1953, usage was sufficiently settled for Willis Ware (in a discussion of ENIAC and von Neumann's subsequent Princeton computer project) to be able to write simply: 'what we now know as the "stored program machine"'.[210] As for the centrality of the stored-program concept, in their writings from the period Williams and Kilburn repeatedly highlighted the concept's key position. For example, Kilburn said in 1949: 'When a new instruction is required from the table of instructions stored in the main storage tube, *S*, a "prepulse" initiates the standard sequence of events'.[211]

Similar examples can be multiplied endlessly from books and articles published on both sides of the Atlantic. That electronic computers could usefully edit their own stored programs was basic common knowledge. The 1959 textbook *Electronic Digital Computers* said:

> [I]t is at once apparent that instructions may be operated upon by circuitry of the same character as that used in processing numerical information. Thus, as the computation progresses, the machine may be caused to modify certain instructions in the code that it is following.[212]

We believe that the useful and well-known term 'stored program' is reasonably clear and precise. Like many terms, however, it will certainly benefit from some careful logical analysis, and this we now offer. Looking back over the early years of the computer's history, as outlined in Sect. 3, at least six different *programming*

---

[207]'The Turing-Wilkinson Lecture Series (1946–7)', in Copeland et al., *Alan Turing's Electronic Brain*, p. 465.

[208]Turing, 'Proposed Electronic Calculator', p. 392.

[209]Williams, F. C., Kilburn, T. 'Electronic Digital Computers', *Nature*, vol. 162, no. 4117 (1948), p. 487.

[210]Ware, W. H. 'The History and Development of the Electronic Computer Project at the Institute for Advanced Study', RAND Corporation report P-377, Santa Monica, 10 March 1953, p. 5.

[211]Kilburn, T. 'The Manchester University Digital Computing Machine', in Williams, M. R., Campbell-Kelly, M. eds *The Early British Computer Conferences* (Los Angeles: Tomash, 1989), p. 138.

[212]Smith, C. V. L. *Electronic Digital Computers* (New York: McGraw Hill, 1959), p. 31.

paradigms can be distinguished. We denote these paradigms **P**1, **P**2, **P**3, **P**4, **P**5 and **P**6. To borrow Turing's onion-skin metaphor, the stored-program concept, like an onion, consists of a number of layers or levels. **P**3, **P**4, **P**5 and **P**6 are four such layers.

4.1 Paradigm P1

Programming consists of physically rewiring the machine, e.g. by means of plugs and switches. Colossus is a leading exemplar of this paradigm, and also ENIAC as originally designed. Since a signal can travel much faster along a wire than through, say, a vacuum tube circuit, this form of programming can lead to an inherently very fast machine—as in the case of Turing's Bombe, for example. **P**1's leading disadvantage is setup time. By 1945, the Colossus group were considering the use both of tape and punched cards to ease setup difficulties.[213] Eckert underlined the impoverished nature of **P**1 in reflections on the ENIAC: 'It was obvious that computer instructions could be conveyed in a numerical code, and that whatever machines we might build after the ENIAC would want to avoid the setup problems that our hastily built first try ENIAC made evident'.[214]

4.2 Paradigm P2

The main advantage of **P**2 over **P**1 is the ease of setting up for a new job and the corresponding reduction in setup time. Instructions are expressed in the form of numbers, or other symbols, and these numbers or symbols are stored in a memory medium such as tape or punched cards. The processes used in writing and reading the instructions are not of the same kind as those used in executing them (an echo of Gandy's statement, quoted above). Exemplars of this paradigm are Babbage's Analytical Engine, whose instructions were pre-punched into cards, and Aiken's ASCC and Zuse's Z1, Z2, Z3 and Z4, whose instructions were pre-punched into tape. This form of programming is read-only and the computer does not edit the instructions as the program runs.

We shall call machines programmed in accordance with **P**1 or **P**2 'program-controlled' machines, in order to set off **P**1 and **P**2 from **P**3–**P**6.

---

[213]Good, I. J., Michie, D., Timms, G. *General Report on Tunny*, Bletchley Park, 1945 (National Archives/Public Record Office, Kew; document reference HW 25/4 (vol. 1), HW 25/5 (vol. 2)), p. 331. A digital facsimile of *General Report on Tunny* is available in *The Turing Archive for the History of Computing* <http://www.AlanTuring.net/tunny_report>.

[214]Eckert, 'The ENIAC', p. 531.

4.3 Paradigm P3

Instructions are expressed in the form of numbers, or other symbols, and these numbers or symbols are stored in a (relatively fast) read-only memory. As with **P**2, the operations used in executing the instructions are not available to edit the instructions, since the memory is read-only. Writing the instructions into the memory may be done by hand—e.g. by means of a keyboard or hand-operated setting switches. The main advantage of **P**3 over **P**2 is that with this form of memory, unlike tape or cards, instructions or blocks of instructions can be read out repeatedly, without any need to create multiple tokens of the instructions in the memory (so avoiding copying blocks of cards or punching blocks of instructions over and again on the tape).

An exemplar of this paradigm is the modified form of ENIAC here called 'ENIAC-1948'. In order to simplify the setup process, ENIAC was operated from 1948 with instructions stored in a 'function table', a large rack of switches mounted on a trolley.[215] (Eckert recorded that this hand-switched, read-only storage system, essentially a resistance network, was based on ideas he learned from RCA's Jan Rajchman.[216]) The switch trolleys offered a slow but workable read-only memory for coded instructions. Richard Clippinger from Aberdeen Ballistic Research Laboratory (where ENIAC was transferred at the end of 1946) was responsible for ENIAC's transition to this new mode of operation. Clippinger explained:

> I discovered a new way to program the ENIAC which would make it a lot more convenient. ... I became aware of the fact that one could get a pulse out of the function table, and put it on the program trays, and use it to stimulate an action. This led me to invent a way of storing instructions in the function table.[217]

It seems, though, that Clippinger had reinvented the wheel. Mauchly stated that Eckert and he had previously worked out this idea.[218] Referring to Clippinger's rediscovery of the idea, Mauchly said: '[P]eople have subsequently claimed that the idea of such stored programs were [sic] quite novel to others who were at Aberdeen'. Eckert emphasized the same point: 'In Aberdeen, Dr. Clippinger later "rediscovered" these uses of the function tables'.[219] However, it can at least be said that Clippinger reduced the idea to practice, with the assistance of Nick Metropolis,

---

[215]Jennings said that ENIAC ran in this mode from April 1948, but Goldstine reported a later date: 'on 16 September 1948 the new system ran on the ENIAC'. Jennings Bartik, *Pioneer Programmer*, p. 120; Goldstine, *The Computer from Pascal to von Neumann*, p. 233.

[216]Presper Eckert interviewed by Christopher Evans in 1975 ('The Pioneers of Computing: an Oral History of Computing', Science Museum: London).

[217]Richard Clippinger interviewed by Richard R. Mertz in 1970 (Computer Oral History Collection, Archives Center, National Museum of American History, Smithsonian Institution, Washington, D.C.), p. I-I-11.

[218]John Mauchly interviewed by Christopher Evans in 1976 ('The Pioneers of Computing: an Oral History of Computing', Science Museum: London).

[219]Eckert, 'The Eniac', p. 529.

Betty Jean Jennings, Adele Goldstine, and Klari von Neumann (von Neumann's wife).[220]

In the secondary literature, this idea of using the switch trolleys to store instructions is usually credited to von Neumann himself (e.g. by Haigh in his recent von Neumann-boosting work).[221] But Clippinger related that 'When Adele Goldstine noted that I had evolved a new way to program the ENIAC, she quickly passed the word along to von Neumann'.[222] According to Clippinger and Jennings, what von Neumann contributed, in the course of discussions with Clippinger and others, was a more efficient format for the stored instructions, a one-address code that allowed two instructions to be stored per line in the function table (replacing Clippinger's previous idea of using a three-address code).[223]

There is a strong tradition in the literature for calling this paradigm 'stored program', and we follow that tradition here. Nick Metropolis and Jack Worlton said that ENIAC-1948 was 'the first computer to operate with a read-only stored program'.[224] Mauchly also referred to the switch trolley arrangement as involving 'stored programs' (in the above quotation). In passing, we note that we would not object *very* strongly to the suggestions that **P3** be termed 'stored-program *in the weak sense*' or 'stored-program in the *minimal* sense', or even as transitional between **P2** and genuine stored programming. However, the important point is that the major difference between **P3** and **P4–P6** should be marked somehow; and so long as the distinction is clearly enough drawn, it hardly matters in the end which words are used. Later in this section, a systematic notation is developed that brings out the minimal and somewhat anomalous status of **P3**.

4.4 Paradigm P4

Instructions in the form of numbers or other symbols are stored in a memory medium, in such a way that the processes used in reading and writing these instructions are of the same kind as those used in executing them. In other words, the processes used in executing the instructions can potentially 'get at' the instructions. The pre-eminent exemplar of this paradigm is the universal Turing machine as Turing described it in 1936.

---

[220] Clippinger interviewed by Mertz, p. I-I-14; Metropolis, N., Worlton, J. 'A Trilogy on Errors in the History of Computing', *Annals of the History of Computing*, vol. 2 (1980), pp. 49–59 (p. 54).

[221] Haigh, '"Stored Program Concept" Considered Harmful', p. 242. Also Goldstine, *The Computer from Pascal to von Neumann*, p. 233.

[222] Clippinger interviewed by Mertz, p. I-I-12. See also Metropolis and Worlton, 'A Trilogy on Errors in the History of Computing', p. 54.

[223] Clippinger interviewed by Mertz, pp. I-I-12, I-I-13; Jennings Bartik, *Pioneer Programmer*, pp. 11–12, 113.

[224] Metropolis and Worlton, 'A Trilogy on Errors in the History of Computing', p. 54.

In the context of electronic machines, use of **P**4 enables the computer to access its instructions at electronic speed, since the access operations are themselves electronic—so avoiding the 'instruction bottleneck' arising when instructions are supplied from some slower-than-electronic medium, such as punched tape.

4.5 Paradigm P5

P4 contains the potential for instruction editing. In **P**5 and **P**6 that potential is realized. The instruction editing made possible by **P**4 conforms to one or other of two logical types: instruction editing without the creation of new instructions, and editing that does produce new instructions. In **P**5 only the first occurs, whereas in **P**6 the second occurs. In **P**5, the processes used in executing instructions are also used to edit the instructions themselves, by adding, manipulating or deleting symbols, in order to *mark* or unmark portions of an instruction.

Again the pre-eminent exemplar is Turing's universal machine of 1936. Turing introduced a number of subroutines for inserting marker symbols into instructions, and for operating on marked up segments of an instruction in various ways; subsequently one of the subroutines would delete the markers, leaving the instructions exactly as they were originally found. For example, Turing's routines *kom* and *kmp* place markers showing the next instruction to be obeyed. His routine *sim* marks up the next instruction with various other symbols, and *inst* copies marked portions of the instruction to other locations on the tape, finally deleting the marker symbols from the instruction.[225]

4.6 Paradigm P6

P6 is but a very short step away from **P**5. In **P**5, the editing of instructions is limited to the insertion, manipulation and deletion of symbols that function as markers, while in **P**6 the editing processes from time to time delete and insert symbols in such a way as to produce a *different instruction*. **P**6 first appeared in the historical record in 1945. 'Proposed Electronic Calculator' and 'First Draft of a Report on the EDVAC' both describe this programming paradigm (although, as noted above, von Neumann initially protected symbols other than address bits from editing, only lifting this restriction in later publications).

In **P**6, the instructions that are edited may be those of the program that is actually running, as with address modification on the fly, or the modifications may be made to the instructions of some other program stored in memory. These different cases are here referred to as 'reflexive' and 'non-reflexive' editing respectively. Section 3

[225]Turing, 'On Computable Numbers', pp. 71–72 in *The Essential Turing*.

noted that in 1945 both Turing and Zuse farsightedly mentioned the idea of one program editing another. The importance of non-reflexive editing of stored programs was in fact widely recognized from computing's earliest days (further disproof of Haigh's claim that the stored program concept had only a 'fairly obscure' existence until the late 1970s). In one of computing's earliest textbooks, published in 1953, Alec Glennie wrote:

> It has been found possible to use the Manchester machine to convert programmes written in a notation very similar to simple algebraic notation into its own code. ... [I]nstructions take the form of words which, when interpreted by the machine, cause the correct instructions in the machine's code to be synthesized. ...[226]

In his 1959 textbook *Electronic Digital Computers*, Charles Smith wrote:

> This ability of the machine to modify a set of instructions to refer to various sets of operands makes it possible so to compress the instruction codes of many problems that they can be held, along with partial results and required data, in the limited internal memory of computers that have no relatively fast secondary or external memory.[227]

The key differences between **P3**, **P4**, **P5** and **P6** can be summarized as follows:

P3. This paradigm is: *read-only stored instructions*. In a notation designed to make the differences between **P3**, **P4**, **P5** and **P6** transparent ('*S*-notation'), **P3** is *S0*, a step beneath the unmarked case *S*.

P4. **P4** is simply *S*: *stored instructions* potentially accessible to editing by the processes used to execute the instructions. Actually making use of this potential leads on to **P5** and **P6**.

P5. This paradigm is: *stored instructions/editing/no instruction change*. **P5** is *S/E/=* ('=' signifying no change of instruction). **P5** includes editing markers in a way that involves applying numerical operations, such as addition and subtraction, to the markers. For example, having marked an instruction with the marker '1010', the mechanism may repeatedly read out that instruction, each time subtracting 1 from the marker, stopping this process when the marker becomes 0000.

P6. This paradigm is: *stored instructions/editing/instruction change*. **P6** is *S/E/ΔI* ('*ΔI*' signifying instruction change). Where it is helpful to distinguish between reflexive and non-reflexive editing, *S/E/ΔI/SELF* is written for the former and *S/E/ΔI/OTHER* for the latter. *S/E* is sometimes used to refer to any or all of *S/E/=*, *S/E/ΔI/SELF* and *S/E/ΔI/OTHER*. For ease of reference, the *S*-notation just presented is summarized in Figure 1.[228]

Important logical features of **P2**, **P3**, **P4**, **P5**, and **P6**, as well as an important forensic principle, can be brought out by means of a short discussion of a memory model we call 'Eckert disk memory'. Eckert considered this form of internal

[226]Glennie, A. E. 'Programming for High-Speed Digital Calculating Machines', ch. 5 of Bowden, B. V. ed. *Faster Than Thought* (London: Sir Isaac Pitman & Sons, 1953), pp. 112–113.

[227]Smith, *Electronic Digital Computers*, p. 31.

[228]Copeland developed the *S*-notation in discussion with Diane Proudfoot.

S0	Read-only stored instructions.
S	The stored instructions are accessible to editing by the processes used to execute the instructions.
S/E	Editing of the stored instructions actually occurs—in any of the following modes:
S/E/=	As in Turing's 1936 paper, the stored instructions are edited, by the addition or removal of marker symbols, but the instructions are not changed ('=' signifying no change of instruction).
S/E/ΔI	The stored instructions are changed during editing ('*ΔI*' signifying instruction change). *S/E/ΔI* divides into two types:
S/E/ΔI/SELF	The program alters some of its own instructions as it runs.
S/E/ΔI/OTHER	The program alters the instructions of another stored program.

Fig. 1 The *S*-notation

memory in 1944, but soon abandoned the idea in favour of his mercury delay line, a technology he had had more experience with, and which, moreover, he thought would offer faster access times.[229] Eckert described the disk memory in a typewritten note dated January 1944.[230] He said 'The concept of general internal storage started in this memo'—but in fact that concept appeared in Zuse's writings in 1936, as Sect. 6 explains.[231]

Eckert disk memory consists of a single memory unit containing a number of disks, mounted on a common electrically-driven rotating shaft (called the time shaft). As Eckert described it, the memory unit formed part of a design for a desk calculating machine, an improved and partly electronic form of 'an ordinary mechanical calculating machine', he said.[232] Some of the disks would have their edges engraved with programming information expressed in a binary code. As the disk rotated, the engravings would induce pulses in a coil mounted near the disk. These pulses would initiate and control the operations required in the calculation (addition, subtraction, multiplication, division). Eckert described this arrangement as 'similar to the tone generating mechanism used in some electric organs'.[233]

The engraved disks offered permanent storage. Other disks, made of magnetic alloy, offered volatile storage. The edges of these disks were to be 'capable of being magnetized and demagnetized repeatedly and at high speed'.[234] Pulses would

[229] Eckert interviewed by Evans.

[230] The note is included in Eckert, 'The ENIAC', pp. 537–539.

[231] Eckert, 'The ENIAC', p. 531.

[232] Eckert, 'The ENIAC', p. 537.

[233] Eckert, 'The ENIAC', p. 537.

[234] Eckert, 'The ENIAC', p. 537.

be written to the disk edges and subsequently read. (This idea of storing pulses magnetically on disks originated with Perry Crawford.[235]) Eckert explained that the magnetic disks were to be used to store not only numbers but also function tables, such as sine tables and multiplication tables, and the numerical combinations required for carrying out binary-decimal-binary conversion.

A 1945 report written by Eckert and Mauchly, titled 'Automatic High Speed Computing: A Progress Report on the EDVAC', mentioned this 1944 disk memory. Eckert and Mauchly said: 'An important feature of this device was that operating instructions and function tables would be stored in exactly the same sort of memory device as that used for numbers'.[236] This statement is true, but requires some careful unpacking.

The storage of programming information on the engraved disks is an exemplification of paradigm **P**2. As with punched tape, the processes used in writing the engraved instructions (etching) and reading them (induction) are manifestly of a different kind from the processes used in executing the instructions. The processes used in execution are incapable of editing the engraved instructions. It is true that instructions are stored *in the same memory device* as numbers—i.e. the disk memory unit itself—but this device has subcomponents and, as we shall see, the subcomponents conform to different programming paradigms. Conceivably Eckert's intention was to control the rotation of the engraved disk in such a way that instructions or blocks of instructions could be read out repeatedly, in which case the engraved disk would conform to paradigm **P**3 rather than **P**2; however, Eckert did not mention this possibility in the memo.

Eckert also considered using a magnetic alloy disk for instruction storage, saying 'programming may be of the temporary type set up on alloy discs or of the permanent type on etched discs'.[237] Storing the programming information on an alloy disk exemplifies paradigm **P**4, since the same read and write operations that are used to fetch and store numbers during execution are also used to write the binary instructions onto the disk and to read them for execution.

When the instructions are stored on one of these magnetic alloy disks, the possibility arises of editing the instructions. For example, spaces could be left between instructions where temporary markers could be written to indicate the next instruction (**P**5); or, indeed, the instructions themselves might be edited, to produce different instructions (**P**6). However, Eckert mentioned none of this; the idea of editing instructions was simply absent from his memo. Of course, it is a short step conceptually from simply storing the instructions on one of these alloy disks to editing them. Nevertheless, it would be a gross mistake to say that Eckert's memo entertains either *S/E/=* or *S/E/*ΔI, since there is nothing in the document to indicate that editing of instructions was envisaged. Eckert might have been aware

[235] Eckert interviewed by Evans.

[236] Eckert and Mauchly, 'Automatic High Speed Computing: A Progress Report on the EDVAC', p. 2.

[237] Eckert, 'The ENIAC', p. 538.

of these possibilities at the time, or he might not. When interpreting historical documents, there is an obvious forensic principle that should be in play, to the effect that ideas not actually mentioned in the document must not be projected into it by the interpreter. Evidence internal to Eckert's 1944 memo enables us to state that he proposed *S* at this time, but there is no evidence in the document that his thinking extended to *S/E/=* or *S/E/*Δ*I*. Similarly, although Turing's 'On Computable Numbers' used *S/E/=*, and *S/E/*Δ*I* is but a very short step from *S/E/=*, it would be a mistake to ascribe *S/E/*Δ*I* to Turing's 1936 document. Turing might very well have realized in 1935–1936 that his machine's instructions could be edited to produce new instructions, simply by applying the editing processes not only to symbols marking the instructions but also to the symbols of the instructions themselves; however, there is no evidence of *S/E/*Δ*I* to be found in the actual document. Eckert's 1944 memo describes *S*, nothing more; and Turing's 1936 paper describes *S/E/=*, nothing more.

Armed with this forensic principle and these various distinctions, let us now turn to a consideration of Zuse's work.

5 Zuse and the Concept of the Universal Machine

In 1948, in a lecture to the Royal Society of London, Max Newman defined *general purpose* computers as 'machines able without modification to carry out any of a wide variety of computing jobs'.[238] Zuse had undoubtedly conceived the idea of a digital, binary, program-controlled general-purpose computer by 1936. In a patent application dating from April of that year he wrote:

> The present invention serves the purpose of automatically carrying out frequently recurring calculations, of arbitrary length and arbitrary construction, by means of calculating machinery, these calculations being composed of elementary calculating operations. ... A prerequisite for each kind of calculation that is to be done is the preparation of a calculating plan ... The calculating plan is recorded in a form that is suitable for the control of the individual devices, e.g. on a punched paper tape. The calculating plan is scanned by the machine, section by section, and provides the following details for each calculating operation: the identifying numbers of the storage cells containing the operands; the basic type of calculation; the identifying numbers of the cell storing the result. The calculating plan's fine detail [*Angaben*] automatically triggers the necessary operations.[239]

Concerning his use of the binary system, Zuse said:

> [O]ne can ignore human habits and choose the simplest number system. Leibniz previously recognized ... the system with base 2 to be the simplest system. This obviously holds for the case of the calculating machine as well.[240]

[238]Newman, M. H. A. 'A Discussion on Computing Machines', *Proceedings of the Royal Society of London*, Series A, vol. 195 (1948), pp. 265–287 (p. 265).

[239]Zuse, Patent Application Z23139, April 1936, pp. 1–2.

[240]Zuse, Patent Application Z23139, April 1936, p. 8.

Is there any evidence that Zuse went further than this in his thinking, to reach the point of formulating the concept of a *universal* computer, independently of Turing? A patent application dating from 1941 contained a passage that might be taken to suggest an affirmative answer. Our view, though, is that the correct answer is negative. In the 1941 application, Zuse first explained that

> New [to the art] is the combination of elements in such a way that orders are given to the whole system from a scanner ... The calculating unit A is connected with the storage unit C so that the calculating unit's results can be transferred to any arbitrary cell of the storage unit, and also stored numbers can be transferred to the individual organs [*Organe*] of the calculating unit.[241] P is the plan unit together with the scanner. It is from here that the calculating unit's operating keys are controlled, as well as the selection unit, Pb, which connects up the requisite storage cells with the calculating unit.[242]

About the arrangement just described Zuse claimed:

> By means of such a combination it is possible, in contrast to currently existing devices, to calculate every arbitrary formula [*Formel*] composed from elementary operations.[243]

This statement might be thought to parallel the Church-Turing thesis that every *effective* calculation can be done by the universal machine, and so to embody an independent notion of universality.[244]

We argue that this is not so. Zuse's 1936 and 1941 patent applications described an automatic numerical calculator—a very advanced form of relay-based desk calculator, programmable and capable of floating-point calculations, and yet compact enough and cheap enough to be stationed permanently beside the user, probably an engineer. Zuse's machine was, in a sense, a personal computer. Rojas said: 'As early as 1935 or so, Zuse started thinking about programmable mechanical calculators specially designed for engineers. His vision at the time, and in the ensuing years, was not of a large and bulky supercomputer but of a desktop calculating machine.'[245] Zuse's desktop calculator added, subtracted, multiplied, and divided. There is no trace in Zuse's 1936 and 1941 descriptions of his machine of Turing's grand vision of a universal machine able not only to calculate, but also to solve non-numerical problems, to learn, and to reproduce the behaviour of a wide range of different forms of physical apparatus. Not until his work on the *Plankalkül* did Zuse consider 'steps in the direction of symbolic calculations, general programs for relations, or graphs, as we call it today, chess playing, and so on'.[246]

---

[241]Von Neumann also spoke of 'specialized organs' for addition, multiplication, and so on; von Neumann, 'First Draft of a Report on the EDVAC', p. 182 in Stern, *From ENIAC to UNIVAC*.

[242]Zuse, Patent Application Z391, 1941, p. 4.

[243]Zuse, Patent Application Z391, 1941, p. 4.

[244]Copeland, B. J. 'The Church-Turing Thesis', in Zalta, E. (ed.) *The Stanford Encyclopedia of Philosophy*, http://plato.stanford.edu/entries/church-turing/.

[245]Rojas, R., Darius, F., Göktekin, C., Heyne, G. 'The Reconstruction of Konrad Zuse's Z3', *IEEE Annals of the History of Computing*, vol. 27 (2005), pp. 23–32 (p. 23).

[246]Zuse, 'Some Remarks on the History of Computing in Germany', p. 625.

The 'five operations: addition, subtraction, multiplication, division and finding the square root, as well as translating from decimal to binary and back' are the basis of Zuse's numerical calculator.[247] In his patent application dated December 1936, he gave a detailed account of the implementation of the operations in his *Relaistechnik* (relay technology).[248] These are the 'elementary operations' [*elementaren Rechenoperationen*], each of which, he explained, in fact reduces to the first, addition, with subtraction reducing to addition of the complement.[249] When Zuse asserted that his machine could 'calculate every arbitrary formula that is composed from elementary operations', there is no reason to think he was envisioning a machine capable of carrying out any and every conceivable algorithm (if given enough memory); and in fact every reason to think he was claiming simply that his machine could, in principle, carry out any arbitrary combination of the *elementaren Rechenoperationen*.

In later life, Zuse was given to saying that his early calculating machines were universal computers. For example, talking about Z1, Z2 and Z3 in a 1968 interview, he stated 'Machines up to the Z3 are universal'; and in 1980 he wrote that Z1–Z4 'were universal computers'.[250] However, we encountered nothing in the documents that we examined from the period 1936–1945 to make us think that Zuse arrived independently at Turing's concept of a universal machine. Indeed, so far as we can tell, Zuse's later pronouncements about universality seem in fact to concern the idea of a general-purpose machine rather than a universal machine. The first statement just quoted was made in connection with contrasting Z1–Z3 with various 'specialized machines', the several versions of a 'specialized calculator' that he built for the German aircraft industry during the war (see Sect. 2).[251] In the second quoted statement, Zuse immediately added 'for numerical calculations'. The phrase '*general-purpose* computers for numerical calculations' makes perfect sense.

In 1997, Rojas argued that Z3 is universal, since Z3 can be programmed to simulate the working of the universal Turing machine, and so with Z3 'one can, in principle, do any computation that any other computer with a bounded memory can perform'.[252] Of course, this does not show that Zuse himself had the concept of universality (and nor did Rojas suggest that it does). The crucial step in Rojas's proof was to establish that the universal Turing machine can be simulated by a program occupying a single (finite) loop of punched tape and containing only Zuse's *elementare Rechenoperationen* of addition, subtraction, multiplication, division, and square root. Rojas concluded: 'Zuse's Z3 is, therefore, at least in principle,

[247] Zuse, Patent Application Z391, 1941, p. 9.

[248] Zuse, Patent Application Z23624, December 1936.

[249] Zuse, Patent Application Z23139, April 1936, p. 12. Square rooting is not mentioned in Z23139 but is dealt with in Z391.

[250] Zuse interviewed by Merzbach; Zuse, 'Some Remarks on the History of Computing in Germany', p. 614; and Zuse makes the same claim in his interview with Evans.

[251] Zuse interviewed by Merzbach.

[252] Rojas, 'How to Make Zuse's Z3 a Universal Computer', p. 53.

as universal as today's computers that have a bounded addressing space'.[253] He admitted, though, that his programming approach 'greatly slows down the computations' of Z3 and that 'the large loop of punched tape ... would pose extraordinary and most likely unsolvable mechanical difficulties'.[254] In fact, it is something of an understatement to speak of a 'large' loop of tape.

Rojas's universality proof for Z3 was the first of a genre: there are now a number of logical results showing that ancient computers, never designed to be universal, are so in principle. For example, Benjamin Wells argued in 2003 that a cluster of ten Colossi can implement a universal Turing machine—and there *were* ten Colossi in 1945, nine in the Newmanry at Bletchley Park and another in Flowers' factory.[255] In a later paper Wells conjectured that 'a single late Colossus Mark 2 endowed with an appropriate tape punch and controller' can implement a universal Turing machine—specifically Alastair Hewitt's 2-state, 3-symbol machine, proposed in 2007.[256] Another proof in the Rojas genre, this time concerning Babbage's Analytical Engine, was offered by Leif Harcke. Harcke described 'a program for the Engine that simulates a small [4-state, 6-symbol] universal Turing machine'.[257] This appears to vindicate Turing's claim, doubted by some, that the 'Analytical Engine was a universal digital computer'.[258]

Universality, while not ubiquitous, turns out to be more widespread than might have been expected. It is even possible that Colossus's largely electromechanical predecessor at Bletchley Park (a machine known simply as 'Heath Robinson', after William Heath Robinson, the Rube Goldberg of British cartoon artists) was universal: Wells conjectures that the construction he used to prove Colossus to be universal could apply to Robinson also, because of its functional similarity to Colossus (although, given the current lack of complete knowledge of the Robinson's hardware, this lies beyond the bounds of formal proof). Indeed Marvin Minsky showed long ago that in the context of a suitable architecture, simply an ability to multiply and divide by 2 and 3 leads to universality (his proof concerns program machines with registers that are capable of holding arbitrarily large numbers).[259] In

[253]Rojas, 'How to Make Zuse's Z3 a Universal Computer', p. 51.

[254]Rojas, 'How to Make Zuse's Z3 a Universal Computer', p. 53.

[255]Wells, B. 'A Universal Turing Machine Can Run on a Cluster of Colossi', *American Mathematical Society Abstracts*, vol. 25 (2004), p. 441.

[256]Wells, B. 'Unwinding Performance and Power on Colossus, an Unconventional Computer', *Natural Computing*, vol. 10 (2011), pp. 1383–1405 (p. 1402); Hewitt, A. 'Universal Computation With Only 6 Rules', http://forum.wolframscience.com/showthread.php?threadid=1432.

[257]Harcke, L. J. 'Number Cards and the Analytical Engine', manuscript (Copeland is grateful to Wells for sending him a copy of this unpublished proof). Wells found a lacuna in Harcke's proof but he believes this to be harmless; Wells says 'A memory management limitation can be overcome by seamlessly incorporating virtual memory, as Harcke agrees'.

[258]Turing, A. M. 'Computing Machinery and Intelligence', p. 455 in *The Essential Turing*.

[259]Minsky, M. L. *Computation: Finite and Infinite Machines* (Englewood Cliffs: Prentice-Hall, 1967), p. 258.

the light of Minsky's theorem, it would have been rather curious had Z3 turned out *not* to be universal.

It is an undeniable feature of all these universality proofs for early machines that the proofs tell us nothing at all about these ancient computers as they actually existed and were actually used. Despite Wells' proof that the Colossi were universal, Flowers' actual machines were very narrowly task-specific. Jack Good related that Colossus could not even be coaxed to carry out long multiplication. This extreme narrowness was no defect of Colossus: long multiplication was simply not needed for the cryptanalytical processing that Colossus was designed to do. Wells' result, then, teaches us a general lesson: even the seemingly most unlikely devices can sometimes be proved to be universal, notwithstanding the actual historical limitations of the devices.

Similar remarks apply to Rojas's result about Zuse's machine. His proof tells us nothing at all about the machine as it was used, and viewed, within its own historical context, and nothing at all about its scope and limits as a practical computer. Nevertheless, these results are certainly not without interest. As Wells put it:

> Colossus *was* the first functioning electronic universal machine. The Analytical Engine, Colossus, and Z3 were all universal. This has nothing to do with the intentions or writings of Babbage, Flowers, or Zuse—it is an objective property of their creations.

6 Zuse and the Stored-Program Concept

Zuse presented instructions in two different formats, a high-level human-friendly format, and the low-level form punched into the programming tape. The high-level format described in his April 1936 patent application was a three-address code, with each instruction consisting of four components: (1) the address of the cell of the store containing the first operand; (2) the address in the store of the second operand; (3) the operation code, which Zuse wrote as 'Add.', 'Subt.', 'Mult.', and so on; and (4) the address to which the result was to be stored.[260] In his 1941 patent application, this three-address code was replaced by a single-address code. The instructions were of the following forms: Read from cell *n*; Store in cell *n*; + ; − ; × ; Output result.[261] The numerical output was displayed on banks of lights.[262]

Zuse's machine-level code corresponded to the punch holes in the program tape.[263] Each of six positions falling on a straight line across the width of the tape could be punched or blank. Every instruction was represented by eight of these lines of punch-holes; eight *Felder*, in Zuse's terminology. The two-dimensional array of holes making up a *Feld* operated the appropriate relays directly. The first two

---

260 Zuse, Patent Application Z23139, April 1936, p. 4.

261 Zuse, Patent Application Z391, 1941, p. 5.

262 Zuse, 'Some Remarks on the History of Computing in Germany', p. 618.

263 Zuse, Patent Application Z391, 1941, p. 40.

Felder indicated whether the instruction was a read-from-store, write-to-store, or an order to the calculating unit. Subsequent *Felder* were used to specify addresses, and also a calculating operation in the case of an order to the calculating unit. As well as punching instructions on the tape, Zuse also made provision for punching numbers, such as $\sqrt{2}$, π, g, and other frequently-used constants. A special tag indicated whether the next group of *Felder* contained an instruction or number, and in the latter case the tag's punch-pattern switched the scanner temporarily to number mode.

As Zuse described the fundamental working of the machine in 1936:

> For each operation, the machine's work routines consist of 4 phases [*Takten*], in accordance with the calculating plan: 1.) and 2.) transfer of the operands into the calculating unit, 3.) calculating operation, 4.) transfer of the result into the storage unit.[264]

In the 1941 design, groups of four machine-code instructions implemented this fetch-operate-store cycle.

Zuse's separation of the addressable store and the calculating unit is possibly what Schmidhuber was referring to when he said that, in the April 1936 patent application, Zuse 'described what is commonly called a "von Neumann architecture" (re-invented in 1945)'. However, this idea in fact goes back to Babbage's Analytical Engine, with its separate Store and Mill.[265] In the 1941 patent application, Zuse himself emphasized that the coupling of an individual storage unit and counting unit was 'well known', citing as prior art 'counting units attached to a drum'.[266]

The machine that Zuse described in detail in his 1936 patent applications falls fairly and squarely into programming paradigm **P2**. The punching and scanning operations used to implant the instructions on the tape and to read them into the machine are not of the same kind as the operations used to execute the instructions, viz store, read from store, add, subtract, and multiply.

However, after describing his machine's four-phase operation, Zuse gave a list of potential refinements and extensions, reproduced here in full:

> If a number remains the same for the next operation in the calculating device, the phases 'store the result' and 'bring over the 1st operand for the next calculation' can be omitted. In this way redundant phases are avoided.
> By installing two connections between the storage unit and calculating device, so enabling numbers to be transferred back and forth, phases can be nested.
> Several calculating units, storage units, distributors, scanners, punches etc. can be installed and thus several operations carried out at the same time.
> Frequently used numbers, such as $\sqrt{2}$, π, and g, can be made permanently available in fixed number storages.
> The scanner and punch for the initial values and result can be replaced by setting and reading devices.

---

264 Zuse, Patent Application Z23139, April 1936, p. 5.

265 Bromley, A. 'Charles Babbage's Analytical Engine, 1838', *Annals of the History of Computing*, vol. 4 (1982), pp. 196–217.

266 Zuse, Patent Application Z391, 1941, p. 3.

> The calculating plan can be stored too, whereby the orders are fed to the control devices in synchrony with the calculation.
> Correspondingly, calculating plans can be stored in a fixed form if the machine is to carry out the same calculation often.[267]

Beyond those two brief sentences, noting the possibility of storing the calculating plan, Zuse said nothing more about the matter in his patent application. In particular, he did not say where or how the plan would be stored. If storage was to be in some unit logically equivalent to Eckert's engraved disks, then the device that Zuse was describing still conforms to paradigm **P2**. If, however, he meant that the calculating plan, expressed in binary code, was to be placed in the addressable relay store, together with the initial values, and whatever numbers were transferred to the store from the calculating unit as the calculation progressed—and it seems reasonable enough to interpret him in this way, since he mentions no other kind of storage apart from the addressable store and the punched tape—then Zuse can reasonably be taken to be suggesting *S0* programming, albeit very briefly. The 1938 documents examined later in this section tend to support this interpretation. If Zuse was indeed thinking of *S0* programming, however, there is little in the 1936 document to indicate that his thinking went beyond *S0* at this time (we discuss his 'two connections between the storage unit and calculating unit' below). In particular there is no evidence to suggest that the further steps involved in *S/E* were in his mind.

Schmidhuber's claim that in 1936, Zuse described 'a "von Neumann architecture" ... with program and data in modifiable storage' is immensely misleading. What Zuse described in 1936, in great detail, was a **P2** architecture. In two brief sentences he mentioned in passing the possibility of storing the program, but gave no architectural detail whatsoever. Moreover, far from offering further development in his 1941 patent application of this idea of storing the program, it is not even mentioned there. 'The calculating plan has the form of punched paper tape', Zuse stated in 1941.[268]

It is not so surprising that in his 1941 design Zuse did not pursue his idea of placing binary coded instructions in the relay store, nor implement the idea in Z3 or Z4. Any speed differential between the relay-based calculating unit and the tape mechanism was not so great as to create an instruction bottleneck, and the internal storage of instructions would use up precious cells of the relay store.

Nevertheless, Zuse did return to the program storage idea: half a dozen handwritten pages in a 1938 workbook extended his cryptic suggestion of 1936. The entries are dated 4 and 6 June and are a mixture of labeled diagrams and notes in the Stolze-Schrey shorthand system. Zuse's shorthand was transcribed into ordinary German by the *Gesellschaft für Mathematik und Datenverarbeitung* (Society for Mathematics and Data Processing) during 1977–1979.[269] In these pages, Zuse

---

[267]Zuse, Patent Application Z23139, April 1936, pp. 6–7.

[268]Zuse, Patent Application Z391, 1941, p. 40.

[269]Both workbook and transcription are in the Deutsches Museum Archiv, NL 207/01949. We are grateful to Matthias Röschner of the Deutsches Museum for information.

introduced a special 'plan storage unit' [*Planspeicherwerk*]. This was of the same type as the main relay store. The plan storage unit was coupled with a 'read-out unit' [*Abfühlwerk*] and these two units are shown in his diagrams as replacing the punched tape for supplying instructions to the calculating unit. As the notes progressed, Zuse dropped the distinction between the plan storage unit and the main relay store, using the main store for plan storage.

Zuse's principal focus in these notes was to develop a 'simpler way' of dealing with 'plans with iterating or periodical parts' (as mentioned previously, if a block of instructions punched on tape needs to be iterated, the instructions are punched over and over again, a clumsy arrangement). In addition to the plan storage unit, Zuse introduced a number of supplementary units to assist with program management and control. He named programs by a 'plan identification number' [*Plan-Nummer*], and to the plan storage unit he added a 'setting unit' [*Einstellwerk*] and a 'plan selection unit' [*Planwählwerk*]. These three units together with the read-out unit made up the 'plan unit' [*Planwerk*]. When a plan identification number was written into the setting unit, the plan selection unit would progressively select the lines of the store containing the identified plan, allowing the read-out unit to deliver the plan's instructions one by one.

Next Zuse introduced the idea of a numbered 'subplan' and of a 'self-controlling' calculating unit. He distinguished between what he called 'outer orders' [*äussere Befehle*] and 'inner orders' [*innere Befehle*]. He wrote: 'Outer orders control the work unit [*Arbeitswerk*], inner orders control the order unit [*Befehlswerk*]'. The work unit consists of the selection unit, the storage unit, and the operation unit (calculating unit). Unfortunately Zuse did not explain the term 'order unit' [*Befehlswerk*], which occurs only once in Zuse's notes and is presumably different from the plan unit [*Planwerk*].

In the default case, the outer orders are delivered sequentially from the plan storage unit. Inner orders, however, can trigger the operation of a 'counting unit' [*Zählwerk*]. There are two of these counting units, E_0 for the main plan and E_1 for the subplans. Zuse's diagram shows the plan storage unit supplying inner orders to the counting units. These inner orders appear to be essentially parameters. The arrival of a parameter from the plan storage unit at one of the counting units toggles the unit into action. Zuse wrote: 'Inner orders trigger changes in the relevant counting unit, whereby E_1 (subplan unit) carries on counting from the desired Nr and E_0 carries on counting from where it stopped last.' While E_0 is toggled on, the plan selection unit causes the plan storage unit to stream outer orders from the main plan to the work unit; and while E_1 is toggled on, the plan storage unit streams outer orders from the subplan. The subplan is selected by writing the subplan identification number into the setting unit.

Finally, Zuse described an arrangement whereby the plan storage unit was able to place the return address (the number of the instruction that will be read out of the plan storage unit once a subplan has been completed) into a special storage unit for instruction numbers. Zuse also explained that, by using additional counting units, all of them controlled by parameters from the plan storage unit, multiple nesting of subplans [*Mehrfachverschachtelung*] could be achieved.

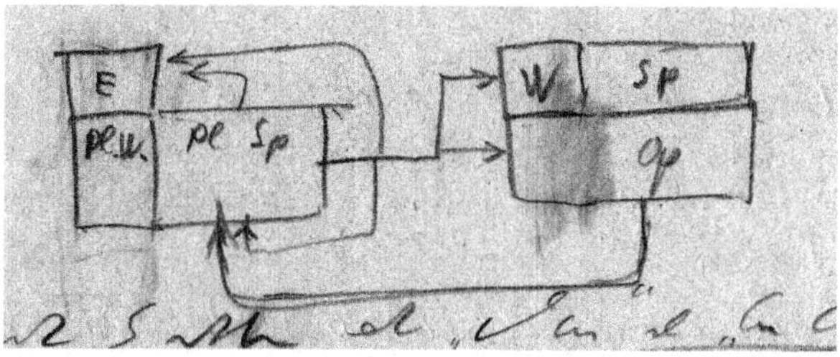

Fig. 2 Transferring information [*Angaben*] from the 'work unit' to the 'plan unit'. *E* is the setting unit, *Pl.W.* is the plan selection unit, *Pl Sp* is the plan storage unit, *W* is the selection unit, *Sp* is the storage unit, and *Op* is the calculating unit. *Credit: Deutsches Museum (Nachlass Konrad Zuse, Bestellnr. NL 207/0717)*

These arrangements of Zuse's are reminiscent of ENIAC-1948—and, like ENIAC-1948, exemplify **P**3. Instruction storage is read-only and Zuse is describing *S0*, not *S*. In Zuse's metaphor, the difference between **P**3 and **P**4 is the existence, in **P**4, not only of 'a controlling line going from left to right, but also from right to left'.[270] The effects of the arrangements described so far, Zuse said, can be achieved equally with punched tape. He emphasized that the whole setup described up to this point in the notes 'only serves the purpose of setting out rigid plans with iterating or periodical parts in a simpler way and it can be replaced in every case by a punched paper tape'.

Then, right at the end of the entry for June 4, Zuse adopted a new tack, in a brief section headed 'Living Plans' [*Lebende Pläne*]. He distinguished what he called living plans from 'rigid plans' [*starre Pläne*]. In two sketch diagrams and a mere 14 words of German he expounded the idea of allowing the work unit to write to the plan unit. Zuse heavily underlined the shorthand for 'Rück-Koppelung'—*feedback* between the work unit and the plan unit. This was a development of his pithy 1936 remark, quoted above, that by 'installing two connections between the storage unit and calculating device, so enabling numbers to be transferred back and forth, phases can be nested'. In this brief section of his 1938 notes, Zuse added a control line leading from 'right to left'. His diagram, showing an arrowed line leading back from the work unit to the plan unit, is captioned 'Transferring information [*Angaben*] from the "work unit" to the "plan unit"' (see Fig. 2).

So Zuse appears to have made the jump from **P**3 to **P**4, from *S0* to *S*. In his 1936 patent application, the two-way connection was suggested only as a means of permitting the nesting of phases [*Takten*], but in the 1938 notes, with his talk of 'living plans', Zuse seems to have had more in mind than nesting *Takten*.

In a final diagram, he swept away the distinction between the plan storage unit and the main store altogether, amalgamating the two. This showed two-way traffic between the work unit and the general-purpose store, with one arrowed line leading

[270]Zuse, 'Some Remarks on the History of Computing in Germany', p. 616.

from the store to the work unit, and a second arrowed line leading back from the work unit to the store.

How was the *potential* implicit in *S*—implicit in the bi-directional connection—to be used? About this Zuse wrote tantalizingly little. All he said, in his June 6 entry in the workbook, headed 'Dependent and independent feedback', was this:

> Independent feedback = independent from the initial details (note: initial details [*Ausgangsangaben*] = input values [*Eingabewerte*]) serves only to enable the plan to be represented in a more compact form, and for this to unfold as it runs. Plans with independent feedback are still said to be rigid.
>
> Dependent feedback = actual living plans. Effect [*Einfluss*] of the details [*Angaben*] calculated [*errechneten*], thus also of the initial details [*Ausgangsangaben*] on the sequence of events [*Ablauf*] in the computation [*Rechnung*].

These comments are altogether too cryptic for an interpreter to be certain what Zuse meant. We offer the following speculative example of a 'living plan', which causes the machine to select a subroutine for itself, on the basis of calculations from input values, and also to calculate from the input how many times to run the subroutine. We select this particular example in order to make the idea of a living plan vivid. The example derives from Turing's work on note-playing subroutines, as described in his *Programmers' Handbook for Manchester Electronic Computer Mark II*.[271] The example conforms well to Zuse's 1976 description of his early ideas: 'instructions stored independently and special units for the handling of addresses and subroutines'.[272]

Consider a simple piece of software for playing musical notes. The program takes as input values (a) a number *n*, and (b) the name of a musical note. In the case we will describe, the note is C_4, middle C (the subscript indicating the octave). These input values cause the computer to play the note of C_4, the number *n* functioning to tell the computer how long to hold the note. The computer plays the note by outputting a stream of suitable pulses (a stream of 1s separated by appropriate numbers of 0s) to an attached amplifier and loudspeaker. The actual details of how the sound is produced are not relevant to the example. The key point is that the note is generated by a subroutine consisting of a loop of instructions; running the loop continuously sends the correct stream of pulses to the loudspeaker. Each note-playing subroutine has a subplan identification number. *n* is the timing number, causing the computer to hold the specified note for *n* seconds.

Inputting the note name C_4 (in binary, of course) causes the calculating unit to calculate a subplan identification number from the binary input, by means of a fixed formula. The calculating unit then feeds this identification number back to the plan storage unit, and thence the number is transferred to the plan selection unit (by some

[271]Turing, A. M. *Programmers' Handbook for Manchester Electronic Computer Mark II* Computing Machine Laboratory, University of Manchester, no date, circa 1950; a digital facsimile is in *The Turing Archive for the History of Computing* at www.AlanTuring.net/programmers_handbook. See also Copeland, B. J., Long, J. 'Electronic Archaeology: Turing and the History of Computer Music', in Floyd, J., Bokulich, A. (eds) *Philosophical Explorations of the Legacy of Alan Turing* (New York: Springer, 2016).

[272]Zuse, 'Some Remarks on the History of Computing in Germany', p. 616.

mechanism that Zuse left unspecified). Next, the calculating unit divides n by the time taken to obey a single outer instruction (assumed to be a constant), and feeds this parameter m back to E_1, the counting unit for subplans. The parameter's arrival has the effect of toggling E_1 on, with the result that the subplan whose identification number is in the plan selection unit starts to run. After m steps, control passes back to the main program, at which point the loudspeaker has played C_4 for n seconds.

This example of **P4** programming, which involves no editing of instructions, appears to us to illustrate the principles described by Zuse in his 1938 notes. While it cannot be ruled out that Zuse might have had in mind editing an inner instruction in the course of the unspecified steps leading to the delivery of the subplan identification number and the parameter m to their respective units, he certainly did not say so. Zuse in fact gave only the vaguest description of the inner order to E_1, saying merely that these orders were of the form: '"continue E_1" Nr ...'. More importantly, we find no trace of evidence in these notes that Zuse was thinking of *S/E* (either *S/E/*Δ*I* or *S/E/*=) in the case of outer orders.

7 Concluding Remarks

To conclude, we reprise the main events in the stored-program concept's early history. We also present translations of key remarks from Zuse's 1945 manuscript 'Der Plankalkül', which, unlike his 1936 and 1938 documents, did describe instruction editing.

The stored-program story began in 1936 (see the timeline in Fig. 3), when in 'On Computable Numbers' Turing published a description of *S/E/*=: storage of instructions, with editing, but no creation of different instructions. *S/E/*Δ*I*, editing that does lead to new instructions, is very obviously available in the setup Turing described in 'On Computable Numbers'; but he did not mention *S/E/*Δ*I* in 1936. In the absence of evidence, it is impossible to say whether or not he noticed this intriguing possibility at the time.

Also in 1936, in an unpublished and subsequently withdrawn application for a patent, Zuse gave a bare mention of *S0*, the bottom rung of the hierarchy of levels comprising the stored-program concept. He also introduced, very briefly, a proto-version of his idea of two-way traffic between the storage unit and the calculating unit, limiting the scope of the idea to nesting *Takten*. In 1938, in a few sheets of handwritten personal notes, Zuse developed his brief 1936 suggestion of placing instructions in his addressable relay store. In these notes he initially described a sequence of architectures involving *S0*, and then, by adding *Rück-Koppelung*—feedback from the work unit to the plan unit—he apparently described a calculating machine that implements *S*. Zuse therefore seems to have been the first to sketch (in barest outline) a practical architecture in which *S* could be achieved. However, he did little at this time to tap the potential of *S*. Not even his 'living plans' can be said to involve *S/E/*= or *S/E/*Δ*I*: any such claim would go well beyond the evidence present in the document.

Fig. 3 Timeline: early history of the stored-program concept[a]

1936

Turing describes *S* and *S/E/=*

Zuse describes *S0*

1938

Zuse describes *S*

1944

Eckert describes *S0* and *S*

1945

Von Neumann describes *S* and restricted *S/E/ΔI/SELF*

Zuse describes *S/E/ΔI/OTHER*

Turing describes unrestricted *S/E/ΔI/SELF* and *S/E/ΔI/OTHER*

1948

ENIAC-1948 implements *S0*

Manchester Baby implements *S*

S/E/ΔI/SELF described at Manchester and possibly implemented

1949

Expanded Baby implements *S/E/ΔI/SELF* using the B-tube (index register)

The Cambridge EDSAC and Eckert-Mauchly BINAC run stored programs, closely followed by MIT's Whirlwind and the Australian CSIR Mk1[a]

[a]Wilkes, M. V. *Memoirs of a Computer Pioneer* (Cambridge, Mass.: MIT, 1985), p. 142; Lukoff, H. *From Dits to Bits: A Personal History of the Electronic Computer* (Portland, Oregon: Robotics Press, 1979), p. 84; Woodger, M. 'Stored-program Electronic Digital Computers' (handwritten

Zuse then turned his back on the stored-program idea. His 1941 Z3 and 1945 Z4 used punched tape to control the computation and are exemplars of the **P**2 programming paradigm. Much later, reflecting on his early work, Zuse said he had felt that implementing the *Rück-Koppelung* of his 1938 notes 'could mean making a contract with the devil'.[273] 'When you make feedback from the calculating unit to the program, nobody is able to foresee what will happen', he explained.[274] Zuse emphasized that Z1–Z4 had 'no feedback in the program unit'.[275]

When Zuse began developing his *Plankalkül* language, however, he did return to the stored-program concept. In 1945, the *annus mirabilis* of the stored-program concept, Zuse wrote his five-chapter manuscript 'Der Plankalkül'. Unpublished and little-known, this document is the shadowy third member of a momentous trilogy of reports from 1945, each developing the stored-program concept in its own way—the others being, of course, von Neumann's 'First Draft of a Report on the EDVAC' and Turing's 'Proposed Electronic Calculator'.

In a section of 'Der Plankalkül' titled 'Repetition plans', Zuse said:

> [T]he orders contained in the repetition part [of the plan] are subject to continuous changes that result from the repetitions themselves.[276]

A 'variation instruction' [*Variationsvorschrift*] produces 'variations of the plan' [*die Variationen des Planes*].[277] There is, however, no 'contract with the devil': Zuse's examples of repetition plans seem, as with his 1938 notes, to involve *S* but no editing of instructions within the meaning of the act.

'Der Plankalkül' does clearly set out the concept of *S/E/*Δ*I/OTHER*, describing the automatic 'calculating of calculating plans'. Zuse wrote:

> The modifying of calculating plans is a function of the plan variables, which can consist of variable operation symbols, plan symbols [introduced so that 'subplans occurring inside a calculating plan can be modified'], structure symbols ['calculating plans of the same structure can adopt different meanings by modification of the algebraic dimension'], or other values. ...
>
> The use of such variable structure symbols plays an important role when the task is to represent calculating plans for storage in calculating machines in as compact a form as possible, in order to capture far-reaching variation with little extra expenditure. ...
>
> In the cases discussed, calculating plans are modified by the simple insertion of variable symbols. However, modifications of an essentially more complicated nature are possible. One can move from this kind of calculating plan to free calculating plans ...[278]

note in the Woodger Papers, Science Museum, Kensington, London); McCann, D., Thorne, P. *The Last of the First. CSIRAC: Australia's First Computer* (Melbourne University Press, 2000), p. 2.

273 Zuse, 'Some Remarks on the History of Computing in Germany', p. 616.

274 Zuse interviewed by Evans.

275 Zuse, 'Some Remarks on the History of Computing in Germany', p. 616.

276 Zuse, 'Der Plankalkül' (manuscript), p. 32.

277 Zuse, 'Der Plankalkül' (manuscript), p. 32.

278 Zuse, 'Der Plankalkül' (manuscript), pp. 23, 24, 25.

Free calculating plans, seemingly the same as or at any rate similar to the 'living plans' of Zuse's 1938 workbook, are those in which 'the actual variables [*eigentlichen Variablen*] have an effect on the course of the calculation'.[279] In 'Der Plankalkül', Zuse described the automatic calculation of free calculating plans, noting that this process might determine 'only parts' of the plan, leaving other details to be filled in by the machine as the plan actually runs.[280] The *Plankalkül* and Zuse's early ideas about compilation are topics for a more detailed discussion in a further article.

Turing also described *S/E/*Δ*I/OTHER* in 1945 and, of course, Turing and von Neumann both described *S/E/*Δ*I/SELF* in that year (in their respective reports on the ACE and the EDVAC). Von Neumann restricted the scope of editing to the instruction's address bits, whereas Turing described unrestricted editing of instructions. It was not until his later papers, from 1946 onwards, that unrestricted *S/E/*Δ*I/SELF* appeared in von Neumann's work.

Jumping back a year in order to mention developments at the Moore School: in 1944 Eckert described a disk memory device involving both *S0* and *S*, but there is no evidence that he was considering *S/E/*Δ*I* or even *S/E/=* in connection with this device. Around the same time, he and Mauchly conceived the idea of storing instructions in the ENIAC's function tables, an idea later reduced to practice at the Ballistic Research Laboratory, by Clippinger and others, in 1948. Storing instructions in ENIAC-1948s read-only function tables was an example of the **P3** programming paradigm, achieving *S0* but not *S*.

S and *S/E/*Δ*I/SELF* were first implemented at Manchester, in Max Newman's Computing Machine Laboratory. *S* was achieved in June 1948, but initially the Baby was used without instruction editing, as a surviving laboratory notebook shows.[281] During the summer, though, Kilburn added a kind of halfway house, a relative control transfer, achieved by adding a number from the store to the address of the next instruction (or subtracting the number from the address), this address being stored in a control tube.[282] Soon the editing of instructions was considered useful enough for Williams and Kilburn to add special hardware for it. The new hardware was known simply as the *B*-tube, the accumulator already being named the *A*-tube and the control tube the *C*-tube. In modern terms, the *B*-tube was an index register. Kilburn explained:

> Instructions can, of course, be modified by the normal processes ... in the same way as numbers, but this is often inconvenient and wasteful of time and storage space. Therefore each instruction ... is preceded by a single digit called the b digit. If $b = 0$, the content of

[279] Zuse, 'Der Plankalkül' (manuscript), p. 25.

[280] Zuse, 'Der Plankalkül' (manuscript), p. 31.

[281] Tootill, 'Digital Computer—Notes on Design & Operation'.

[282] Kilburn interviewed by Copeland, July 1997.

line *B0* of *B* (normally zero) is added into the present instruction ... before this instruction is used. If $b = 1$, the content of line *B1* of *B* is used in the same manner.[283]

Baby. Tom Kilburn is on the left, Freddie Williams on the right. *Credit: University of Manchester School of Computer Science*

In the Manchester laboratory notebook previously mentioned (compiled by engineer Geoff Tootill), the earliest dated appearance of the *B*-tube idea was in an entry for 13 October 1948, giving coded instructions for the operations that Kilburn described in the above quotation.[284] The original idea had emerged some weeks earlier, during a discussion between Newman, Williams, Kilburn and Tootill.[285] It arose initially (Williams and Kilburn explained) as 'a convenient means of shifting the effect of a whole block of instructions by a constant amount, while leaving others

[283]Kilburn, T. 'The University of Manchester Universal High-Speed Digital Computing Machine', *Nature*, vol. 164, no. 4173 (1949), pp. 684–7 (p. 687).

[284]Tootill, 'Digital Computer—Notes on Design & Operation', list of the machine's instructions dated 13/10/48. There are also two undated references to instructions involving the B-tube a few pages earlier. Kilburn included the same B-tube instructions in his 'Code for Charge Storage Computer', a list of the machine's instructions that is dated 30 November 1948. (Copeland is grateful to Simon Lavington for sending him a copy of Kilburn's document, seemingly a lecture handout. Lavington included a retype of the document in his 'Computer Development at Manchester University', in Metropolis, Howlett and Rota, *A History of Computing in the Twentieth Century*, p. 439.)

[285]Kilburn interviewed by Copeland, July 1997.

that were not B-modified unaffected'.[286] A young assistant, Dai Edwards, was given the job of developing the new piece of hardware.[287]

The B-tube was probably being used to run engineering test programs by March 1949 and was ready for routine use in April 1949.[288] It was part of an extensive upgrade, completed in April, that transformed the computer from a small demonstration model into a usable machine: the upgrade also included a magnetic drum memory, improved CRT storage, a hardware multiplier, and an increase in word length to 40 bits.[289] Manchester was well ahead of the field, with a number of other pioneering computer projects in the UK, US and Australia succeeding in running stored programs later in 1949 (see the timeline in Fig. 3).

Was the less convenient method of instruction modification that Kilburn mentioned—not involving the B-tube—ever actually implemented during the period June 1948–April 1949? This is a tantalizing and important question, since the first implementation of instruction editing was a key moment in the history of computing. The method was important enough for Turing to discuss it in detail in his *Programmers' Handbook*: he described 'the formation of an instruction in the accumulator by addition, the copying of this instruction into the list of instructions, and the subsequent obeying of it' (adding, 'This is a rather clumsy process').[290] Turing gave several examples of small programs using this method of instruction editing, in sections of the *Handbook* describing what he called the 'reduced machine'. The 'reduced machine' is by and large the Manchester computer as it existed before April 1949.

Thus it is certainly possible that Manchester's first use of instruction editing preceded the arrival of the B-tube. But if so, no record appears to have survived. Therefore the question of precisely when instruction editing was first implemented—one of computer science's most historic dates—remains open.

Acknowledgments Copeland is grateful to the following institutions for supporting this research: University of Canterbury, New Zealand; University of Queensland, Australia; Federal Institute of Technology (ETH), Zurich, Switzerland; and Det Informationsvidenskabelige Akademi,

[286] Williams, F. C., Kilburn, T. 'The University of Manchester Computing Machine', in Bowden, *Faster Than Thought*, p. 122.

[287] Lavington, S. H. *A History of Manchester Computers* (Manchester: NCC Publications 1975), p. 12.

[288] Tootill, 'Digital Computer—Notes on Design & Operation', note dated 27/3/49 inserted within an older note dated 15/7/48; Dai Edwards in interview with Simon Lavington, 6 May 2015. (Copeland is grateful to Lavington for making the transcript of this interview available.)

[289] Williams and Kilburn, 'The University of Manchester Computing Machine', pp. 121–122; Edwards in interview with Lavington, 6 May 2015; Kilburn, T. Tootill, G. C., Edwards, D. B. G., Pollard, B. W. 'Digital Computers at Manchester University', *Proceedings of the Institution of Electrical Engineers*, vol. 77 (1953), pp. 487–500.

[290] Turing, *Programmers' Handbook for Manchester Electronic Computer Mark II*, pp. 19–20. A typo has been corrected: in Turing's original, the second occurrence of 'instruction' was typed 'instructions'.

Copenhagen University, Denmark. Copeland and Sommaruga are grateful to Herbert Bruderer, Brian Carpenter, Martin Davis, Bob Doran, Simon Lavington, Teresa Numerico, Diane Proudfoot, and Benjamin Wells for helpful comments on a draft of this chapter, and to the Deutsches Museum for providing digital facsimiles of various original documents by Zuse.

Part II
Generalizing Turing Computability Theory

Theses for Computation and Recursion on Concrete and Abstract Structures

Solomon Feferman

Abstract The main aim of this article is to examine proposed theses for computation and recursion on concrete and abstract structures. What is generally referred to as Church's Thesis or the Church-Turing Thesis (abbreviated **CT** here) must be restricted to concrete structures whose objects are finite symbolic configurations of one sort or another. Informal and principled arguments for **CT** on concrete structures are reviewed. Next, it is argued that proposed generalizations of notions of computation to abstract structures must be considered instead under the general notion of algorithm. However, there is no clear general thesis in sight for that comparable to **CT**, though there are certain wide classes of algorithms for which plausible theses can be stated. The article concludes with a proposed thesis **RT** for recursion on abstract structures.

Keywords Church's Thesis • Church-Turing Thesis • Computation • Algorithm • Recursion • Concrete structures • Abstract structures

1 Introduction

The concepts of recursion and computation were closely intertwined from the beginning of the efforts early in the 1930s to obtain a conceptual analysis of the informal notion of *effective calculability*. I provide a review of those efforts in Sect. 2 as background to the remainder of this article, but I have nothing new to add here to the extensive historical and analytical literature.[1] It is generally agreed that the conceptual analysis of effective calculability was first provided most convincingly by Turing [3]. Not long before that Church [4] had proposed identifying effective calculability with the Herbrand-Gödel notion of general recursiveness, soon enough proved equivalent to Turing computability among other suggested explications.

[1]Gandy [1] is an excellent introductory source to those developments; cf. also [2].

S. Feferman (✉)
Department of Mathematics, Stanford University, Stanford, CA, USA
e-mail: feferman@stanford.edu

G. Sommaruga, T. Strahm (eds.), *Turing's Revolution*,
DOI 10.1007/978-3-319-22156-4_4

Curiously, the subject of effective computation came mostly to be called *Recursion Theory*, even though that is only one of the possible forms of its development.[2]

In his influential book, *Introduction to Metamathematics*, Kleene [6] baptized the statement that every effectively calculable function is general recursive as *Church's Thesis*. He went on to baptize as *Turing's Thesis* the statement that "every function which would naturally be regarded as computable is computable ... by one of his machines"[3] and to say that this is equivalent to Church's Thesis, since the general recursive functions are exactly the same as the (total) functions computable by Turing machines. This led Kleene to speak further on in his book in ambiguous terms of the *Church-Turing Thesis*. Among workers in recursion theory it is common to take Church's Thesis and Turing's Thesis to be equivalent for the same reason as given by Kleene, and to follow him in referring to them without distinction as Church's Thesis or as the Church-Turing Thesis; I shall also use '**CT**' ambiguously as an abbreviation for either of these.[4]

My main concern in this article is to examine proposed theses for computation and recursion on both concrete and abstract structures. By *concrete structures* I mean those given by sets of finite symbolic configurations (e.g., finite strings, trees, graphs, hereditarily finite sets, etc.) together with the appropriate tests and operations for their transformation. In Sect. 3 I review some efforts to "prove" **CT** for computation on concrete structures that began with Gandy [8] and were later pursued by Sieg and by Dershowitz and Gurevich among others (references below). The approaches in question proceed by isolating basic properties of the informal notion of effective calculability or computation in axiomatic form and proving that any function computed according to those axioms is Turing computable. It is not my aim here to argue for one or another of these approaches but rather to emphasize what they hold in common (and what is usually taken for granted), namely that *in whatever way one understands* ***CT****, there is no calculation without representation*, i.e. it is a necessary ingredient of whatever constitutes effective calculability that one operates only on finite symbolic configurations by means that directly transform such configurations into new ones using appropriate tests along the way.

Beginning in the late 1950s a number of generalizations of the notions of computation and recursion were made to a great variety of *abstract structures*, including general first-order (or algebraic) structures, finite-type structures over the natural numbers, and structures of sets. These developments witnessed impressive success in obtaining various analogues of leading results from classical recursion

---

[2]Soare in his articles [2, 5] has justifiably made considerable efforts to reconfigure the terminology of the subject so as to emphasize its roots in the notion of computation rather than recursion, for example to write 'c.e.' for 'computably enumerable' in place of 'r.e.' for 'recursively enumerable', but they do not seem to have overcome the weight of tradition.

[3]Kleene's wording derives from Turing [3, p. 249]: "No attempt has yet been made to show that the 'computable' numbers include all numbers *which would naturally be regarded as computable.*" [Italics mine]

[4]Cf. Copeland [7] for an introductory article on the Church-Turing thesis.

theory. Nevertheless, the point of my emphasis on concrete structures as a *sine qua non* for **CT** is to raise questions about proposed generalizations of it to abstract structures that have been suggested. In particular, I shall concentrate in Sect. 4 on general theories of "computation" on first-order structures that descend from Friedman [9] via his adaptation of the Shepherdson-Sturgis register machine approach (equivalently, the Turing machine approach) on the one hand, and that of Tucker and Zucker [10, 11] via "While" schemata on the other. I shall argue (as Friedman already did) that the proposed generalizations are more properly examined under the concept of *algorithmic procedures*, which are meaningful for abstract structures (or "data types") since algorithms are independent of the means by which individual data items and the operations on them may be represented. As will be explained at the end of Sect. 4, substantial efforts have been made to answer the question, "What is an algorithm?," leading to quite different conclusions, among them by Moschovakis [12, 13] and Gurevich [14]. In view of the controversy, it is perhaps premature to propose a general associated version of **CT** for algorithms, though wide classes of algorithms may be candidates for such. In particular, Tucker and Zucker (op. cit.) have formulated a thesis for algebraic algorithmic procedures on abstract structures that deserves special attention under that heading.

Finally, by comparison with these, Sect. 5 reviews a general notion of recursive definition applied to suitable second-order structures, and a general thesis **RT** related to such is proposed for consideration.

Before turning to this material, one may well ask: What purpose is served by the questions of formulation and justification of these theses? We should not expect a single answer in each case, let alone a single overall answer; here are a few. In the case of **CT** itself, we had the historical pressure on the negative side to demonstrate the effective unsolvability of the *Entscheidungsproblem* in logic and, subsequently, many more examples of that in mathematics. On the positive side, it was and continues to be used to provide a solid foundation to informal proofs of computability, i.e. to justify "proofs by Church's Thesis." (And for Gödel, it was used to bolster his conviction that mind is not mechanical, via the incompleteness theorems and the identification of formal systems in their most general form with Turing machines.) Moving on to the proposed extension of notions of computability to abstract structures, a prime structure of interest is that of the real numbers (and relatedly the complex numbers), the arena of numerical analysis (aka scientific computation). Among the concerns here are to understand its limits and to provide a unified foundation. Moreover, as we shall see, a crucial issue is to separate conceptually algebraic methods from analytical ones. In addition, the latter are relevant to questions as to whether and in what sense the laws of physics are mechanical.

More generally, it is important to separate and refine concepts that are often confused, namely those of *computation procedure*, *algorithmic procedure* and *recursive definition* that are the primary concern of this article. Those who are philosophically inclined should find interest in these case studies in conceptual analysis (aka explication of informal concepts) both settled and unsettled. Finally, it may be hoped that satisfactory progress on these questions would help lead to the same on concepts of *feasibility* in these various areas, concepts that are wholly untouched here.

2 Recursion and Computation on the Natural Numbers, and the Church-Turing Thesis

The question as to which forms of definitions of functions on the natural numbers are finitistically acceptable was an important issue in the development of Hilbert's program in the 1920s. In the views of the Hilbert school this certainly included the primitive recursive functions (going back to Dedekind), but was not limited to such once Ackermann produced his example of a non-primitive recursive function. Formally, that was given by a nested double recursion, but it could also be analyzed as a recursion on the ordinals up to ω^2. Hilbert's unsupported claim [15] to have demonstrated the Continuum Hypothesis by means of his proof theory involved the use of functions given by higher transfinite recursions; it would thus seem that Hilbert counted them among the finitistically acceptable functions, though no clear conditions were given as to what would make them so.

A more specific analysis of which functions of the natural numbers are definable by recursive means came about as a result of a single exchange of correspondence between Herbrand and Gödel in 1931. Herbrand wrote first after receiving a copy of Gödel's paper on the incompleteness of arithmetic that happened to make essential technical use of the schemata for primitive recursive functions. Concerning the contents of that letter, in 1963 Jean van Heijenoort asked Gödel about the exchange in connection with his [16] formulation of the notion of general recursive function, the idea for which he had credited in part to Herbrand. In his response to van Heijenoort, Gödel said that he could not locate the correspondence in his papers but that the formulation in the 1934 notes was exactly as had been proposed to him three years prior to that by Herbrand. The two letters in the exchange continued to be missing in the following years until they were found finally by John W. Dawson, Jr. in 1986 in the course of his cataloguing of the Gödel *Nachlass* at the Institute for Advanced Study in Princeton. As explained in detail in Dawson [17] and Sieg [18, 19] it turned out that Gödel misremembered some essential points.[5]

Herbrand's letter to Gödel informally proposed a characterization of the extent of the finitistically acceptable functions in terms of recursive definitions via a single formal system of arithmetic with quantifier-free induction and axioms for a sequence

[5]For convenience I shall only refer to Sieg [18] in the following, though there is considerable overlap with Dawson [17].

of functions f_n satisfying three conditions on successive quantifier free axioms for the f_n, of which the main one is:

> We must be able to show, by means of intuitionistic [finitary] proofs, that with these axioms it is possible to compute the value of the functions univocally for each specified system of values of their arguments. (cf. [18], p. 6)

Among examples, Herbrand lists the recursion equations for addition and multiplication, Gödel's schemata for primitive recursion, Ackermann's non-primitive recursive function, and functions obtained by diagonalization. Gödel's reply, though friendly, was critical on a number of points, among which Herbrand's proposal that finitism could be encompassed in a single formal system.[6]

In the notes for his 1934 lectures on the incompleteness theorems at the IAS, Gödel returned to Herbrand's proposal in the last section under the heading "general recursive functions." He there recast Herbrand's suggestion to define a new function φ in terms of "known" functions $\psi_1, \ldots, \psi_k$ and possibly φ itself. The main requirement is now taken to be that for each set of natural numbers $k_1, \ldots, k_l$ there is one and only one m such that $\varphi(k_1, \ldots, k_l) = m$ is a derived equation, where the rules of inference for the notion of derivation involved are simply taken to be those of the equation calculus, i.e. substitution of numerals for variables and substitution of equals for equals (cf. [16], p. 26).

Before turning to this bridge to the conceptual analysis of effective calculability, note that there are two features of Gödel's definition of general recursive function that make it ineffective in itself, namely (given "known" effectively calculable functions $\psi_1, \ldots, \psi_k$) one can't decide whether or not a given system of equations E has (i) the uniqueness property, namely that for each sequence of arguments $k_1, \ldots, k_l$ there is at most one m such that $\varphi(k_1, \ldots, k_l) = m$ is derivable from E, nor that E has (ii) the existence property, namely that there is at least one such m for each such sequence of arguments. In Kleene's treatments of general recursion elaborating the Herbrand-Gödel idea, beginning in 1936 and ending in Kleene [6] Ch. XI, one may use *any* system of equations E to determine a partial recursive function, simply by taking $\varphi(k_1, \ldots, k_l)$—when defined—to be the value m given by the least derivation (in a suitable primitive recursive coding) ending in some value for the given arguments. It is of course still undecidable whether the resulting function is total. It finally remained for Kleene to give the most satisfactory general formulation of effective recursion on the natural numbers via his Recursion Theorem (op. cit., p. 348). But this requires the notion of partial recursive functional to which we shall return in Sect. 5 below.

Now, not only did Gödel in 1934 misremember the details of Herbrand's 1931 formulation, he made a crucial conceptual shift there from the question of characterizing the totality of finitistically acceptable functions to that of characterizing the totality of functions given by a "finite computation" procedure, despite his clear

[6] Gödel was not long after to change his mind about that; cf. Gödel [70].

reservations both about the possibility of such and of general recursion in particular as a prime candidate. Church was bolder:

> We now define [sic!] the notion ... of an effectively calculable function of positive integers by identifying it with the notion of recursive function of positive integers (or of a λ-definable function of positive integers). [4, p. 356]

Church had previously proposed to identify the effectively calculable functions with the λ-definable ones, but by 1936 he could depend on their co-extensiveness with the general recursive functions established in Kleene [20, 21]. It was Kleene [22] who baptized Church's "definition" of effectively calculable function as 'Thesis I' and then as Church's Thesis in Kleene [6, p. 300]. See Gandy [1, pp. 76 ff] for Church's full statement of the Thesis (*qua* definition) and the initial arguments that he made in its favor, among which the equivalence of "two such widely different and ... equally natural definitions of effective calculability"; Gandy calls that *the argument by confluence of ideas.*[7]

This argument was soon to be extended by the most significant step in the analysis of the concept of effectively calculable function on the natural numbers, namely that made by Turing [3]. To be noted is that Turing conceives of the calculations as being carried out by an (abstract) human being following a fixed finite set of instructions or routine using a finite set of symbols specified in advance, via entering or deleting the contents of the workspace given by a potentially infinite set of symbol locations or cells. The confluence argument was bolstered by Turing's [24] proof of the equivalence of his notion of computability with that of λ-definability. Church quickly accepted Turing's analysis of effectively calculable as the preferred notion of the three then available. As he wrote in his review of Turing's paper:

> [Turing's notion] has the advantage of making the identification with effectiveness in the ordinary (not explicitly defined) sense evident immediately—i.e., without the necessity of proving preliminary theorems. [25]

Gödel, too, accepted Turing's explication of effective calculability, but by when is not clear. In the unpublished lecture designated **193?* in Gödel [26]—probably prepared in 1938—he describes the notion of general recursive function, and then writes, "[t]hat this really is the correct definition of mechanical computability was established beyond any doubt by Turing" [26, p. 168]; however, he does not refer to proofs of their equivalence as justification for this. The first evidence of his view in print appears briefly at the beginning of his remarks before the 1946 Princeton Bicentennial, where he speaks of "the great importance of the concept of general recursiveness (or Turing's computability)" because of the independence of the concept of calculability in a formal system from the formalism chosen (cf. [27], p. 150). But again he does not take Turing's explanation of effective computability

[7]Church [4, pp. 100–102] had another "step-by-step" argument for the Thesis, but there is a semi-circularity involved; cf. Shagrir [23, p. 224].

to be the primary one. Only later, in his June 1964 Postscript to Gödel [16], does he address the concept on its own terms as follows:

> Turing's work gives an analysis of the concept of "mechanical procedure" (alias "algorithm" or "computation procedure" or "finite combinatorial procedure"). This concept is shown to be equivalent with that of a "Turing machine". (cf. [28], p. 370)

As described in Sect. 1 above, it was Kleene [6, pp. 300 et seq], who led one to talk of Church's Thesis, Turing's Thesis, and then, ambiguously, of the Church-Turing Thesis for the characterization through these equivalences of the effectively calculable functions. In another influential text, Rogers [29, pp. 18 ff], took the argument by confluence as one of the basic pins for **CT** and used that to justify informal proofs "by Church's Thesis".

3 Computation on Concrete Structures and "Proofs" of CT

Beginning with work of Kolmogorov in 1953 (cf. [30]), efforts were made to move beyond the argument by confluence, among others, to more principled arguments for **CT**. The first significant step in that direction was made by Gandy [8] which, together with its successors to be described below, may be construed as following a more axiomatic approach, in the sense that one (cl)aims to isolate basic properties of the informal notion of effective calculability or computation and proves that any function computed according to those conditions is computable by a Turing machine. As one sees by closer inspection, all the axioms used by the work in question take the notion of finiteness for granted, hence may be construed formally as carried out within weak second-order logic, but otherwise there is considerable difference as to how they are formulated.

To begin with, Gandy asserts (as he did again in Gandy [1]) that Turing outlined a proof of the following in his famous paper [3]:

Thesis T What can be calculated by an abstract human being working in a routine way is computable [by a Turing machine].

Actually, such a statement, being non-mathematical, can't be proved, nor did Turing claim to have done so. Rather, what he *did* do in Sect. 9 of the paper was to present three informal arguments as to why his analysis catches everything that "would naturally be regarded as computable;" it is the first of these arguments that leads most directly to the concept of a Turing machine. The argument in question sets out five informal restrictive conditions on the idealized work space and possible actions within it of a human computer. As recast by Sieg [31, 32], in order to proceed to a theorem there are two steps involved: first, Turing's conditions are reformulated more generally in terms of boundedness, locality and determinacy conditions (still at the informal level), and, secondly, those conditions are given a precise mathematical expression for which it can be shown that any function satisfying them is computable by a Turing machine. (Sieg calls the second part a representation theorem.)

By contrast to Thesis T, the aim of Gandy's 1980 article was to argue for the following:

Thesis M What can be calculated by a machine is computable [by a Turing machine].

Again, there is no proof of Thesis M, but rather a two part procedure that eventuates in a definite theorem. The first part consists of Gandy's informal restrictions on what constitute "mechanical devices" [8, pp. 125–126], namely, that he excludes from consideration devices that are "*essentially* analogue machines" and that "[t]he only physical presuppositions made about mechanical devices ... are that there is a lower bound on the linear dimensions of every atomic part of the device and that there is an upper bound (the velocity of light) on the speed of propagation of changes." Furthermore, Gandy assumes that the calculations by a mechanical device are describable in discrete terms and that the behavior of the device is deterministic, though calculations may be carried out in parallel. These restrictions then lead to the formulation of four mathematically precise principles I-IV that express the informal conditions on what constitutes a machine in his sense. The main result of Gandy (1980) is a theorem to the effect that any function calculated by a mechanical device satisfying principles I-IV is computable on a Turing machine.

By the way, in the informal part of his argument for **Thesis** M, Gandy enlarges on the discreteness aspect in a way that is particularly useful for our purposes below.

> Our use of the term "discrete" presupposes that each state of the machine can be adequately described in finite terms. ... [*W*]*e want this description to reflect the actual, concrete, structure of the device in a given state.* On the other hand, we want the form of the description to be sufficiently abstract to apply uniformly to mechanical, electrical or merely notional devices. We have chosen to use hereditarily finite sets; other forms of description might be equally acceptable. We suppose that the labels are chosen for the various parts of the machine—e.g., for the teeth of cog wheels, for a transistor and its electrodes, for the beads and wires of an abacus. Labels may also be used for positions in space (e.g., for squares of the tape of a Turing machine) and for physical attributes (e.g., the color of a bead, the state of a transistor, the symbol on a square). ([8] p. 127, italics mine)

In other words, just as with **Thesis T**, one is working throughout with finite symbolic configurations.

Gandy's case for **Thesis M** was substantially recast by Sieg [31, 32] much as he had done for **Thesis T**. The first part of that work is again that in carrying out effective calculations, the machine is limited by general boundedness, locality and determinacy conditions, but those are now widened to allow acting on given finite configurations in parallel and then reassembling the results into the next configuration. That led Sieg to a statement of new simpler precise principles on mechanisms as certain kinds of discrete dynamical systems for which a representation theorem is proved, i.e. for which it is shown that whatever satisfies those principles "computes" only Turing computable functions.

Yet another approach toward establishing a version of **CT** for certain kinds of mechanisms is that due to Dershowitz and Gurevich [33], entitled "A natural axiomatization of computability and proof of Church's Thesis."[8] The mechanisms in question are called Abstract State Machines (ASMs); fundamental to this work is that of Gurevich [34] in which it is argued that sequential algorithms are captured by Sequential ASMs. Regrettably, the statement by Dershowitz and Gurevich of Church's Thesis is not the one usually understood but rather their Theorem 4.8, according to which "[e]very numeric partial function computed by an arithmetical algorithm is (partial) recursive." And for this, arithmetical algorithms are defined as state transition systems satisfying certain Sequential and Arithmetical Postulates. Unfortunately, the postulates are not fully precise as presented, and so it may be questioned whether one even has a proof there of a definite mathematical theorem.[9] One point to be noted for the following is that the ASM approach takes states to be (abstract) structures having a common fixed finite vocabulary, and a crucial assumption thereto is Postulate III (p. 319) according to which state transitions are "determined by a fixed finite 'glossary' of 'critical terms'." Thus, though the terminology makes it seem otherwise, Abstract State Machines work concretely with suitable finite symbolic configurations in their computational processes; cf. also op. cit., Definition 3.1, p. 321.

The work by Gandy, Sieg, Dershowitz and Gurevich described in the preceding must be valued for taking seriously the task of providing a two part argument for **CT** mediated by axioms of one form or another. However, it may be questioned whether the axioms provided for any of these yet reaches the desired degree of evidence, in other words of being close to compelling on inspection. But what is common to these axiomatic approaches and cannot be denied is that the *sine qua non* of **CT** is that *there is no calculation without representation.* That is, the data with which one works consists of finite symbolic configurations where the symbols (or labels) are drawn from some finite set S given in advance. These represent finite concrete configurations such as finite linear inscriptions by human beings, or mathematical configurations such as finite trees or graphs, or states of various kinds of mechanisms such as described in the quote above from [8, p. 127]. More abstractly, Gandy considered finite symbolic configurations to be themselves represented in the hereditarily finite non-empty sets over the basic set S of symbols, though I think representation in the hereditarily finite non-empty sequences over S would be more appropriate since that gives an order in which things must be read; of course, each can be coded in the other. The operations on finite symbolic configurations must be limited to purely formal transformations following inspections and appropriate tests. Thus the claim here is that a general discussion

[8]Dershowitz and Gurevich [33, p. 305] state that the aim of their work is "to provide a small number of convincing postulates in favor of Church's Thesis"; in that same article, pp. 339–342, they provide a comprehensive survey of the literature sharing that aim, going back to work of Kolmogorov in [30].

[9]Cf. the Postscriptum to Sieg [35] for a detailed critique of this work.

of **CT** as it applies to computation over arbitrary structures *only* makes sense when applied to computation over concrete structures whose elements are finite symbolic configurations of one sort or another and that posit appropriate tests and operations on such.[10]

4 Proposed Generalizations of Theories of Computation and CT to Abstract Structures; Theses for Algorithms

Beginning in the late 1950s, various generalizations were made of theories of computation and recursion theory to abstract structures in general and of certain specific kinds of structures.[11] My concern here is entirely with the foundations of such approaches, not with the results obtained on their basis. The article Feferman [40] contains a more extensive exposition of these matters; the reader is referred to that for more details. What is discussed here—but not there—are proposed generalizations of **CT** to structures that need not be concrete. I shall consider two such below, one (implicitly) due to Blum et al. [41] and the other (explicitly) due to Tucker and Zucker [10]. I shall argue that these are more appropriately to be viewed as theses for algorithms.

An important starting point is the notion of computability on an arbitrary algebraic structure made by Friedman [9] via a generalization of the Shepherdson and Sturgis [42] register machine approach to ordinary recursion theory. By a (first-order) structure or algebra $\mathfrak{A}$ is meant one of the form $\mathfrak{A} = (A, c_1, \ldots, c_j, f_1, \ldots, f_k, R_1, \ldots, R_m)$, where A is a non-empty set, each c_i is a member of A, each f_i is a partial function of one or more arguments from A to A, and each R_i is a (possibly partial) relation of one or more arguments in A. For non-triviality, both k and m are not zero. Of special note is that the test for equality of elements of A is *not* assumed as one of the basic operations; rather, if equality is to be a basic test, that is to be included as one of the relations R_i. A *finite algorithmic procedure* (fap) π on $\mathfrak{A}$ is given by a finite list of instructions among which one is designated as initial and one as terminal. The "machine" has registers $r_0, r_1, r_2, \ldots$, though only a finite number of these are needed for any given "computation", namely those mentioned in π; the register r_0 is reserved for the output. (The r_i may also be thought of as variables.) The fap π may be used to calculate a partial n-ary function f on A^n to A for any n. Given an input $(x_1, \ldots, x_n)$, one enters x_i into register r_i, and proceeds to the initial instruction. The active instructions are: (1) replace the content of one register by that of another; (2) enter one of the c_i in a specified register; (3) enter a value of one of the f_i applied to the contents of specified registers into another

[10] For a systematic treatment of computability on concrete structures see Tucker and Zucker [36].

[11] The literature on generalized recursion theory is very extensive and could use an up-to-date survey. Lacking that, some initial sources can be found in the bibliographies in the works of Barwise [37], Fenstad [38], Sacks [39], and Tucker and Zucker [11].

such; and, finally, (4) test one of the R_i on specified registers and go to designated other instructions depending on the value of the test ("if ... then ... else"). The computation terminates only if the instructions of the form (3) and (4) are defined at each stage where they are called and one eventually lands in the terminal instruction. In that case the content of register r_0 is the value of $f(x_1, \ldots, x_n)$. An n-ary relation R is decidable by a fap π if its characteristic function is computable by π. The class of fap computable partial functions on $\mathfrak{A}$ is denoted by **FAP**($\mathfrak{A}$). Friedman [9] also gives an extensionally equivalent formulation of computability on $\mathfrak{A}$ in terms of generalized Turing machines, as well as one in terms of what he calls effective definitional schemata given by an effective infinite enumeration of definition by cases.

For the structure $\mathfrak{N} = (\mathbb{N}, 0, Sc, Pd, =)$, where $\mathbb{N}$ is the set of natural numbers and Sc and Pd are respectively the successor and predecessor operations (taking $Pd(0) = 0$), **FAP**($\mathfrak{N}$) is equal to the class of partial recursive functions. For general structures $\mathfrak{A}$, Friedman [9] also introduced the notion of *finite algorithmic procedure with counting*, in which certain registers are reserved for natural numbers and one can perform the operations and tests on the contents of those registers that go with the structure $\mathfrak{N}$. Then **FAPC**($\mathfrak{A}$) is used to denote the partial functions on A determined in this way.

The notion of finite algorithmic procedure is directly generalized to many-sorted structures $\mathfrak{A} = (A_1, \ldots, A_n, c_1, \ldots, c_j, f_1, \ldots f_k, R_1, \ldots, R_m)$; each register comes with a sort index limiting which elements can be admitted as its contents. In particular, **FAPC**($\mathfrak{A}$) can be identified with **FAP**($\mathfrak{A}$, $\mathfrak{N}$) where ($\mathfrak{A}$, $\mathfrak{N}$) denotes the structure $\mathfrak{A}$ augmented by that for $\mathfrak{N}$. A further extension of Friedman's notions was made by Moldestad et al. [43, 44], using *stack registers* that may contain finite sequences of elements of any one of the basic domains A_i, including the empty sequence. The basic operations for such a register are to remove the top element of a stack (*pop*) and to add to the contents of one of the registers of type A_i (*push*). This leads to the notion of what is computable by a *finite algorithmic procedures with stacks*, **FAPS**($\mathfrak{A}$), where we take the structure $\mathfrak{A}$ to contain with each domain A_i the domain $A_i{}^*$ of all finite sequences of elements of A_i, and with operations corresponding to pop and push. If we want to be able to calculate the length n of a stack and the jth element of a stack, we need also to have the structure $\mathfrak{N}$ included. This leads to the notion of *finite algorithmic procedure with stacks and counting,* whose computable partial functions are denoted by **FAPCS**($\mathfrak{A}$). In the case of the structure $\mathfrak{N}$, by any one of the usual primitive recursive codings of finite sequences of natural numbers, we have

$$\mathbf{FAP}\,(\mathfrak{N}) = \mathbf{FAPC}\,(\mathfrak{N}) = \mathbf{FAPS}\,(\mathfrak{N}) = \mathbf{FAPCS}\,(\mathfrak{N})\,.$$

Trivially, in general for any structure $\mathfrak{A}$ we have the inclusions,

$$\mathbf{FAP}\,(\mathfrak{A}) \subseteq \mathbf{FAPC}\,(\mathfrak{A}) \subseteq \mathbf{FAPCS}\,(\mathfrak{A})\,, \text{ and}$$
$$\mathbf{FAP}\,(\mathfrak{A}) \subseteq \mathbf{FAPS}\,(\mathfrak{A}) \subseteq \mathbf{FAPCS}\,(\mathfrak{A})\,.$$

It is proved in Moldestad et al. [43] that for each of these inclusions there is a structure $\mathfrak{A}$ which makes that inclusion strict.

An alternative approach to computability over arbitrary algebraic structures is provided in a usefully detailed expository piece, Tucker and Zucker [11] that goes back to their joint work [10]; this uses definition by schemata rather than (so-called) machines. By a *standard structure* $\mathfrak{A}$ is one that includes the structure $\mathfrak{B}$ with domain $\{t, f\}$ and basic Boolean functions as its operations. The Tucker-Zucker notion of computability for standard algebras is given by procedure statements S: these include explicit definition, and are closed under composition, and under statements of the form, *if b then S_1 else S_2*, and *while b do S*, where 'b' is a Boolean term. The set of partial functions computable on $\mathfrak{A}$ by means of these schemata is denoted by **While**($\mathfrak{A}$). Then to deal with computability with counting, Tucker and Zucker simply expand the algebra $\mathfrak{A}$ to the algebra ($\mathfrak{A}$, $\mathfrak{N}$). To incorporate finite sequences for each domain A_i, they make a further expansion of that to suitable $\mathfrak{A}$*. The notions of computability $\mathbf{While}^{\mathfrak{N}}(\mathfrak{A})$ and **While***($\mathfrak{A}$) over $\mathfrak{A}$ are given simply by **While**($\mathfrak{A}$, $\mathfrak{N}$) and **While**($\mathfrak{A}$*), respectively. The following result is stated in Tucker and Zucker [11, p. 487] for any standard algebra $\mathfrak{A}$:

$$\begin{aligned}&\mathbf{While}\,(\mathfrak{A}) = \mathbf{FAP}\,(\mathfrak{A})\,,\ \mathbf{While}^{\mathfrak{N}}\,(\mathfrak{A}) = \mathbf{FAPC}\,(\mathfrak{A})\,,\ \text{and}\\ &\mathbf{While} * (\mathfrak{A}) = \mathbf{FAPCS}\,(\mathfrak{A})\,.\end{aligned}$$

Thus we have a certain robustness (confluence of ideas) for notions of computation on abstract algebraic structures, depending on the choice as to whether or not to include the natural numbers or finite sequences.

The first interesting special non-concrete case to which these notions may be applied is the structure of real numbers; this is of particular significance because it is the principal domain for numerical analysis (aka scientific computation). One approach to the foundations of that subject is given by a model of computation over the reals due to Blum et al. [45]—the BSS model—subsequently worked out at length in the book, Blum et al. [41]. Actually, the model is divided into two cases, the finite dimensional one and the infinite dimensional one. In the first of these, the reals are treated as a purely algebraic structure, namely the ordered field $\mathfrak{R} = \left(\mathbb{R},\ 0,\ 1, +, -, \times, ^{-1}, <\right)$, while in the second case, one also computes with arbitrary finite sequences of reals. According to Friedman and Mansfield [46, p. 298], in the finite dimensional case the BSS computable functions are exactly the same as the **FAP**($\mathfrak{R}$) functions, and in the infinite dimensional case the BSS computable functions are exactly the same as the **FAPS**($\mathfrak{R}$) functions. Moreover, one also has **FAPS**($\mathfrak{R}$) = **FAPCS**($\mathfrak{R}$), because $\mathfrak{N}$ can be embedded in $\mathfrak{R}$ (op. cit., p. 300). Note that the relations of equality and order on the reals are an essential part of the BSS model, as is equality in general for all algebraic structures in that model.

The case made in Blum et al. [41] for the extension of the "classical" notion of computation to computation on the reals via the BSS model is not one argued on its own merits but rather mainly by its claimed requisite applicability. This follows a brief review of theories of computation for concrete structures (the subject of "computer science") as well as Church's Thesis, whose support is bolstered by the confluence of notions. Then the authors say that "[c]ompelling motivation clearly would be required to justify yet a new model of computation" (op. cit., p. 22). And that is claimed to come from a need to give foundations to the subject of numerical analysis:

> A major obstacle to reconciling scientific computation and computer science is the present view of the machine, that is, the digital computer. As long as the computer is seen as a finite or discrete object, it will be difficult to systematize numerical analysis. We believe that the Turing machine as a foundation for real number algorithms can only obscure concepts. Toward resolving the problem we have posed, we are led to expanding the theoretical model of the machine to allow real numbers as inputs. (op. cit., p. 23)

An analogy (pp. 23–24) is made with Newton's problem of reconciling the discrete corpuscular view of matter that he accepted with the mathematics of the calculus that he found necessary to describe bodies in *prima facie* continuous motion; the resolution came via idealized infinitesimal masses.

> Now our suggestion is that the modern digital computer could be idealized in the same way that Newton idealized his discrete universe. ... Moreover, if one regards computer-graphical output such as our picture of the Mandelbrot or Julia sets with their apparently fractal boundaries and asks to describe the machine that made these pictures, one is driven to the idealization of machines that work on real or complex numbers in order to give a coherent explanation of these pictures. For a wide variety of scientific computations the continuous mathematics that the machine is simulating is the correct vehicle for analyzing the operation of the machine itself.
>
> These reasonings give some justification for taking as a model for scientific computation a machine model that accepts real numbers as inputs. (op. cit., p. 24)

What is puzzling in this analogy is that on the BSS model of computation, the relation of order and hence that of equality between real numbers is taken as total and decidable by the idealized machine, and so one is immediately led to discontinuous functions, such as point and step functions. Moreover, the BSS model makes use only of the algebraic structure of the real numbers and nothing that directly reflects its analytic/topological character. So it fails to provide a genuine notion of computation on the real numbers *as such* for which a version of **CT** would be claimed to hold. Nevertheless, as illustrated by a number of leading examples, Blum et al. [41] makes a substantial case that the BSS model provides a proper foundation for the subject of numerical analysis where the basic data is taken to be given by real (or complex) numbers. Why this turns out to be so is a matter to which I shall return at the beginning of the next section.

The question whether there is a sensible generalization of the Church-Turing Thesis to abstract structures is addressed directly by Tucker and Zucker [10, pp. 196 ff] and again in Tucker and Zucker [11, pp. 493 ff]. In the latter it is said that the answer to the question is difficult to explain fully and briefly so that only a sketch is given, and the reader is referred back to the former for more details. Though the later publication is a bit more succinct than the earlier one on this issue, I didn't find that it leaves out any essential points, so I shall use that as the reference in the following.[12] The authors begin with the statement of a "naïve" generalized **CT** for abstract algebras, namely that "[t]he functions that are 'effectively computable' on a many-sorted algebra $\mathfrak{A}$ are precisely the functions that are **While*** computable on $\mathfrak{A}$." This is immediately qualified as follows:

> [T]he idea of effective calculability is complicated, as it is made up from many philosophical and mathematical ideas about the nature of finite computation with finite or concrete elements. For example, its analysis raises questions about the mechanical representation and manipulation of finite symbols; about the equivalence of data representations; and about the formalization of *constituent concepts* such as *algorithm*; *deterministic procedure*; *mechanical procedure*; *computer program*; *programming language*; *formal system*; *machine*; and the functions definable by these entities. ... However, only *some* of these constituent concepts can be reinterpreted or generalized to work in an abstract setting; and hence the general concept, and term, of 'effective computability' does not belong in a generalization of the Church-Turing thesis. In addition, since finite computation on finite data is truly a fundamental phenomenon, it is appropriate to preserve the term with its established special meaning. (Tucker and Zucker [11, p. 494], italics in the original.)

In other words, these authors and I are in complete agreement with the view asserted at the end of the preceding section. Nevertheless, they go on to formulate three versions of a generalized **CT** *not* using the notion of effective calculability, corresponding to the three perspectives of algebra, programming languages, and specification on data types; only the first of these is relevant to the discussion here. Namely:

> **Tucker-Zucker thesis for algebraic computability.** The functions computable by finite deterministic algebraic algorithms on a many-sorted [first-order] algebra $\mathfrak{A}$ are precisely the functions While* computable on $\mathfrak{A}$. (op. cit., p. 495)

This goes back to the work of Tucker [47] on computing in algebraic structures; cf. also Stoltenberg-Hansen and Tucker [48]. Hermann's algorithm for the ideal membership problem in $K[x_1, \dots, x_n]$ for arbitrary fields K is given as a paradigmatic example, but there is no principled argument for this thesis analogous to the work of Gandy, Sieg, Dershowitz and Gurevich described in the preceding section. One may ask, for example, why the natural number structure and arrays are assumed in the Tucker-Zucker Thesis, and why these suffice beyond the structure $\mathfrak{A}$ itself. Moreover, nothing is said about assuming that the equality relation for $\mathfrak{A}$ is to be included in it, even though that is common in algebraic algorithms. Finally, one would like to see a justification of this thesis or possible variants comparable to the ones described for classical **CT**, both informal and of a more formal axiomatic kind.

In any case, the Tucker-Zucker Thesis and supporting examples suggest that *all the notions of computability on abstract first order structures considered in this section should be regarded as falling under a general notion of algorithm.* What distinguishes algorithms from computations is that they are independent of the representation of the data to which they apply but only require how data

[12] In addition, the reference Tucker and Zucker [10] is not as widely available as their year 2000 survey.

is packaged structurally, i.e. they only need consider the data up to structural isomorphism. Friedman was already sensitive to this issue and that is the reason he gave for baptizing his notion using generalized register machines, *finite algorithmic procedures*:

> The difference between [symbolic] configuration computations and algorithmic procedures is twofold. Firstly, in configuration computations the objects are symbols, whereas in algorithmic procedures the objects operated on are unrestricted (or unspecified). Secondly, in configurational computations at each stage one has a finite configuration whose size is not restricted before computation. On the other hand in algorithmic procedures one fixes beforehand a finite number of registers to hold the objects. Thus for some n, at each stage one has at most n objects. (Friedman [9], p. 362).

The general question, "What is an algorithm?" has been addressed by Moschovakis [13] and Gurevich [14] (both under that title), but with very different conclusions.[13] In [14] Sect. 6, Gurevich criticizes Moschovakis' answer on several grounds among which that distributed algorithms do not fall under the latter's central notion of *recursor.* Moreover, even those algorithms that fall under the notion of recursor may do so by losing certain essential aspects of the procedure in question. Whether or not one agrees with all of Gurevich's critiques of Moschovakis' analysis, in my view that is more appropriately to be considered under general theses for recursion that are taken up in the next section. In contrast to Moschovakis, Gurevich asks whether the notion of algorithm can be defined at all; his answer is "yes and no". On the negative side, he writes:

> In our opinion, the notion of algorithm cannot be rigorously defined in full generality, at least for the time being. The reason is that the notion is expanding. Concerning the analogy of algorithms to real numbers, mentioned in sec.1, Andreas Blass suggested a better analogy: algorithms to numbers. Many kinds of numbers have been introduced throughout history: positive integers, natural numbers, rationals, reals, complex numbers, quaternions, infinite cardinals, infinite ordinals, etc. Similarly many kinds of algorithms have been introduced. In addition to classical sequential algorithms, in use from antiquity, we have now parallel, interactive, distributed, real-time, analog, hybrid, quantum, etc. algorithms. New kinds of numbers and algorithms may be introduced. The notions of numbers and algorithms have not crystallized (and maybe never will) to support rigorous definitions. ([14], Sect. 2)

On the positive side he says that even though it is premature to try to propose a general answer to the question, "What is an algorithm?," convincing answers have been given for large classes of such, among which sequential algorithms, synchronous parallel algorithms and interactive sequential algorithms (cf. ibid for references). In particular, the Tucker-Zucker Thesis or something close to it is a plausible candidate for what one might call the *Algebraic Algorithmic Procedures Thesis*. And more generally, it may be possible to distinguish algorithms used in pure mathematics from those arising in applied mathematics and computer science, where such algorithms as "interactive, distributed, real-time, analog, hybrid, quantum, etc." would fall. If there is a sensible separation between the two, Moschovakis'

[13] See also Blass and Gurevich [49].

explanation of what is an algorithm could be justified as being confined to those of pure mathematics. Moreover, his theory leads to the remarkable result that there is a decidable criterion for identity of algorithms in his sense [12].

Clearly, all this requires deeper consideration, and I must leave it at that.

5 Recursion on Abstract Structures

Let us return to the claim of Blum et al. [41] that the BSS model of computation on the reals (and complex numbers) is requisite for the foundations of the subject of scientific computation. That was strongly disputed by Braverman and Cook [50], where the authors argued that the requisite foundation is provided by a quite different "bit computation" model that is *prima facie* incompatible with the BSS model. It goes back to ideas due to Banach and Mazur in the latter part of the 1930s, but the first publication was not made until Mazur [51]. In the meantime, the bit computation model was refined and improved by Grzegorczyk [52] and independently by Daniel Lacombe [53] in terms of a theory of recursively computable functionals. Terminologically, something like "effective approximation computability" is preferable to "bit computability" as a name for this approach in its applications to analysis.

This competing approach was explained in Feferman [40] in rough terms as follows. To show that a real valued function f on a real interval into the reals is computable by effective approximation, given any x in the interval as argument to f, one works *not* with x but rather with an arbitrary *sequential representation of* x, i.e. with a Cauchy sequence of rationals $\langle q_n \rangle_{n \in \mathbf{N}}$ which approaches x as its limit, in order to effectively determine another such sequence$\langle r_m \rangle_{m \in \mathbf{N}}$ which approaches $f(x)$ as limit. The sequences in question are functions from $\mathbb{N}$ to $\mathbb{Q}$, and so what is required is that the passage from $\langle q_n \rangle_{n \in \mathbf{N}}$ to $\langle r_m \rangle_{m \in \mathbf{N}}$ is given by an effective type-2 functional on such functions. Write T for the class of all total functions from $\mathbb{N}$ to $\mathbb{N}$, and P for the class of all partial functions from $\mathbb{N}$ to $\mathbb{N}$. By the effective enumeration of the rational numbers, this reduces the notion of effective approximation computability of functions f on the reals to that of effective functionals F from T to T, and those in turn are restrictions to T of the partial recursive functionals F' (from P to P) whose values on total functions are always total.[14] It may be shown that by the continuity in the recursion theoretic sense of partial recursive functionals we may infer continuity in the topological sense of the functions f on the reals that are effective approximation computable. Thus step functions that are computable in the BSS model are not computable in this sense.

[14]Note that a partial recursive functional F need not have total values when restricted to total arguments.

On the other hand, the exponential function is an example of one that is computable in the effective approximation model that is not computable in the BSS model.[15]

The reader must be referred to Blum et al. [41] and Braverman and Cook [50] for arguments as to which, if either of these, is the appropriate foundation for scientific computation.[16] I take no position on that here, but simply point out that we have been led in a natural way from computation on the reals in the effective approximation sense back to the partial recursive functionals F on partial functions of natural numbers. Now Kleene's principal theorem for such functionals is the "first" Recursion Theorem, according to which each such F has a least fixed point (LFP) f, i.e. one that is least among all partial functions g such that $g = F(g)$ [6, p. 348]. This is fundamental in the following sense: the partial recursive functions and functionals are just those generated by closing under explicit definition and LFP recursion over the structure $\mathfrak{N}$. For, first of all, one immediately obtains closure under the primitive recursive schemata. Then, given primitive recursive $g(\underline{x}, y)$, one obtains the function $f(\underline{x}) \simeq (\mu y)[g(\underline{x}, y) = 0]$ by taking $f(\underline{x}) \simeq h(\underline{x}, 0)$ where $h(\underline{x}, z) \simeq z$ if $(\forall y < z)$ $[g(\underline{x}, y) > 0 \wedge g(\underline{x}, z) = 0]$, else $h(\underline{x}, z')$. It follows that all partial recursive functions (and thence all partial recursive functionals) are obtained by Kleene's Normal Form Theorem.

This now leads one to consider generation of partial functions and functionals by explicit definition and LFP recursion over arbitrary abstract many-sorted structures $\mathfrak{A}$. The development of that idea originates with Platek [63], a PhD thesis at Stanford that, though never published, came to be widely known by workers in the field.

[15]There is a considerable literature on computation on the real numbers under various approaches related to the effective approximation one via Cauchy representations. A more comprehensive one is that given by Kreitz and Weihrauch [54, 55] and Weihrauch [56]; that features surjective representations from a subset of $\mathbb{N}^{\mathbb{N}}$ to $\mathbb{R}$. Bauer [57] introduced a still more general theory of representations via a notion of realizability, that allows one to consider classical structures and effective structures of various kinds (including those provided by domain theory) under a single framework; cf. also Bauer and Blanck [58]. The work of Pour-El surveyed in her article [59] contains interesting applications of the effective approximation approach to questions of computability in physical theory.

[16]Cf. also Blum [60], to which Braverman and Cook [50] responds more directly. Actually, the treatment of a number of examples from numerical analysis in terms of the BSS model that takes up Part II of Blum et al. [41] via the concept of the "condition number" of a procedure in a way brings it in closer contact with the effective approximation model. As succinctly explained to me in a personal communication from Lenore Blum, "[r]oughly, 'condition' connects the BSS/BCSS theory with the discrete theory of computation/complexity in the following way: The 'condition' of a problem instance measures how outputs will vary under perturbations of the input (think of the condition as a normed derivative)." The informative article, Blum [61], traces the idea of the condition number back to a paper by Turing [62] on rounding-off errors in matrix computations from where it became a basic common concept in various guises in numerical analysis. (An expanded version of Blum [61] is forthcoming.) It may be that the puzzle of how the algebraic BSS model serves to provide a foundation for the mathematics of the continuous, at least as it appears in numerical analysis, is resolved by noting that the verification of the algorithms it employs requires in each case specific use of properties of the reals and complex numbers telling which such are "well-conditioned."

The only requirement on a type-2 functional F on partial functions over $\mathfrak{A}$ for it to have a least fixed point is that F be *monotonic increasing*. In addition, for a thorough-going theory of partial recursive functions and functionals over $\mathfrak{A}$, one must use only those F that have themselves been obtained by LFP recursion in terms of previously defined functions and functionals. For that purpose Platek made use of a hierarchy of hereditarily monotonic partial functionals of arbitrary finite type over the domains of $\mathfrak{A}$. That allows one to start not only with given functions over $\mathfrak{A}$ but also given functionals at any level in that hierarchy. On the other hand, Platek showed that in the special case that the initial functionals are of type level ≤ 2, everything of type level ≤ 2 that can be generated via explicit definition and LFP recursion in higher types from that data can already be generated via explicit definition and LFP recursion at type level equal to 2. Platek's approach using the full hierarchy of hereditarily monotonic functionals allowed him to subsume and simplify Kleene's theory of recursion in finite types using hereditarily total functionals as the arguments of partial functionals defined by certain schemata [64].[17] Later, Kechris and Moschovakis [65] showed how to subsume Kleene's theory under LFP recursion at type level ≤ 2 by treating the finite type structure as a many-sorted first-order structure (with infinitely many sorts).

Moschovakis [12, 66] took the LFP approach restricted to functionals of type level ≤ 2 in his explanation of the notion of algorithm over arbitrary structures featuring simultaneous LFP definitions, though those can be eliminated in favor of successive LFP definitions of the above form. Both the Platek and Moschovakis approaches are *extensional*. In order to tie that up both with computation over abstract data types and with Bishop's approach to constructive mathematics, in a pair of papers Feferman [67, 68], I extended the use of LFP schemata to cover *intensional* situations, by requiring each basic domain $\mathrm{A_i}$ of $\mathfrak{A}$ to be equipped with an equivalence relation $=_\mathrm{i}$ that the initial functions and functionals preserve. The resulting partial functions and functionals are called there *Abstract Computation Procedures* (ACPs). They are successively generated over any structure $\mathfrak{A} = (A_0, A_1, \ldots, A_k, F_0, \ldots, F_m)$ by explicit definition and LFP recursion, where each A_i is non-empty and the F_js are constants, partial functions or monotonic increasing partial functionals of type level 2 over the A_i. Also, one of the domains, say A_0, is fixed to be the booleans $\{t, f\}$ and the operations of negation and conjunction are taken among the basic operations; this allows conditional definition. The details of the schemata may be found in Feferman [40].

Let us note two comparisons of ACPs with other approaches. First is the following result due to Xu and Zucker [69]: if $\mathfrak{A}$ is an $\mathfrak{N}$-standard structure with arrays, then **While***($\mathfrak{A}$) = **ACP**($\mathfrak{A}$). Secondly, we have a matchup with the [66] theory of recursors by the result of Feferman [67] Sect. 9 that the ACPs are closed under simultaneous LFP recursion. In the particular case of the structure $\mathfrak{N}$ of natural numbers, the arguments above in connection with Kleene's First Recursion

[17]Platek [63] also used the LFP approach to subsume recursion theory on the ordinals under the theory of recursion in the Sup functional.

Theorem show that the partial functions and functionals generated by the abstract computation procedures are just those that are partial recursive. This shows that the effective approximation approach to computation on the reals is accounted for at the second-order level under **ACP**($\mathfrak{N}$), while the Xu-Zucker result shows that the BSS model is subsumed at the first-order level under **ACP**($\mathfrak{R}$).

Clearly it is apt to use the word 'abstract' in referring to the procedures in question since they are preserved under isomorphism. But given the arguments I have made in the preceding sections, it was a real mistake on my part to use 'computation' as part of their designation, and I very much regret doing so. A better choice would have been simply to call them *Abstract Recursion Procedures*, and I have decided to take this occasion to use 'ARP' as an abbreviation for these, in place of 'ACP', thus **ARP**($\mathfrak{A}$) in place of **ACP**($\mathfrak{A}$). The main point now is to bring matters to a conclusion by using these to propose the following thesis on definition by recursion that in no way invokes the concepts of computation or algorithm.

Recursion Thesis (RT) Any function defined by recursion over a first-order structure $\mathfrak{A}$ (with Booleans) belongs to **ARP**($\mathfrak{A}$).

This presumes an informal notion of being a function f defined by recursion over a first- order structure $\mathfrak{A}$ that is assumed to include the Boolean constants and basic operations. Roughly speaking, the idea for such a definition is that f is determined by an equation $f(x) \simeq E(f, x)$.where E is an expression that may contain a symbol for f and symbols for the initial functions and constants of $\mathfrak{A}$ as well as for functions previously defined by recursion over $\mathfrak{A}$. Now here is the way such a justification for **RT** might be argued. At any given $x = x_0$, $f(x)$ may not be defined by E, for example if $E(f, x) =$ [if $x = x_0$ and $f(x) = 0$ then 1, else 0]. But if $f(x)$ is defined at all by E, it is by use made of values $f(y)$ that are previously defined. Write $y \prec x$ if $f(y)$ is previously defined and its value is used in the evaluation of x; then let f_x be f restricted to $\{y : y \prec x\}$. Thus the evaluation of $f(x)$ is determined by f_x when it is defined, i.e. $f(x) = E(f_x, x)$ for each such x. It may be that $\{y : y \prec x\}$ is empty if $f(x)$ is defined outright in terms of previous functions; in that case x is minimal in the $\prec$ relation. In the case it is not empty, we may make a similar argument for $f(y)$ for each $y \prec x$, and so on. In order for this to terminate, the $\prec$ relation must be well-founded. Next, take F to be the functional given by $F(f, x) = E(f_x, x)$; F is monotonic increasing, because if $f \subseteq g$ then $f_x = g_x$. So F has a LFP g. But F defines our function f by transfinite recursion on $\prec$, so f is a fixed point of F and hence $g \subseteq f$. To conclude that $f \subseteq g$, we argue by transfinite recursion on $\prec$: for a given x, if $f(y) = g(y)$ for all $y \prec x$ then $f(x) = F(f, x) = E(f_x, x) = E(g_x, x) = F(g, x) = g(x)$. Thus f is given by LFP recursion in terms of previously obtained functions in **ARP**($\mathfrak{A}$) and hence itself belongs to **ARP**($\mathfrak{A}$).

The reason for restricting to first-order structures $\mathfrak{A}$ in the formulation of **RT** is so as not to presume the property of monotonicity as an essential part of the idea of definition by recursion. I should think that all this can be elaborated, perhaps in an axiomatic form, but if there is to be any thesis at all for definition by recursion over an arbitrary first-order structure (with Booleans), I cannot see that it would differ in any essential way from **RT**. If there is a principled argument for assuming monotonicity of the functionals in a given second-order structure then we would also have a reasonable extension of **RT** to such.

Acknowledgements I wish to thank Lenore Blum, Andrej Bauer, John W. Dawson, Jr., Nachum Dershowitz, Yuri Gurevich, Grigori Mints, Dana Scott, Wilfried Sieg, Robert Soare, John V. Tucker, and Jeffery Zucker for their helpful comments on an early draft of this article. Also helpful were their pointers to relevant literature, though not all of that extensive material could be accounted for here.

References

1. R.O. Gandy, The confluence of ideas in 1936, in *The Universal Turing Machine. A Half-Century Survey*, ed. by R. Herken (Oxford University Press, Oxford, 1988), pp. 55–111
2. R.I. Soare, The history and concept of computability, in *Handbook of Computability Theory*, ed. by E. Griffor (Elsevier, Amsterdam, 1999), pp. 3–36
3. A. Turing, On computable numbers, with an application to the Entscheidungsproblem, in *Proceedings of the London Mathematical Society. Series 2,* vol. 42 (1936–1937), pp. 230–265; a correction, *ibid.* 43, 544–546
4. A. Church, An unsolvable problem of elementary number theory. Am. J. Math. **58**, 345–363 (1936)
5. R.I. Soare, Computability and recursion. Bull. Symb. Log. **2**, 284–321 (1996)
6. S.C. Kleene, *Introduction to Metamathematics* (North-Holland, Amsterdam, 1952)
7. B.J. Copeland, The Church-Turing Thesis (Stanford Encyclopedia of Philosophy 2002), http://plato.stanford.edu/entries/church-turing/
8. R.O. Gandy, Church's thesis and principles for mechanisms, in *The Kleene Symposium*, ed. by J. Barwise, H.J. Keisler, K. Kunen (North-Holland, Amsterdam, 1980), pp. 123–145
9. H. Friedman, Algorithmic procedures, generalized Turing algorithms, and elementary recursion theory, in *Logic Colloquium '69*, ed. by R.O. Gandy, C.M.E. Yates (North-Holland, Amsterdam, 1971), pp. 361–389
10. J.V. Tucker, J.I. Zucker, *Program Correctness over Abstract Data Types* (CWI Monograph, North-Holland, Amsterdam, 1988)
11. J.V. Tucker, J.I. Zucker, Computable functions and semicomputable sets on many-sorted algebras, in *Handbook of Logic in Computer Science,* vol. 5, ed. by S. Abramsky, et al. (Oxford University Press, Oxford, 2000), pp. 317–523
12. Y.N. Moschovakis, The formal language of recursion. J. Symb. Log. **54**, 1216–1252 (1989)
13. Y.N. Moschovakis, What is an algorithm? in *Mathematics Unlimited 2001 and Beyond*, ed. by B. Engquist, W. Schmid (Springer, Berlin, 2001), pp. 919–936
14. Y. Gurevich, What is an algorithm? in *SOFSEM 2012: Theory and Practice of Computer Science*, ed. by M. Bielikova et al. (Springer, Berlin, 2012), LNCS, vol. 7147, pp. 31–42
15. D. Hilbert, Über das Unendliche. Math. Ann. **95**, 161–190 (1926); English translation in van Heijenoort (ed.), *From Frege to Gödel. A Source Book in Mathematical Logic, 1879–1931* (Harvard University Press, Cambridge, 1967), pp. 367–392
16. K. Gödel, On undecidable propositions of formal mathematical systems, (mimeographed lecture notes by S.C. Kleene and J.B. Rosser); reprinted with revisions in M. Davis (1965) (ed.), in *The Undecidable. Basic Papers on Undecidable Propositions, Unsolvable Problems, and Computable Functions* (Raven Press, Hewlett, 1934), pp. 39–74, and in Gödel (1986), pp. 346–371
17. J.W. Dawson Jr., Prelude to recursion theory: the Gödel-Herbrand correspondence, in *First International Symposium on Gödel's Theorems*, ed. by Z.W. Wolkowski (World Scientific, Singapore, 1993), pp. 1–13

18. W. Sieg, Introductory note to the Gödel-Herbrand correspondence, in K. Gödel, *Collected Works. Vol. V, Correspondence H-Z*, ed. by S. Feferman et al. (Oxford University Press, Oxford, 2003), pp. 3–13
19. W. Sieg, Only two letters: the correspondence between Herbrand and Gödel. Bull. Symb. Log. **11**, 172–184 (2005)
20. S.C. Kleene, General recursive functions of natural numbers. Mathematische Annalen **112**, 727–742 (1936)
21. S.C. Kleene, λ-definability and recursiveness. Duke Math. J. **2**, 340–353 (1936)
22. S.C. Kleene, Recursive predicates and quantifiers. Trans. Am. Math. Soc. **53**, 41–73 (1943)
23. O. Shagrir, Effective computation by humans and machines. Minds Mach. **12**, 221–240 (2002)
24. A. Turing, Computability and λ-definability. J. Symb. Log. **2**, 153–163 (1937)
25. A. Church, Review of Turing (1936–37). J. Symb. Logic **2**, 42–43 (1937)
26. K. Gödel, *Collected Works. Vol. III, Unpublished Essays and Lectures*, ed. by S. Feferman et al. (Oxford University Press, New York, 1995)
27. K. Gödel, *Collected Works. Vol. II, Publications 1938–1974*, ed. by S. Feferman et al. (Oxford University Press, New York, 1990)
28. K. Gödel, *Collected Works. Vol. I, Publications 1929–1936*, ed. by S. Feferman et al. (Oxford University Press, New York, 1986)
29. H. Rogers, *Theory of Recursive Functions and Effective Computability* (McGraw-Hill Publ. Co., New York, 1967)
30. A.N. Kolomogorov, V.A. Uspenski, On the definition of an algorithm. Uspehi Math. Nauk. 8, 125–176 (1953); Am. Math. Soc. Transl. 29, 217–245 (1963)
31. W. Sieg, Calculations by man and machine: Conceptual analysis, in *Reflections on the Foundations of Mathematics. Essays in Honor of Solomon Feferman,* ed. by W. Sieg, R. Sommer, C. Talcott. Lecture notes in logic, vol. 115 (Association for Symbolic Logic, A. K. Peters, Ltd., Natick, 2002), pp. 390–409
32. W. Sieg, Calculations by man and machine: Mathematical presentation, in *Proceedings of the Cracow International Congress of Logic, Methodology and Philosophy of Science* (Synthese Series, Kluwer Academic Publishers, Dordrecht, 2002), pp. 245–260
33. N. Dershowitz, Y. Gurevich, A natural axiomatization of computability and proof of Church's thesis. Bull. Symb. Log. **14**, 299–350 (2008)
34. Y. Gurevich, Sequential abstract state machines capture sequential algorithms. ACM Trans. Comput. Log. **1**, 77–111 (2000)
35. W. Sieg, Axioms for computability: do they allow a proof of Church's thesis? in *A Computable Universe – Understanding and Exploring Nature as Computation*, ed. by H. Zenil (World Scientific, Singapore, 2013), pp. 99–123
36. J.V. Tucker, J.I. Zucker, Abstract versus concrete computability: The case of countable algebras, in *Logic Colloquium '03*, ed. by V. Stoltenberg-Hansen, J. Väänänen. Lecture notes in logic, vol. 24 (Association for Symbolic Logic, A. K. Peters, Ltd., Wellesley, 2006), pp. 377–408
37. J. Barwise, *Admissible Sets and Structures* (Springer, Berlin, 1975)
38. J. Fenstad, *General Recursion Theory. An Axiomatic Approach* (Springer, Berlin, 1980)
39. G.E. Sacks, *Higher Recursion Theory* (Springer, Berlin, 1990)
40. S. Feferman, About and around computing over the reals, in *Computability. Turing, Gödel, Church, and Beyond*, ed. by B.J. Copeland, C.J. Posy, O. Shagrir (MIT, Cambridge, 2013), pp. 55–76
41. L. Blum, F. Cucker, M. Shub, S. Smale, *Complexity and Real Computation* (Springer, New York, 1997)
42. J.C. Shepherdson, H.E. Sturgis, Computability of recursive functions. J. Assoc. Comput. Mach. **10**, 217–255 (1963)
43. J. Moldestad, V. Stoltenberg-Hansen, J.V. Tucker, Finite algorithmic procedures and inductive definability. Mathematica Scandinavica **46**, 62–76 (1980)
44. J. Moldestad, V. Stoltenberg-Hansen, J.V. Tucker, Finite algorithmic procedures and inductive definability. Mathematica Scandinavica **46**, 77–94 (1980)

45. L. Blum, M. Shub, S. Smale, On a theory of computation and complexity over the real numbers: NP-completeness, recursive functions and universal machines. Bull. Am. Math. Soc. **21**, 1–46 (1989)
46. H. Friedman, R. Mansfield, Algorithmic procedures. Trans. Am. Math. Soc. **332**, 297–312 (1992)
47. J.V. Tucker, Computing in algebraic systems, in *Recursion Theory, Its Generalizations and Applications*, ed. by F.R. Drake, S.S. Wainer (Cambridge, University Press, Cambridge, 1980)
48. V. Stoltenberg-Hansen, J.V. Tucker, Computable rings and fields, in *Handbook of Computability Theory*, ed. by E. Griffor (Elsevier, Amsterdam, 1999), pp. 363–447
49. A. Blass, Y. Gurevich, Algorithms: a quest for absolute definitions. Bull. EATCS **81**, 195–225 (2003)
50. M. Braverman, S. Cook, Computing over the reals: foundations for scientific computing. Not. Am. Math. Soc. **51**, 318–329 (2006)
51. S. Mazur, Computable analysis. Rozprawy Matematyczne **33**, 1–111 (1963)
52. A. Grzegorczyk, Computable functionals. Fundamenta Mathematicae **42**, 168–202 (1955)
53. D. Lacombe, Extension de la notion de fonction récursive aux fonctions d'une ou plusieurs variables réelles, I, II, III, *Comptes Rendus de l'Académie des Science Paris*, **240**, 2470–2480 (1955) (**241**, 13–14, **241**, 151–155)
54. C. Kreitz, K. Weihrauch, A unified approach to constructive and recursive analysis, in *Computation and Proof Theory*, ed. by M. Richter, et al. Lecture notes in mathematics, vol. 1104 (1984), pp. 259–278
55. C. Kreitz, K. Weihrauch, Theory of representations. Theor. Comput. Sci. **38**, 35–53 (1985)
56. K. Weihrauch, *Computable Analysis* (Springer, New York, 2000)
57. A. Bauer, *The Realizability Approach to Computable Analysis and Topology*, PhD Thesis, Carnegie Mellon University, 2000; Technical Report CMU-CS-00-164
58. A. Bauer, J. Blanck, Canonical effective subalgebras of classical algebras as constructive metric completions. J. Universal Comput. Sci. **16**(18), 2496–2522 (2010)
59. M.B. Pour-El, The structure of computability in analysis and physical theory, in *Handbook of Computability Theory*, ed. by E. Griffor (Elsevier, Amsterdam, 1999), pp. 449–471
60. L. Blum, Computability over the reals: where Turing meets Newton. Not. Am. Math. Soc. **51**, 1024–1034 (2004)
61. L. Blum, Alan Turing and the other theory of computation, in *Alan Turing: His Work and Impact*, ed. by Cooper and van Leeuwen (Elsevier Pub. Co., Amsterdam, 2013), pp. 377–384
62. A. Turing, Rounding-off errors in matrix processes. Q. J. Mech. Appl. Math. **1**, 287–308 (1948); reprinted in Cooper and van Leeuwen (2013), pp. 385–402
63. R.A. Platek, *Foundations of Recursion Theory*, PhD Dissertation, Stanford University, 1966
64. S.C. Kleene, Recursive functionals and quantifiers of finite type I. Trans. Am. Math. Soc. **91**, 1–52 (1959)
65. A. Kechris, Y. Moschovakis, Recursion in higher types, in *Handbook of Mathematical Logic*, ed. by J. Barwise (North-Holland Pub. Co., Amsterdam, 1977), pp. 681–737
66. Y.N. Moschovakis, Abstract recursion as a foundation for the theory of recursive algorithms, in *Computation and Proof Theory* ed. by M.M. Richter, et al. Lecture notes in computer science, vol. 1104 (1984), pp. 289–364
67. S. Feferman, A new approach to abstract data types, I: informal development. Math. Struct. Comput. Sci. **2**, 193–229 (1992)
68. S. Feferman, A new approach to abstract data types, II: computability on ADTs as ordinary computation, in *Computer Science Logic*, ed. by E. Börger, et al. Lecture notes in computer science, vol. 626 (1992), pp. 79–95
69. J. Xu, J. Zucker, First and second order recursion on abstract data types. Fundamenta Informaticae **67**, 377–419 (2005)
70. K. Gödel, The present situation in the foundations of mathematics (unpublished lecture), in Gödel (1995), pp. 45–53

Generalizing Computability Theory to Abstract Algebras

J.V. Tucker and J.I. Zucker

Abstract We present a survey of our work over the last four decades on generalizations of computability theory to many-sorted algebras. The following topics are discussed, among others: (1) abstract *v* concrete models of computation for such algebras; (2) computability and continuity, and the use of many-sorted topological partial algebras, containing the reals; (3) comparisons between various equivalent and distinct models of computability; (4) generalized Church-Turing theses.

Keywords Computability and continuity • Computability on abstract structures • Computability on the reals • Generalized church-turing thesis • Generalized computability

AMS Classifications: 03D75, 03D78

1 Introduction

Since 1936, most of the development of computability theory has focused on sets and functions of strings and natural numbers—though computability theory immediately found applications, first in logic and algebra and then pretty much everywhere. Indeed, recall that when Turing created his model of computation on strings [38], he applied it to solve a problem involving real numbers. The applications were via codings or representations of other data.

Why would we want to generalize classical computability theory from strings and natural numbers to arbitrary abstract algebras? How should we do it?

J.V. Tucker
Department of Computer Science, Swansea University, Singleton Park, Swansea SA2 8PP, Wales, UK
e-mail: J.V.Tucker@swansea.ac.uk

J.I. Zucker (✉)
Department of Computing and Software, McMaster University, Hamilton, ON, Canada L8S 4K1
e-mail: zucker@mcmaster.ca

G. Sommaruga, T. Strahm (eds.), *Turing's Revolution*,
DOI 10.1007/978-3-319-22156-4_5

Computability theory is a general theory about what and how we compute. At its heart is the notion of an algorithm that processes data. Today, algorithms and data abound in all parts of our professional, social and personal lives. Data are composed of numbers, texts, video and audio. We know that, being digital, this vast range of data is coded or represented by bitstrings (or strings over a finite alphabet). However, the algorithms that make data useful are created specifically for a high-level, independent model of the data close to the use and users of the data. Therefore, computability theory cannot be content with bitstrings and natural numbers.

Now, the algorithms are designed to schedule sequences of basic operations and tests on the data in order to accomplish a task. They are naturally high-level and their level of abstraction is defined by the nature of the basic operations and tests. In fact, a fundamental observation is this:

What an algorithm knows of the data it processes is precisely determined by what operations and tests on the data it can employ.

This is true of strings and natural numbers, of course. This observation leads to the fundamental idea of an *abstract data type* in computer science. Mathematically, an implementation of a data type is modelled by an algebraic structure with operations and relations, and an abstract data type is modelled by a class of algebraic structures. The signature of an algebra is a syntactic interface to the operations and tests. The abstract data type can be specified by giving a signature and a set of axioms that the operators and relations must satisfy. The algebraic theory of abstract data types is a theory about all data, now and in the future.

What about the models of computation? Thanks to computer science, there is an abundance of computation models, practical and theoretical. Even within the theory of computation, the diversity is daunting: so many motivations, intuitions and technical developments.[1] In this chapter we will give an introduction to *one* abstract model of computation for an arbitrary abstract algebra. Our account will be rather technical and very quick, but it will contain what we think are the key ideas that can launch a whole mathematical theory and sustain its application. The model is a simple form of imperative programming, being an idealised programming language for manipulating data in a store using the constructs of assignments, sequencing, conditionals and iteration. By concentrating on this model the reader will be able to explore and benchmark other models, however complicated or obscure their origins.

To make some sense of the jungle of models of computation, we discuss, in Sect. 2, two basic types of model, abstract and concrete. The distinction is invaluable when we meet other models, and applications to specific data types, later in Sect. 8. The first part of this chapter introduces data types modelled by *algebras* (Sect. 3) and the *imperative* programming model (Sect. 4), and covers *universality* (Sect. 5), and *semicomputability* (Sect. 6). Here we consider only the case where algebras have *total* operations and tests. Next (Sect. 7) we look closely at data types with

[1]At one time it was possible for us to investigate most of the mathematical models, and compare and classify them [29, 31]!

continuous operations, such as the real numbers. At this point the theory deepens. New questions arise and there are a number of changes to the imperative model, not least the use of algebras that have *partial* operations and tests and the need for *nondeterministic* constructs. Finally (Sect. 8) we take a quick look other models and propose some *generalizations of the Church-Turing thesis* to abstract many-sorted algebras.

2 On Generalizing Computability Theory

By a *computability theory* we mean a theory of functions and sets that are definable using a model of computation. By a *model of computation* we mean a theoretical model of some general method of calculating the value of a function or of deciding, or enumerating, the elements of a set. We allow the functions and sets to be constructed from any kind of data. Thus, classical computability theory on the set $\mathbb{N}$ of natural numbers is made up of many computability theories (based upon Turing machines, recursive definitions, register machines, etc.).

We divide computability theories into two types:

In an *abstract computability theory* the computations are *independent* of all the representations of the data. Computations are *uniform* over all representations and are necessarily isomorphism invariant. Typical of abstract models of computation are models based on abstract ideas of *program*, *equation*, *recursion scheme*, or *logical formula*.

In a *concrete computability theory* the computations are *dependent* on *some* data representation. Computations are *not* uniform, and different representations can yield different results. Computations are not automatically isomorphism invariant. Typical of concrete models of computation are those based on concrete ideas of *coding*, *numbering*, or *data representations* using numbers or functions.

Now in computer science, it is obvious that a computation is fundamentally dependent on its data. By a *data type* we mean

(1) data,
together with
(2) some primitive operations and tests on these data.
Often we also have in mind the ways these data are
(3) axiomatically specified, and
(4) represented or implemented.

To choose a computation model, we must think carefully about what forms of data the user may need, how we might model the data in designing a system—where some high level but formal understanding is important—and how we might implement the data in some favoured programming language.

We propose the working principle:

Any computability theory should be focused equally on the data types and the algorithms.

Now this idea may be difficult to appreciate if one works in one of the classical computability theories of the natural numbers, for data representations rarely seem to be an important topic there. Although the translation between Turing machines and register machines involves data transformations, these can be done on an ad hoc basis.

However, representations are *always* important. Indeed, representations are a subject in themselves. This is true even in the simple cases of discrete data forming countable sets and structures. From the beginning there has been a great interest in comparisons between different kinds of numberings, for example in Mal'cev's theory of computability on sets and structures [18]. The study of different notions of reduction and equivalence of numberings, and the space of numberings, has had a profound influence on the theory and has led to quite remarkable results, such as Goncharov's Theorem and its descendants[2] [5, 24]. These notions also include the idea of invariance under computable isomorphisms, prominent in computable algebra, starting in [10]. In the general theory of computing with enumerated structures, there was always the possibility of computing relative to a reasonable numbering that was standard in some sense. For example, earlier work on the word problem for groups, such as [19], and on computable rings and fields, such as [22], was not concerned with the choice of numberings.

We see the importance of representations even more clearly when computing with continuous data forming uncountable sets. For example, in computing with real numbers, it has long been known that if one represents the reals by infinite decimal expansions then one cannot even compute addition (consider, e.g., $0.333\cdots + 0.666\ldots$). But if one chooses the Cauchy sequence representations then a great deal is computable [3].

In general, what is the relationship between abstract and concrete computability models with a common set of data D?

Let $\boldsymbol{AbstComp}_A(D)$ be the set of functions on the data set D that are computable in an abstract model of computation associated with a structure A containing D.

Let $\boldsymbol{ConcComp}_R(D)$ be the set of functions on D that are computable in a concrete model of computation with representation R.

- **Soundness**: An abstract model of computation $\boldsymbol{AbstComp}_A(D)$ is *sound* for a concrete model of computation $\boldsymbol{ConcComp}_R(D)$ if

$$\boldsymbol{AbstComp}_A(D) \subseteq \boldsymbol{ConcComp}_R(D).$$

[2]For example: for all n there exists a computable algebra with precisely n inequivalent computable numberings.

- **Adequacy**: An abstract model of computation $\boldsymbol{AbstComp}_A(D)$ is *adequate* for a concrete model of computation $\boldsymbol{ConcComp}_R(D)$ if

$$\boldsymbol{ConcComp}_R(D) \subseteq \boldsymbol{AbstComp}_A(D).$$

- **Completeness**: An abstract model of computation $\boldsymbol{AbstComp}_A(D)$ is *complete* for a concrete model of computation $ConcComp_R D$ if it is both sound and adequate, i.e.,

$$\boldsymbol{AbstComp}_A(D) = \boldsymbol{ConcComp}_R(D).$$

As an example for the l.h.s. here, let us take the data set $D = \mathbb{R}$, the set of reals, the structure $A = \mathcal{R}_\mathsf{p}^N$, the partial algebra $\mathcal{R}_\mathsf{p}$ of reals defined below in Sect. 7, with the naturals adjoined, and $\boldsymbol{AbstComp}_A(D) = \boldsymbol{While}^*\ (\mathcal{R}_\mathsf{p}^N)$, the set of functions on $\mathbb{R}$ definable by the ***While*** programming language with arrays over $\mathcal{R}_\mathsf{p}^N$ defined in Sect. 5. For the r.h.s., take a standard enumeration $\alpha : \mathbb{N} \to \mathbb{Q}$ of the rationals, which generates a representation $\overline{\alpha}$ of the computable reals (as described in Sect. 7.1.1 below), and let $\boldsymbol{ConcComp}_{\overline{\alpha}}(\mathbb{R})$ be the corresponding "$\overline{\alpha}$-tracking" model. Then our abstract model is sound, but not adequate, for this concrete model:

$$\boldsymbol{While}^*(\mathcal{R}_\mathsf{p}) \subsetneqq \boldsymbol{ConcComp}_{\overline{\alpha}}(\mathbb{R}).$$

On the other hand, if we take for our abstract model over $\mathcal{R}_\mathsf{p}$ the non-deterministic ***WhileCC**** (***While*** + "countable choice" + arrays) language, and further replace "computability" by "approximable computability" [33, 34], then we obtain completeness (see Sect. 7.3 below):

$$\boldsymbol{WhileCC}^* - \boldsymbol{approx}(\mathcal{R}_\mathsf{p}) = \boldsymbol{ConcComp}_{\overline{\alpha}}(\mathbb{R}). \tag{1}$$

3 *While* Computation on Standard Many-Sorted Total Algebras

We will study a number of high level imperative programming languages based on the "while" construct, applied to a many-sorted signature Σ. We give semantics for these languages relative to a total Σ-algebra A, and define the notions of *computability, semicomputability* and *projective semicomputability* for these languages on A. Much of the material is taken from [31].

We begin by reviewing basic concepts: many-sorted signatures and algebras. Next we define the syntax and semantics of the ***While*** programming language. Then we extend this language with special programming constructs to form two new languages: $\boldsymbol{While}^N$ and $\boldsymbol{While}^*$.

3.1 Basic Concepts: Signatures and Partial Algebras

A many-sorted signature Σ is a pair $\langle \textbf{\textit{Sort}}(\Sigma), \textbf{\textit{Func}}\,(\Sigma)\rangle$ where

(a) $\textbf{\textit{Sort}}(\Sigma)$ is a finite set of basic types called *sorts* $s, s', \ldots$.
(b) $\textbf{\textit{Func}}\,(\Sigma)$ is a finite set of basic *function symbols*

$$F\colon s_1 \times \cdots \times s_m \to s \qquad (m \geq 0).$$

The case $m = 0$ gives a *constant symbol*; we then write $F\colon \;\to s$.

A *product type* has the form $s_1 \times \cdots \times s_m$ $(m \geq 0)$, where $s_1, \ldots, s_m$ are sorts. We write $u, v, \ldots$ for product types. A *function type* has the form $u \to s$, where u is a product type.

A Σ-algebra A has, for each Σ-sort s, a non-empty set A_s, the carrier of sort s, and for each Σ-function symbol $F\colon s_1 \times \cdots \times s_m \to s$, a (basic) function

$$F^A : A^u \to A_s$$

where $u = s_1 \times \cdots \times s_m$, and $A^u = A_{s_1} \times \cdots \times A_{s_m}$.

We write $\Sigma(A)$ for the signature of an algebra A.

A Σ-algebra is called *total* if all the basic functions are total; it is called *partial* in the absence of such an assumption. Sections 3–6 will be devoted to total algebras. In Sect. 7 we will turn to a more general theory, with partial algebras.

Example 3.1.1 (Booleans) The signature $\Sigma(\mathcal{B})$ of the booleans is

signature	$\Sigma(\mathcal{B})$
sorts	bool
functions	true, false: $\to$ bool,
	not: bool $\to$ bool
	or, and: bool$^2 \to$ bool

The algebra $\mathcal{B}$ of booleans contains the carrier $\mathbb{B} = \{\mathtt{t}, \mathtt{f}\}$ of sort bool, and the standard interpretations of the constant and function symbols of $\Sigma(\mathcal{B})$.

Note that for a structure A to be useful for computational purposes, it should be susceptible to testing, which means it should contain the carrier $\mathbb{B}$ of booleans and the standard boolean operations; in other words, it should contain the algebra $\mathcal{B}$ as a retract. Such an algebra A is called *standard*. All the examples of algebras discussed below will be standard.

Example 3.1.2 (Naturals) The signature $\Sigma(\mathcal{N})$ of the naturals is

signature	$\Sigma(\mathcal{N})$
import	$\Sigma(\mathcal{B})$
sorts	nat
functions	$0\colon \ \to$ nat,
	suc: nat $\to$ nat
	$\mathsf{eq_N}, \mathsf{less_N}$: nat^2 $\to$ bool

The algebra $\mathcal{N}$ of naturals consists of the carrier $\mathbb{N} = \{0, 1, 2, \ldots\}$ of sort nat, the carrier $\mathbb{B}$ of sort bool, and the standard constants and functions $0_{\mathsf{N}}\colon \ \to \mathbb{N}$, $\mathsf{suc_N} : \mathbb{N} \to \mathbb{N}$, and $\mathsf{eq_N}, \mathsf{less_N}\colon \mathbb{N}^2 \to \mathbb{B}$ (as well as the standard boolean operations).

We will use the infix notations "=" and "<" for "$\mathsf{eq_N}$" and "$\mathsf{less_N}$", and also use "$\vee$" and "$\wedge$" for the boolean operations "or" and "and".

Example 3.1.3 (Total Algebra of Reals) The signature $\Sigma(\mathcal{R}_t)$ of the total algebra of reals is:

signature	$\Sigma(\mathcal{R}_t)$
import	$\Sigma(\mathcal{B})$
sorts	real
functions	$0, 1\colon \ \to$ real,
	$+, \times$: real^2 $\to$ real,
	$-$: real $\to$ real,
	$\mathsf{eq_R}, \mathsf{less_R}$: real^2 $\to$bool

(We will study a partial algebra of reals in Sect. 7.) The algebra $\mathcal{R}_t$ of reals has the carrier $\mathbb{R}$ of sort real, as well as the imported carrier $\mathbb{B}$ of sort bool with the boolean operations, the real constants and operations $(0, 1, +, \times, -)$, and the (total) boolean-valued functions $\mathsf{eq_R}\colon \mathbb{R}^2 \to \mathbb{B}$. and $\mathsf{less_R}\colon \mathbb{R}^2 \to \mathbb{B}$. Again, we will use the infix notations "=" and "<" for these.

Definition 3.1.4 (Minimal Carriers; Minimal Algebra) Let A be a Σ-algebra, and s a Σ-sort.

(a) A is *minimal at s* (or the carrier A_s is *minimal in* A) if A_s is generated by the closed Σ-terms of sort s.
(b) A is *minimal* if it is minimal at every Σ-sort.

To take two examples:

- Every N-standard algebra (see Sect. 3.3) is minimal at sorts bool and nat.
- The algebra $\mathcal{R}_t$ of reals (Example 3.1.3) is not minimal at sort real.

3.2 Syntax and Semantics of Σ-Terms

For a signature Σ, the set $\boldsymbol{Tm}(\Sigma)$ of Σ-terms is defined from Σ-variables $\mathrm{x}^s, \ldots$ of sort s (for all Σ-sorts s) by

$$t^s ::= \mathrm{x}^s | F(t_1^{s_1}, \ldots, t_m^{s_m})$$

where F is a Σ-function symbol of type $s_1 \times \cdots \times s_m \to s$.

We write $t : s$ to indicate that t is a Σ-term of sort s, and more generally, $t : u$ to indicate that t is a tuple of terms of product type u. We also write $b, \ldots$ for boolean Σ-terms, i.e. Σ-terms of sort bool.

We turn to the semantics of terms.

A *state* over an algebra A is a family $\langle \sigma_s \mid s \in \boldsymbol{Sort}(\Sigma) \rangle$ of functions σ_s: $\boldsymbol{Var}_s \to A_s$ (where $\boldsymbol{Var}_s$ is the set of variables of sort s). Let $\boldsymbol{State}(A)$ be the set of states on A. We will write $\sigma(\mathrm{x})$ for $\sigma_s(\mathrm{x})$ where $\mathrm{x} : s$. Also, for a tuple $\mathrm{x} \equiv (\mathrm{x}_1, \ldots, \mathrm{x}_m)$, we write $\sigma[\mathrm{x}]$ for $(\sigma(\mathrm{x}_1), \ldots, \sigma(\mathrm{x}_m))$.

Let σ be a state over A, and for some Σ-product type u, let $\mathrm{x} \equiv (\mathrm{x}_1, \ldots, \mathrm{x}_n) : u$ and $a = (a_1, \ldots, a_n) \in A^u$ (for $n \geq 1$). We define the *variant* $\sigma\{\mathrm{x}/a\}$ to be the state over A formed from σ by replacing its value at x_i by a_i for $i = 1, \ldots, n$.

For a term $t : s$, we will define the function

$$[\![t]\!]^A : \boldsymbol{State}(A) \to A_s$$

where $[\![t]\!]^A \sigma$ is the value of t in A at state σ.

The definition of $[\![t]\!]^A \sigma$ is by structural induction on Σ-terms t:

$$\begin{aligned} [\![\mathrm{x}]\!]^A \sigma &= \sigma(\mathrm{x}) \\ [\![F(t_1, \ldots, t_m)]\!]^A \sigma &= F^A([\![t_1]\!]^A \sigma, \ldots, [\![t_m]\!]^A \sigma) \end{aligned} \tag{2}$$

3.3 Adding Counters: N-Standard Signatures and Algebras

A signature Σ is *N-standard* if (1) it is standard (see Sect. 3.1), and (2) it contains the standard signature of naturals (Example 3.1.2), i.e., $\Sigma(\mathcal{N}) \subseteq \Sigma$.

Given an N-standard signature Σ, a Σ-algebra A is *N-standard* if it is an expansion of $\mathcal{N}$, i.e., it contains the carrier $\mathbb{N}$ with the standard arithmetic operations.

N-standardness is clearly very useful in computation, with the presence of counters, and the facility for enumerations and numerical coding.

Any standard signature Σ can be "N-standardised" to a signature Σ^N by adjoining the sort nat and the operations 0, suc, eq_N and less_N. Correspondingly, any standard Σ-algebra A can be N-standardised to an algebra A^N by adjoining the carrier $\mathbb{N}$ together with the corresponding arithmetic and boolean functions on $\mathbb{N}$.

3.4 Adding Arrays: Algebras A^* of Signature Σ^*

Given a standard signature Σ, and standard Σ-algebra A, we expand Σ and A in two stages: (1) N-standardise these to form Σ^N and A^N, as in Sect. 3.3; and (2) define, for each sort s of Σ, the carrier A_s^* to be the set of *finite sequences* or *arrays* a^* over A_s, of "starred sort" s^*.

The resulting algebras A^* have signature Σ^*, which extends Σ^N by including, for each sort s of Σ, the new starred sorts s^*, and certain new function symbols to read and update arrays. Details are given in [30, 31].

We conclude this section with a very useful syntactic conservativity theorem, which says that every Σ^*-term with sort in Σ is effectively semantically equivalent to a Σ-term. This theorem will be used later for proving universality for ***While**** computations by a ***While***^{N} procedure (Theorem 3*) and deriving a strong form of Engeler's Lemma (Theorem 8).

Theorem 1 (Σ^*/Σ Conservativity for Terms) *For every Σ-sort s, every Σ^*-term t of sort s without any variables of starred sort is effectively semantically equivalent*[3] *to a Σ-term.*

4 The *While* Programming Language

Note that we will use "≡" to denote syntactic identity between two expressions.

We define ***Stmt***(Σ) to be the class of ***While***(Σ)-statements $S, \ldots$ generated by:

$$S ::= \mathsf{skip} \mid \mathsf{x} := t \mid S_1 \,; S_2 \mid \mathsf{if}\ b\ \mathsf{then}\ S_1\ \mathsf{else}\ S_2\ \mathsf{fi} \mid \mathsf{while}\ b\ \mathsf{do}\ S_0\ \mathsf{od}$$

where the variable x and term t have the same Σ-sort.

Proc(Σ) is the class of ***While***(Σ)-procedures $P, \ldots$, of the form:

$$P \;\equiv\; \mathsf{proc\ in\ a\ out\ b\ aux\ c\ begin}\ S\ \mathsf{end} \tag{3}$$

where S is the body, and and a, b and c are tuples of (distinct) input, output and auxiliary variables respectively.

If $\mathsf{a} : u$ and $\mathsf{b} : v$, then P is said to have type $u \to v$, written $P\colon u \to v$.

We turn to the semantics of statements and procedures.

The meaning $[\![S]\!]^A$ of a statement S is a partial state transformer[4] on an algebra A:

$$[\![S]\!]^A : \textbf{\textit{State}}(A) \rightharpoonup \textbf{\textit{State}}(A).$$

[3] I.e., we can effectively find a Σ-term t' such that $[\![t']\!]^A\sigma = [\![t]\!]^{A^*}\sigma$ for all Σ-algebras A and states σ over A^* (or A).

[4] "$\rightharpoonup$" denotes a partial function.

Its definition is standard [30, 31] and lengthy, and so we omit it. Briefly, it is based on defining the *computation sequence* of states from S starting in a state σ, or rather the n-th component of this sequence, by a primary induction on n, and a secondary induction on the size of S.

Next, given a procedure (3) of type $u \to v$, its meaning is a partial function $P^A : A^u \rightharpoonup A^v$ defined as follows. For $a \in A^u$, let σ be any state on A such that $\sigma[\mathtt{a}] = a$, and $\sigma[\mathtt{b}]$ and $\sigma[\mathtt{c}]$ are given preassigned default values. Then

$$P^A(a) \simeq \begin{cases} \sigma'[b] & \text{if } [\![S]\!]^A\sigma \downarrow \sigma' \quad \text{(say)} \\ \uparrow & \text{if } [\![S]\!]^A\sigma \uparrow. \end{cases}$$

Here "$\simeq$" means that the two sides either both converge to the same value, or both diverge ("Kleene equality" [15, Sect. 63]).

We are also using the notation $[\![S]\!]^A\sigma \downarrow$ to mean that evaluation of $[\![S]\!]^A$ at σ halts or converges; $[\![S]\!]^A\sigma \downarrow \sigma'$ that it converges to σ', and $[\![S]\!]^A\sigma \uparrow$ that it diverges.

It is worth noting that the semantics of ***While***(Σ) procedures is invariant under Σ-isomorphism.

Modifications in these semantic definitions (equations (2) in Sect. 3.2, and (3) in Sect. 4) required for partial algebras will be indicated in Sect. 7 (Remark 7.3.1).

4.1 *While, While^N and While* Computability*

A (partial) function f on A is ***While*** *computable* if $f = P^A$ for some ***While*** procedure P.

Consider now the ***While*** programming language over Σ^N and Σ^*.

A ***While***$^N(\Sigma)$ procedure is a ***While***(Σ^N) procedure in which the input and output variables have sorts in Σ. However the *auxiliary* variables may have sort nat.

Similarly, a ***While***$^*(\Sigma)$ procedure is a ***While***(Σ^*) procedure in which the input and output variables have sorts in Σ. However the *auxiliary* variables may have starred sorts.

A function f on A is ***While***^{N} (or ***While****) *computable* if $f = P^A$ for some ***While***^{N} (or ***While****) procedure P.

We write ***While***(A), ***While***$^N(A)$ and ***While***$^*(A)$ for the classes of functions ***While***, ***While***^{N} and ***While**** computable on A.

Remarks 4.1.1

(a) Clearly, if A is N-standard, then ***While***^{N} computability coincides with ***While*** computability on A.
(b) Because of the effective enumeration of the set $\mathbb{N}^*$, ***While***^{N} and ***While**** computability coincide with ***While*** computability on $\mathcal{N}$, which is in turn equivalent to classical *partial recursiveness* over $\mathbb{N}$.

(c) ***While**** computability will be the basis for a generalized Church-Turing Thesis, as we will see later (Sect. 8.2). On the other hand, $\boldsymbol{While}^N$ computability is useful for representing the syntax of ***While*** programming within the formalism, by means of coding (Sect. 5).

5 Representations of Semantic Functions; Universality

We examine whether the ***While*** programming language is a so-called *universal model* of computation. This means answering questions of the following form. Let A be a standard Σ-algebra.

> *Is there a universal* **While** *procedure* $U_{\mathsf{proc}} \in \boldsymbol{Proc}(\Sigma)$ *that can compute all the* **While** *computable functions on* A*?*

To this end we need the techniques of numerical codings (Gödel numberings) and symbolic computations on terms. More accurately, for this to be possible, we need the sort nat, and so we will consider the possibility of representing the syntax of a standard Σ-algebra A (not in A itself, but) in its N-standardisation A^N, or (failing that) in the array algebra A^*. We will see that

> *For any given* Σ*-algebra A, there is a universal* **While** *procedure over A if, and only if, there is a* **While** *program for term evaluation over A.*

Consequently, since term evaluation is always ***While**** computable on A, we have

> *For any* Σ*-algebra A there is a universal* **While*** *program and universal* **While*** *procedure over A.*

Thus, for any algebra A our ***While**** model of computation is universal.

Hence, if the Σ-algebra A has a ***While*** program to compute term evaluation, then

$$\boldsymbol{While}^*(A) \;=\; \boldsymbol{While}^N(A).$$

5.1 Numbering of Syntax

We assume given families of effective numerical codings of the syntactic classes E with which we deal, i.e., 1–1 maps $\boldsymbol{code}: E \to \mathbb{N}$, with $\ulcorner e \urcorner = \boldsymbol{code}(e)$ denoting the code of the expression $e \in E$. Further, we assume standard effective numberings of sets such as $\mathbb{N}^*$, $\mathbb{Q}^2$, etc. Hence we assume that we can primitive recursively simulate all operations involved in processing the syntax of the programming language.

By "effective(ly)", we will mean *effective* in the codes of the syntactic or mathematical objects referred to.

We will use the notation

$$\ulcorner \boldsymbol{Tm} \urcorner \;=\; \{\ulcorner t \urcorner \mid t \in \boldsymbol{Tm}\},$$

etc., for sets of codes of syntactic expressions.

5.2 Representation of States

We will be interested in the representation of various semantic functions on syntactic classes such as $\boldsymbol{Tm}(\Sigma)$, $\boldsymbol{Stmt}(\Sigma)$ and $\boldsymbol{Proc}(\Sigma)$ by functions on A or A^*, and in the computability of these representing functions. These semantic functions have states as arguments, so we must first define a representation of states.

Let x be a u-tuple of program variables. A state σ on A is *represented* (relative to x) by a tuple of elements $\boldsymbol{a} \in A^u$ if $\sigma[\mathrm{x}] = \boldsymbol{a}$.

The *state representing function*

$$\boldsymbol{Rep}^A_{\mathrm{x}} : \boldsymbol{State}(A) \;\to\; A^u$$

is defined by

$$\boldsymbol{Rep}^A_{\mathrm{x}}(\sigma) \;=\; \sigma[\mathrm{x}].$$

5.3 Representation of Term Evaluation; Term Evaluation Property

Let x be a u-tuple of variables. Let $\boldsymbol{Tm}_{\mathrm{x}} = \boldsymbol{Tm}_{\mathrm{x}}(\Sigma)$ be the class of all Σ-terms with variables among x only, and for all sorts s of Σ, let $\boldsymbol{Tm}_{\mathrm{x},s} = \boldsymbol{Tm}_{\mathrm{x},s}(\Sigma)$ be the class of such terms of sort s.

The *term evaluation function on A relative to* x

$$\boldsymbol{TE}^A_{\mathrm{x},s} : \boldsymbol{Tm}_{\mathrm{x},s} \times \boldsymbol{State}(A) \;\to\; A_s,$$

defined by

$$\boldsymbol{TE}^A_{\mathrm{x},s}(t, \sigma) \;=\; [\![t]\!]^A \sigma,$$

is *represented* by the function

$$\boldsymbol{te}^A_{\mathrm{x},s} : \ulcorner \boldsymbol{Tm}_{\mathrm{x},s} \urcorner \times A^u \;\to\; A_s$$

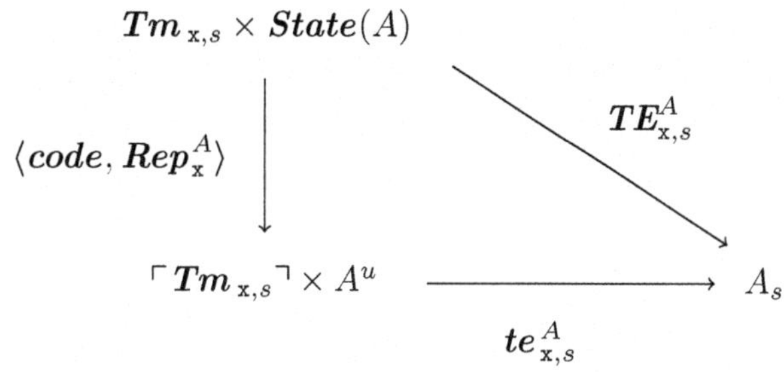

Fig. 1 Term evaluation representing function

defined by

$$te^A_{x,s}(\ulcorner t \urcorner, a) = [\![t]\!]^A \sigma,$$

where σ is any state on A such that $\sigma[x] = a$, in the sense that the diagram in Fig. 1 commutes.

We will be interested in the computability of this term evaluation representing function.

Definition 5.3.1 (Term Evaluation) The algebra A has the *term evaluation property (TEP)* if for all x and s, the term evaluation representing function $te^A_{x,s}$ is ***While*** computable on A^N.

Many well-known varieties (i.e., equationally axiomatisable classes of algebras) have (uniform versions of) the TEP. Examples are: semigroups, groups, and associative rings with or without unity. This follows from the *effective normalisability* of the terms of these varieties. In the case of rings, this means an effective transformation of arbitrary terms to polynomials.

Thus, for example, the algebra $\mathcal{R}_t$ of reals has the TEP.

Theorem 2 *The term evaluation representing function on A^* is **While** computable on A^*.*

Corollary 5.3.2 *The term evaluation representing function on A is **While*** computable on A^N.*

Recall the Definition 3.1.4 of minimal carriers.

Corollary 5.3.3

(a) *If A is minimal at s, then there is a **While*** computable enumeration (or listing) of the carrier A_s, i.e., a surjective total mapping*

$$enum^A_s : \mathbb{N} \to A_s,$$

*which is **While*** computable on A^N.*

(b) *If in addition A has the TEP, then $enum^A_s$ is also **While** computable on A^N.*

Theorem 3 (Universality Characterization Theorem for *While* Computations) *The following are equivalent:*

(i) *A has the TEP;*
(ii) *For all Σ-product types u, v, there is a* ***While***(Σ^N) *procedure*

$$\mathsf{Univ}_{u \to v} : \mathsf{nat} \times u \to v$$

which is universal for $\boldsymbol{Proc}_{u \to v}$ *on A, in the sense that for all* $P \in \boldsymbol{Proc}_{u \to v}$ *and* $\boldsymbol{a} \in A^u$,

$$\mathsf{Univ}^A_{u \to v}(\ulcorner P \urcorner, a) \simeq P^A(a).$$

Using the Σ^*/Σ conservativity theorem (Theorem 1), we can strengthen the above theorem, so as to construct a universal ***While***(Σ^N) procedure for ***While**** computation,

Theorem 3* (Universality Characterization Theorem for *While Computations)** *The following are equivalent.*

(i) *A has the TEP.*
(ii) *For all Σ-product types u, v, there is a* ***While***(Σ^N) *procedure*

$$\mathsf{Univ}_{u \to v} : \mathsf{nat} \times u \to v$$

which is universal for $\boldsymbol{Proc}^*_{u \to v}$ *on A, in the sense that for all* $P \in \boldsymbol{Proc}^*_{u \to v}$ *and* $\boldsymbol{a} \in A^u$,

$$\mathsf{Univ}^A_{u \to v}(\ulcorner P \urcorner, a) \simeq P^A(a).$$

We conclude that there are universal ***While***^{N} procedures for ***While**** computation on $\mathcal{R}_t$.

6 Concepts of Semicomputability

We want to generalize the notion of *recursive enumerability* to many-sorted algebras. There turn out to be many non-equivalent ways to do this.

The primary idea is that a set is ***While*** semicomputable if, and only if, it is the domain or halting set of a ***While*** procedure; and similarly for ***While***^{N} and ***While**** semicomputability.

This concept satisfies the standard closure properties (under finite union and intersection) and also Post's Theorem:

A set is computable if, and only if, it and its complement are semicomputable.

The second idea of importance is that of a *projection* of a computable or semicomputable set. set. In classical computability theory on $\mathbb{N}$, the class of semicomputable sets is closed under projections, but this is not true in the general case of algebras, as we will see. Projective semicomputability is strictly more powerful (and less algorithmic) than semicomputability.

We will also characterize the semicomputable sets as the sets definable by some effective countable disjunction of boolean valued terms. This result, first observed by E. Engeler, has a number of interesting applications.

Definition 6.0.4

(a) R is ***While*** *computable on* A if its characteristic function is.
(b) R is ***While*** *semicomputable* on A if it is the halting set on A of some ***While*** procedure P, i.e., $R = \{a \in A^u \mid P^A(a) \downarrow\}$.

Examples 6.0.5 We will have need for the notation $\mathcal{R}_t^o$ to indicate the algebra $\mathcal{R}_t$ without the "<" operation.

(a) On the naturals $\mathcal{N}$ the ***While*** semicomputable sets are precisely the recursively enumerable sets, and the ***While*** computable sets are precisely the recursive sets.
(b) On $\mathcal{R}_t^o$ the set of *naturals* (as a subset of $\mathbb{R}$) is ***While*** semicomputable, being the halting set of the following procedure:

```
is_nat ≡ proc in x : real
            begin while not x = 0
                     do x := x − 1 od
            end
```

(c) Similarly, the set of *integers* is ***While*** semicomputable on $\mathcal{R}_t^o$.
(d) However, the sets of naturals and integers are ***While*** *computable* on $\mathcal{R}_t$, as can be easily seen.
(e) The set of *rationals* is ***While*** semicomputable on $\mathcal{R}_t^o$. (*Exercise*. Hint: Prove this first for $\mathcal{R}_t^{oN}$.)

6.1 Merging Two Procedures; Closure Theorems

The classical "merge" theorems generalize:

Theorem 4 *The union and intersection of two* **While** *semicomputable relations of the same type are again* **While** *semicomputable.*

In the case of union, if we assume that (1) A is N-standard, and (2) A has the TEP, then the construction of the "merge" of the two characteristic procedures, i.e., interleaving their steps to form the new procedure, simply follows the classical proof for computation on $\mathbb{N}$. Failing this, the construction of the merge procedure (by structural induction on the pair of statements) is quite challenging. (The tricky case is where both are "while" statements.)

If R is a relation on A of type u, we write the *complement of* R as $R^{\mathsf{c}} = A^u \backslash R$.

Theorem 5 (Post's Theorem for *While* Semicomputability) *For any relation R on A*

$$R \text{ is } \textbf{\textit{While}} \text{ computable} \iff R \text{ and } R^{\mathsf{c}} \text{ are } \textbf{\textit{While}} \text{ semicomputable.}$$

Note that the proofs of the above two theorems depend strongly on the totality of A. (See Remark 7.3.2.)

Another useful closure result, applicable to N-standard structures, is:

Theorem 6 (Closure of *While* Semicomputability Under $\mathbb{N}$-Projections) *Suppose A is N-standard. If $R \subseteq A^{u \times \mathsf{nat}}$ is* ***While*** *semicomputable on A, then so is its $\mathbb{N}$-projection $\{x \in A^u \mid \exists n \in \mathbb{N}\, R(x, n)\}$.*

To outline the proof: From a procedure P which halts on R, we can effectively construct another procedure which halts on the required projection. Briefly, for input x, we search by "dovetailing" for a number n such that P halts on (x, n).

We can generalize this theorem to the case of an A_s-projection for any *minimal carrier* A_s (recall Definition 3.1.4), provided A has the TEP.

Corollary 6.1.1 (Closure of *While* Semicomputability Under Projections off Minimal Carriers) *Suppose A is N-standard and has the TEP. Let A_s be a minimal carrier of A. If $R \subseteq A^{u \times s}$ is* ***While*** *semicomputable on A, then so is its projection off A_s.*

Note that Corollary 6.1.1 is a many-sorted version of (part of) Theorem 2.4 of [9], cited in [23]. The minimality condition (a version of Friedman's Condition III) means that *search* in A_s is *computable* (or, more strictly, *semicomputable*) provided A has the TEP. Thus in minimal algebras, many of the results of classical recursion theory carry over, e.g.,

- the domains of semicomputable sets are closed under projection (as above)
- a semicomputable relation has a computable selection function
- a function with semicomputable graph is computable [9, Theorem 2.4].

If, in addition, there is *computable equality* at the appropriate sorts, other results of classical recursion theory carry over, e.g.,

- the range of a computable function is semicomputable [9, Theorem 2.6].

6.2 Projective While semicomputability and Computability

A set $R \subseteq A^u$ is *projectively* **While** *semicomputable* (or *computable*) on A iff R is a projection of a **While** semicomputable (or computable) set on A, i.e., for some product types u and v,

$$\forall x \in A^u \left[x \in R \iff \exists y \in A^v : (x, y) \in R'\right].$$

where R' is a semicomputable (or computable) subset of $A^{u \times v}$.

We note that although the emphasis in this subsection is on projective *semi*-computability, the concept of projective *computability* will be used in our formulation of a generalized Church-Turing thesis for specifiability (in Sect. 8).

In this connection we note further that

(1) The concepts of projective ***While**** semicomputability and projective ***While**** computability coincide, by the projective equivalence theorem (Theorem 10 below).
(2) Projective ***While***$^{(*)}$ semicomputability is, in general, a broader concept than ***While***$^{(*)}$ semicomputability (Sect. 6.7).

We do, however, have closure of semicomputability in the case of $\mathbb{N}$-projections, i.e., existential quantification over $\mathbb{N}$, as we saw in Theorem 6. Further, we have from Corollary 6.1.1:

Theorem 7 *Suppose A is N-standard and minimal and has the TEP. Then on A*

$$\textit{projective } \textbf{While} \textit{ semicomputability} \;=\; \textbf{\textit{While}} \textit{ semicomputability.}$$

6.3 While* Semicomputability

A relation R on A is ***While**** *semicomputable* if it is the halting set of some ***While***(Σ^*) procedure on A^*.

Again, we have Post's Theorem for ***While**** computability and semicomputability, and again, we have closure of ***While**** semicomputability under $\mathbb{N}$-projections, and projections off minimal carriers.

Note that we do not have to assume the TEP for the latter (cf. Corollary 6.1.1), since the term evaluation representing function is always ***While**** computable.

Example 6.3.1 The subalgebra relation[5]:

$$\textbf{\textit{subalg}}^A(x, y) \iff x \text{ is in the subalgebra of } A \text{ generated by } y$$

[5]We are suppressing sort superscripts here.

is ***While***** semicomputable on A. This follows from ***While***** computability of term evaluation on A^N (Corollary 5.3.2).

6.4 Projective While* Semicomputability

A relation R on A is said to be *projectively* ***While***** *computable* (or *semicomputable*) on A if R is a projection of a ***While***(Σ^*) computable (or semicomputable) relation on A^*.

Theorem 7 can be re-stated for ***While***** semicomputability:

Theorem 7* *Suppose A is a minimal. Then on A*

$$\textit{projective } \textbf{\textit{While}}^{*} \textit{ semicomputability} \;=\; \textbf{\textit{While}}^{*} \textit{ semicomputability.}$$

Note again that the TEP does not have to be assumed here (cf. Theorem 7). Also we are using the fact that if A is minimal then so is A^*.

Example 6.4.1 In $\mathcal{N}$, the various concepts we have listed: ***While***, ***While***^{N} and ***While***** semicomputability, as well as projective ***While***, ***While***^{N} and ***While***** semicomputability, all reduce to *recursive enumerability* over $\mathbb{N}$.

In general, however, projective ***While***** semicomputability is strictly broader than projective ***While***^{N} semicomputability. In other words, projecting along starred sorts is stronger than projecting along simple sorts or nat. (Intuitively, this corresponds to existentially quantifying over a finite, but unbounded, sequence of elements.) An example to show this will be given below.

We do, however, have the following equivalence:

$$\textit{projective } \textbf{\textit{While}}^{*} \textit{ semicomputability} = \textit{projective } \textbf{\textit{While}}^{*} \textit{ computability.}$$

This is the projective equivalence theorem for ***While***** (Theorem 10).

6.5 Computation Trees; Engeler's Lemma

For any ***While*** statement S over Σ, we can define a (possibly infinite) *computation tree* for S. The construction is by structural induction on S. Details are given in [31].

Using this and the Σ/Σ^* conservativity theorem (Theorem 1), we can prove the following. Let R be a relation on A.

Theorem 8 (Engeler's Lemma for *While Semicomputability)** *R is* ***While***** *semicomputable over A iff R can be expressed as an effective countable disjunction of booleans over* Σ.

If, moreover, A has the TEP, then we can say more:

Theorem 9 (Semicomputability Equivalence Theorem) *Suppose A has the TEP. Then the following assertions are equivalent:*

(i) *R is* ***While***^{N} *semicomputable on A;*
(ii) *R is* ***While**** *semicomputable on A.*
(iii) *R can be expressed as an effective countable disjunction of booleans over Σ.*

The step (i)⇒(ii) is trivial, and (ii)⇒(iii) is just Engeler's Lemma for ***While****. The new step (iii)⇒(i) follows from an analysis of the coding of an effective infinite disjunction.

Corollary 6.5.1 *Suppose A has the TEP. Then the following are equivalent:*

(i) *R is* ***While**** *computable on A;*
(ii) *R is* ***While***^{N} *computable on A.*

This follows from the above theorem, and Post's Theorem for ***While***^{N} and ***While****.

6.6 *Projective Equivalence Theorem for While**

The following theorem uses Engeler's Lemma, and the ***While**** computability of term evaluation.

Theorem 10 (Projective Equivalence Theorem) *The following are equivalent:*

(i) *R is projectively* ***While**** *semicomputable on A;*
(ii) *R is projectively* ***While**** *computable on A.*

We can strengthen the theorem with a third equivalent clause, if we add an assumption about computability of equality in A.

First we must define certain syntactic classes of formulae over Σ.

Let ***Lang***$^* = $ ***Lang***(Σ^*) be the first order language with equality over Σ^*. We are interested in special classes of formulae of ***Lang****.

Formulae of ***Lang**** are formed from the atomic formulae by means of the propositional connectives and universal and existential quantification over variables of any Σ^*-sort.

Definition 6.6.1 (Classes of Formulae of *Lang*(Σ^*))

(a) An *atomic formula* is an equality between a pair of terms of the same Σ^*-sort.
(b) A *bounded quantifier* has the form '$\forall k < t$' or "$\exists k < t$", where $t : \mathsf{nat}$.
(c) An *elementary formula* is one with only bounded quantifiers.
(d) A Σ_1^* *formula* is formed by prefixing an elementary formula with existential quantifiers only.
(e) An *extended* Σ_1^* *formula* is formed by prefixing an elementary formula with a string of existential quantifiers and bounded universal quantifiers (in any order).

We can show that an extended Σ_1^* formula is equivalent to a Σ_1^* formula over Σ. Hence we will use the term "Σ_1^*" to denote (possibly) extended Σ_1^* formulae.

We can now re-state the projective equivalence theorem in the presence of equality.

Theorem 10$^=$ (Projective Equivalence Theorem for Σ^* with Equality) *Suppose Σ has an equality operator at all sorts. Then the following are equivalent:*

(i) *R is projectively **While*** semicomputable on A;*
(ii) *R is projectively **While*** computable on A;*
(iii) *R is Σ_1^* definable.*

6.7 Semicomputability and Projective Semicomputability on $\mathcal{R}_t$

We apply some of the above ideas and results to the algebra $\mathcal{R}_t$. Details can be found in [31, Sects. 6.2, 6.3][6]

We begin again with a restatement of the semicomputability equivalence theorem (Theorem 9), for the particular case of $\mathcal{R}_t$.

Theorem 11 (Semicomputability for $\mathcal{R}_t$) *Suppose $R \subseteq \mathbb{R}^n$ $(n = 1, 2, \ldots)$. Then the following are equivalent:*

(i) *R is **While**^{N} semicomputable on $\mathcal{R}_t$,*
(ii) *R is **While*** semicomputable on $\mathcal{R}_t$,*
(iii) *R can be expressed as an effective countable disjunction*

$$x \in R \Longleftrightarrow \bigvee_i b_i(x)$$

where each $b_i(x)$ is a finite conjunction of equations and inequalities of the form

$$p(x) = 0 \qquad \textit{and} \qquad q(x) > 0,$$

where p, q are polynomials in $x \equiv (x_1, \ldots, x_n) \in \mathbb{R}^n$, with coefficients in $\mathbb{Z}$.

We also have:

Theorem 12 *In $\mathcal{R}_t$, the following three concepts coincide for subsets of $\mathbb{R}^n$:*

(i) ***While**^{N} semicomputability,*
(ii) ***While*** semicomputability,*
(iii) *projective **While**^{N} semicomputability.*

[6]The notation in [31] is unfortunately not completely consistent with the present notation: $\mathcal{R}$ and $\mathcal{R}^<$ in [31] correspond (resp.) to $\mathcal{R}_t^0$ and $\mathcal{R}_t$ here.

The proof of equivalence between (iii) and the other two concepts follows from the fact that semialgebraic sets are closed under projection, which in turn follows from Tarski's quantifier-elimination theorem for real closed fields [16, Chap. 4].

Interestingly, in the algebra $\mathcal{R}_t^o$ (i.e., $\mathcal{R}_t$ without the order relation "<"), where Tarski's theorem fails, one can find an example of a relation (namely, "<" itself!) which is projectively ***While*** semicomputable, but not ***While*** (or ***While****) semicomputable.

On the other hand (returning to $\mathcal{R}_t$) the three equivalent concepts of semicomputability given in Theorem 12 differ from a fourth:

(iv) *projective* ***While**** *semicomputability*,

as we now show.

Example 6.7.1 (A set which is projectively ***While**** *semicomputable, but not projectively* ***While***^{N} *semicomputable)* In order to prepare for this example, we must first enrich the structure $\mathcal{R}_t$. Let $E = \{e_0, e_1, e_2, \ldots\}$ be a sequence of reals such that

$$\text{for all } i,\ e_i \text{ is transcendental over } \mathbb{Q}(e_0, \ldots, e_{i-1}).$$

We define $\mathcal{R}_t^E$ to be the algebra $\mathcal{R}_t$ augmented by the set E as a separate sort E, with the embedding $j : E \to \mathbb{R}$ in the signature, thus:

algebra	$\mathcal{R}_t^E$
import	$\mathcal{R}_t$
carriers	E
functions	$j : E \to \mathbb{R}$

We write $\overline{E} \subset \mathbb{R}$ for the real algebraic closure of $\mathbb{Q}(E)$.

It is easy to see that $\overline{E}$ is projectively ***While**** semicomputable in $\mathcal{R}_t^E$. (In fact, $\overline{E}$ is the projection on $\mathbb{R}$ of a ***While*** semicomputable relation on $\mathbb{R} \times E^*$.) We must show that, on the other hand, $\overline{E}$ is not projectively ***While***^{N} semicomputable in $\mathcal{R}_t^E$.

Briefly, we proceed by showing that if $F \subseteq \overline{E}$ is any projectively ***While***^{N} semicomputable set in $\mathcal{R}_t^E$, then, using Engeler's Lemma and Tarski's theorem, we can show that F cannot equal $\overline{E}$.

The proof further shows that $\overline{E}$ (although a projection on $\mathbb{R}$ of a ***While*** semicomputable relation on $\mathbb{R} \times E^*$) is not a projection of a ***While***^{N} semicomputable relation in $\mathcal{R}_t^E$. In fact, it can be shown (still using Engeler's Lemma) that $\overline{E}$ is not even a projection of a ***While**** semicomputable relation on $\mathbb{R}^n \times E^m$ (for any $n, m > 0$). Thus to define $\overline{E}$, we must project off the *starred sort* E^*, or (in other words) existentially quantify over a *finite, but unbounded* sequence of elements of E.

7 Computation on Topological Partial Algebras

When one considers the relation between abstract and concrete models, a number of intriguing problems appear. We will explain them by considering a series of examples based on the data type of real numbers. Then we formulate our strategy for solving these problems. The picture for topological algebras in general will be clear from our examples with the reals.

7.1 Abstract Versus Concrete Data Types of Reals; Continuity; Partiality

7.1.1 Abstract and Concrete Data Types of Reals

To compute on $\mathbb{R}$ with an abstract model of computation, we have only to select an algebra A in which $\mathbb{R}$ is a carrier set. Abstract computability on $\mathbb{R}$ is then computability on A, and we may apply the general theory of computable functions on many-sorted algebras outlined in the previous sections.

By contrast, to compute on $\mathbb{R}$ with a concrete model of computation (say the tracking model), we first take a standard enumeration of the rationals $\alpha : \mathbb{N} \to \mathbb{Q}$, which in turn yields a representation $\overline{\alpha} : C \to \mathbb{R}_\mathsf{c}$ that maps the set $C \subset \mathbb{N}$ of codes of effective fast Cauchy sequences of rationals onto the computable reals $\mathbb{R}_\mathsf{c} \subset \mathbb{R}$. With this natural number representation, computable functions on $\mathbb{R}$ are investigated by means of their (classically computable) *$\overline{\alpha}$-tracking functions* on $\mathbb{N}$ [33, 34].

7.1.2 Continuity

Computations with real numbers involve infinite data. Computations are finite processes that approximate in some way infinite processes. The topology of $\mathbb{R}$ defines a process of approximation for infinite data; the functions on the data that are *continuous* in the topology are exactly the functions that can be approximated to any desired degree of precision. This suggests a *continuity principle*:

$$\textit{computability} \implies \textit{continuity}. \tag{4}$$

For abstract models, we assume the algebra A that contains $\mathbb{R}$ is a *topological algebra*, i.e., one in which the basic operations are continuous in its topologies. This implies, in turn, that *all* computable functions will be continuous. As it turns out, the class of functions that can be exactly abstractly computed is, in general, quite limited—"approximate" computations are also necessary [30].

In the concrete models, on the other hand, continuity of computable functions is a consequence of the the Kreisel-Lacombe-Tseitin Theorem [17, 26, 27].

Thus, in both abstract and concrete approaches, an analysis of basic concepts leads to the continuity principle.

7.1.3 Partiality

In computing with an abstract model on A we assume A has some boolean-valued functions to test data. For example, in computing on $\mathbb{R}$ we need the functions

$$\mathsf{eq_R} : \mathbb{R}^2 \to \mathbb{B} \qquad \text{and} \qquad \mathsf{less_R} : \mathbb{R}^2 \to \mathbb{B}. \tag{5}$$

This presents a problem, since total continuous boolean-valued functions on the reals, being continuous functions from a connected space $\mathbb{R}^n$ to a discrete space $\mathbb{B}$, must be constant. Furthermore, in consequence, we can show that the "while" and "while"-array computable functions on connected total topological algebras are precisely the functions explicitly definable by terms over the algebra [30]. This demands the use of partiality for such functions.

To study the full range of real number computations, we must therefore redefine these tests as *partial* boolean-valued functions. This has interesting effects on the theory of computable functions in the areas of nondeterminism and many-valuedness, as we will see.

We turn to some examples to illustrate these features.

7.2 Examples of Nondeterminism and Many-Valuedness

We look at two examples of computing functions on $\mathbb{R}$.

Example 7.2.1 (Nonzero Selection Function) Define the function

$$\boldsymbol{piv} : \mathbb{R}^n \rightharpoonup \{1, \ldots, n\}$$

by

$$\boldsymbol{piv}(x_1, \ldots, x_n) = \begin{cases} \text{some } i : \ x_i \neq 0 & \text{if such an } i \text{ exists} \\ \uparrow & \text{otherwise.} \end{cases} \tag{6}$$

Computation of this nondeterministic ("pivot") function is a crucial step in the Gaussian elimination algorithm for inverting matrices.

Note that (depending on the precise semantics for the phrase "some i" in (6)) $\boldsymbol{piv}$ is *nondeterministic* or (alternatively) *many-valued* on $\boldsymbol{dom}(\boldsymbol{piv}) = \mathbb{R}^n \backslash \{0\}$. Further:

(a) There is no *single-valued* function which satisfies definition (2) in Sect. 3.2 and is *continuous* on $\mathbb{R}^n$ (as can be easily seen).

(b) However there *is* a computable (and hence continuous!) single-valued function

$$\boldsymbol{piv}\colon \mathbf{CS}^n \rightharpoonup \{1,\ldots,n\}$$

(where **CS** is the space of fast Cauchy sequences of rationals) with a simple algorithm. Note however that ***piv*** is *not extensional* on $\mathbf{CS}^n$, in the sense that it cannot be factored through $\mathbb{R}^n$:

$$\begin{array}{ccc} \mathbf{CS}^n & \xrightarrow{\ \boldsymbol{piv}\ } & \\ \kappa^n \downarrow & & \\ \mathbb{R}^n & \overset{?}{\dashrightarrow} & \{1,\ldots,n\} \end{array}$$

where κ is the map from Cauchy sequences to their limits.

In effect, we can regain continuity (for a single-valued function), by foregoing extensionality.

(c) Alternatively, we can maintain continuity *and* extensionality by giving up single-valuedness. For the *many-valued* function

$$\boldsymbol{piv}_{\mathsf{m}}\colon \mathbb{R}^n \to \wp(\{1,\ldots,n\})$$

defined, for $k = 1,\ldots,n$, by

$$k \in \boldsymbol{piv}_{\mathsf{m}}(x_1,\ldots,x_n) \iff x_k \neq 0$$

is *extensional* and *continuous*, where a function $f\colon A \to \mathcal{P}(B)$ is defined to be continuous iff for all open $Y \subseteq B$, $f^{-1}[Y]$ $(=_{df} \{x \in A \mid f(x) \cap Y \neq \emptyset\})$ is open in A.

Note that the complete Gaussian algorithm for inverting matrices is *continuous* and *deterministic* (hence *single-valued*) and *extensional*, even though it contains ***piv*** as an essential component!

Example 7.2.2 (Finding the Root of a Function—Adapted from [39]) Consider the function f_a (Fig. 2),[7] with real parameter a, defined by

$$f_a(x) = \begin{cases} x+a+2 & \text{if } x \leq -1 \\ a-x & \text{if } -1 \leq x \leq 1 \\ x+a-2 & \text{if } 1 \leq x. \end{cases}$$

[7] Figures 2 and 3 are taken by kind permission from [33], © 2004 ACM Inc.

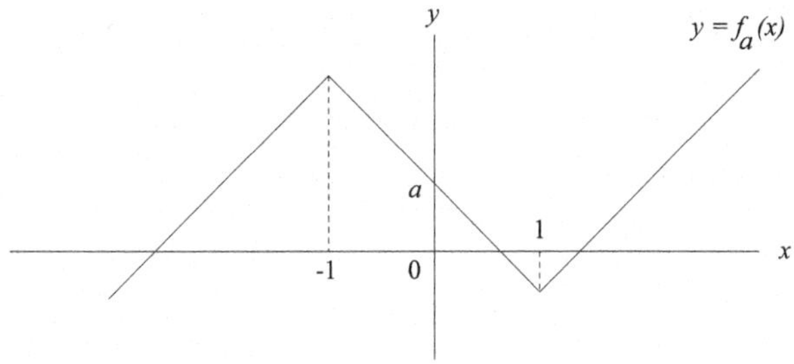

Fig. 2 Function f_a in Example 7.2.2

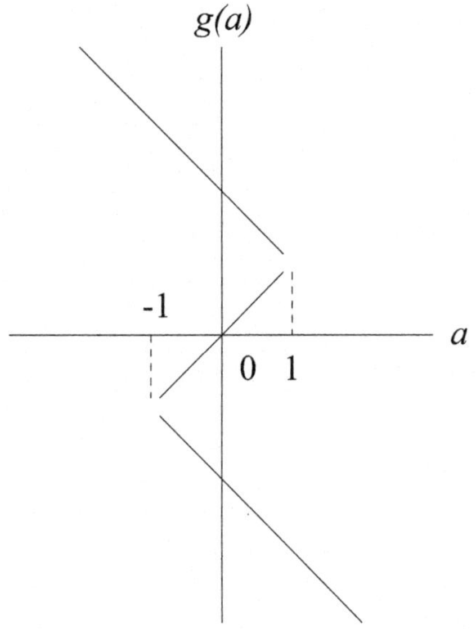

Fig. 3 Non-repeated roots of f_a

This function has either 1 or 3 roots, depending on the size of a. For $a < -1$, f_a has a single (large positive) root; for $a > 1$, f_a has a single (large negative) root; and for $-1 < a < 1$, f_a has three roots, two of which become equal when $a = \pm 1$.

Let g be the (many-valued) function, such that $g(a)$ gives all the non-repeated roots of f_a (Fig. 3). Again we have the situation of the previous examples:

(a) We cannot choose a (single) root of f_a continuously as a function of a.
(b) However, one can easily choose and compute a root of f_a continuously as a function of a *Cauchy sequence representation* of a, i.e., non-extensionally in a.
(c) Finally, $g(a)$, as a *many-valued* function of a, is continuous. (Note that in order to have continuity, we must exclude the repeated roots of f_a, at $a = \pm 1$.)

Other examples of a similar nature abound, and can be handled similarly; for example, the problem of finding, for a given real number x, an integer $n > x$.

7.3 Partial Algebra of Reals; Completeness for the Abstract Model

At the level of *concrete models* of computation, there is no real problem with the issues raised by these examples, since concrete models work only by computations on *representations* of the reals (say by Cauchy sequences).

The real problem arises with the construction of *abstract models* of computation on the reals which should model the phenomena illustrated by these examples, and also correspond, in some sense, to the concrete models.

An immediate problem in this regard is that the total boolean-valued functions eq_R and less_R are not continuous, and hence also (by the continuity principle, Sect. 7.1.2) not (concretely) computable.

We therefore define an N-standard *partial algebra* $\mathcal{R}_\mathsf{p}$ on the reals, formed from the total algebra $\mathcal{R}_\mathsf{t}$ (Example 3.1.3) by *replacing* the total boolean-valued functions eq_R and less_R (Sect. 7.1.3, (5)) by the partial functions

$$\mathsf{eq}_{\mathsf{R,p}}(x,y) \ \simeq \ \begin{cases} \uparrow & \text{if } x = y \\ \mathbb{f} & \text{otherwise,} \end{cases}$$

$$\mathsf{less}_{\mathsf{R,p}}(x,y) \ \simeq \ \begin{cases} \mathbb{t} & \text{if } x < y \\ \mathbb{f} & \text{if } x > y \\ \uparrow & \text{if } x = y. \end{cases}$$

These partial functions (unlike the total versions), *are* continuous, and hence $\mathcal{R}_\mathsf{p}$ (unlike $\mathcal{R}_\mathsf{t}$) is a topological partial algebra. Moreover, these partial functions are concretely computable (by e.g. the tracking model, cf. Sect. 7.1.1).

Then we have the question:

Can such continuous many-valued functions be computed on the abstract data type A containing $\mathbb{R}$ using new abstract models of computation?
If so, are the concrete and abstract models equivalent?

The solution presented in [33] was to take $A = \mathcal{R}_\mathsf{p}^N$, the N-standard extension of $\mathcal{R}_\mathsf{p}$, and then extend the ***While****** programming language over A [31] with a nondeterministic "countable choice" programming construct, so that in the rules of program term formation,

$$\mathsf{choose}\ \mathtt{z} : b$$

is a new term of type nat, where $\mathtt{z}$ is a variable of type nat and b a term of type bool. In addition (calling the resulting language ***WhileCC**** for ***While**** computability with countable choice), ***WhileCC**** *computability* is replaced by ***WhileCC**** *approximability* [33, 34]. We then obtain a *completeness theorem* for abstract/concrete computation, i.e. the equivalence (1) shown at the end of Sect. 2.

Actually (1) was proved in [33] for N-standard metric algebras satisfying some general conditions.

The above considerations lead us to propose the topological partial algebra $\mathcal{R}_p$ as a better basis for abstract models of computation on $\mathbb{R}$ than the (total) algebra $\mathcal{R}_t$—better in the sense of being more faithful to the intuition of computing on the reals.[8]

Remark 7.3.1 (Semantics of Partial Algebras) We briefly indicate the semantics for terms and statements over partial algebras, or rather indicate how the semantics for total algebras given in Sects. 3.2 and 4 can be adapted to partial algebras.

First, the semantics of terms is as given by the Eq. (2) in Sect. 3.2 (with the second "$=$" replaced by "$\simeq$"), using strict evaluation for partial functions (i.e., divergence of any subterm entailing divergence of the term).[9]

Secondly, the semantics of statements is as given in Sect. 4; i.e., the value $[\![S]\!]^A\sigma$ of a statement S at a state σ is the last state in a computation sequence (i.e. a sequence of states) generated by S at σ, provided that the sequence is (well defined and) finite. Otherwise (with an infinite computation sequence) the value *diverges*. The case of partial algebras is similar, except that there are now *two* cases where the value of $[\![S]\!]^A\sigma$ diverges: (1) (as before, *global divergence*) where the computation sequence is infinite, and (2) (a new case, *local divergence*) where the computation sequence is finite, but the last item diverges (instead of converging to a state) because of a divergent term on the right of an assignment statement or a divergent boolean test.

Remark 7.3.2 (Comparison of Formal Results for $\mathcal{R}_p$ and $\mathcal{R}_t$) It would be interesting to see to what extent the results concerning abstract computing on the reals with $\mathcal{R}_t$ detailed in Sects. 3–6 (for example, the merging and closure theorems (Sect. 6.1) and comparisons of various notions of semicomputability and projective semicomputability in $\mathcal{R}_t$ (Sect. 6.7) hold, or fail to hold, in $\mathcal{R}_p$.[10]

It should be noted, in this regard, that the merging procedure used in our proofs of the closure theorems (Theorems 3–5) depend heavily on the totality of the algebra A.

[8]For another perspective on computing with total algebras on the reals, see [2].

[9]As a general rule. For a case where boolean operators with non-strict semantics are appropriate, see [40, Sect. 3].

[10]Some striking results in this connection have been obtained by Mark Armstrong [1], e.g. the failure of closure of ***While***-semicomputable relations under union in $\mathcal{R}_p$ (cf. Theorem 4 in Sect. 6.1).

8 Comparing Models and Generalizing the Church-Turing Thesis

To conclude, we will mention several other abstract approaches to computability on abstract algebras, comment on their comparison, and discuss how to generalize the Church-Turing thesis. These other methods have a variety of technical intuitions and objectives, though they share the abstract setting of an algebraic structure. So let us suppose their common purpose to be the characterization of those functions that are computable in an abstract setting.

8.1 *Abstract Models of Computation*

The computable functions on an abstract algebra can also be characterized by approaches based upon

(1) machine models;
(2) high-level programming constructs;
(3) recursion schemes;
(4) axiomatic methods;
(5) equational calculi;
(6) fixed-point methods for inductive definitions;
(7) set-theoretic methods;
(8) logical languages.

We consider only a couple of these; a fuller survey can be found in [31].

Recursion schemes. Kleene's recursion schemes suggest that we create the class $\boldsymbol{\mu PR}(A)$ of functions $\boldsymbol{\mu PR}$ computable on a standard algebra A, namely those functions definable from the basic operations of A by the application of *composition*, *simultaneous primitive recursion* and *least number search*. We can also extend this to the class $\boldsymbol{\mu PR}^*(A)$ of functions on A definable by the $\boldsymbol{\mu PR}$ operations on A^* (analogous to the definition of the class $\boldsymbol{While}^*(A)$ from $\boldsymbol{While}(A^*)$ in Sect 4.1).

Alternatively, we can define the class $\boldsymbol{\mu CR}(A)$ of functions on A in which simultaneous primitive recursion is replaced by simultaneous *course-of-values recursive schemes*. (Simultaneous recursions are needed because the structures are many-sorted.) Then we have:

Theorem 13 (Recursive Equivalence Theorem) *For any N-standard Σ-algebra A,*

$$\boldsymbol{\mu CR}(A) = \boldsymbol{\mu PR}^*(A) = \boldsymbol{While}^*(A).$$

The question of unbounded memory—A^* versus A—re-appears here in the difference between primitive and course-of-values recursion. This model of computation was created [29] with the needs of equational and logical definability in mind.

Axiomatic Methods. In an axiomatic method one defines the concept of a *computation theory* as a set $\boldsymbol{\Theta}(A)$ of partial functions on an algebra A having some of the essential properties of the set of partial recursive functions on $\mathbb{N}$. To take an example, $\boldsymbol{\Theta}(A)$ can be required to contain the basic algebraic operators and tests of A; be closed under operations such as *composition*; and, in particular, possess an enumeration for which appropriate *universality* and *s-m-n properties* are true. Thus in Sect. 5 we saw that ***While***$^*(A)$ is a computation theory in this sense.

The definition of a computation theory used here is due to Fenstad [7, 8] who takes up ideas from Moschovakis [21]. Computation theory definitions typically require a code set (such as $\mathbb{N}$) to be part of the underlying structure A for the indexing of functions.

The following fact is easily derived from [20] (where register machines are used); see also Fenstad [8, Chap. 0].

Theorem 14 (Minimal Computation Theory) *The set* ***While***$^*(A)$ *of* ***While**** *computable functions on an N-standard algebra A is the smallest set of partial functions on A to satisfy the axioms of a computation theory; in consequence,* ***While***$^*(A)$ *is a subset of every computation theory* $\boldsymbol{\Theta}(A)$ *on A.*

8.2 Generalizing the Church-Turing Thesis

The ***While**** computable functions are a mathematically interesting and useful generalization of the partial recursive functions on $\mathbb{N}$ to abstract many-sorted algebras A and classes $\mathbb{K}$ of such algebras. Do they also give rise to an interesting and useful generalization to A and $\mathbb{K}$ of the Church-Turing thesis, concerning effective computability on $\mathbb{N}$? They do; though this answer is difficult to explain fully and briefly. In this section we will only *sketch* some reasons. The issues are discussed in more detail in [29, 31], as well as in the chapter by Feferman in this volume [6].

First, consider the following naive attempt at a generalization of the Church-Turing thesis.

Thesis 1 (A Naive Generalized Church-Turing Thesis) *The functions "effectively computable" on a many-sorted algebra A are precisely the functions* ***While**** *computable on A.*

Consider now: what can be meant by "effective computability" on an abstract algebra?

The idea of effective computability is inspired by a variety of distinct philosophical and mathematical ideas about the nature of finite computation with finite elements. There are many ways to analyse and formalize the notion of effective calculability, by thinking about concepts such as algorithm; deterministic procedure;

mechanical procedure; computer program; programming language; formal system; machine; device; and, of course, the functions definable by these entities.

The idea of effective computability is invaluable because of the close relationships that exist between its constituent concepts. However, only a few of these constituent concepts make sense in an abstract setting. Therefore *the general concept of "effective computability" does not belong in a generalization of the Church-Turing thesis*. We propose to use the term "effective computation" only to talk about finite computation on finite data.

In seeking a generalization of the Church-Turing thesis, we are trying to make explicit certain primary informal concepts that are formalized by the technical definitions, and hence to clarify the nature and use of the computable functions.

We will start by trying to clarify the nature and use of abstract structures. There are three points of view from which to consider the step from concrete to abstract structures, and hence three points of view from which to consider ***While****** computable functions.

(1) There is *abstract algebra*, which is a theory of calculation based upon the "behaviour" of elements in calculations without reference to their "nature". This abstraction is achieved through the concept of isomorphism between concrete structures; an abstract algebra A can be viewed as "a concrete algebra considered unique only up to isomorphism".
(2) There is the viewpoint of *formal logic*, concerned with the scope and limits of axiomatizations and formal reasonings. Here structures are used to discuss formal systems and axiomatic theories in terms of consistency, soundness, completeness, and so on.
(3) There is *data type theory*, an offshoot of programming language theory, which is about data types that the user may care to define and that arise independently of programming languages. Here structures are employed to discuss the semantics of data types, and isomorphisms are employed to make the semantics independent of implementations. In addition, axiomatic theories are employed to discuss their specifications and implementation.

Data type theory is built upon and developed from the first two subjects: it is our main point of view.

Computation in each of the three cases is thought of slightly differently. In algebra, it is natural to think informally of algorithms built from the basic operations that compute functions and sets in algebras, or over classes of algebras uniformly. In formal logic, it is natural to think of formulae that define functions and sets, and their manipulation by algorithms. In data type theory, we use programming languages to define computations. We return to a consideration of each of these approaches, which, because of its special concerns and technical emphasis, leads to a distinctive theory of computability on abstract structures:

Going first to (1): suppose the ***While****** computable functions are considered with the needs of doing algebra in mind. Then the context of studying algorithms and decision problems for algebraic structures (groups, rings, fields, etc.) leads to a formalization of a generalized Church-Turing thesis tailored to the language and use of algebraists:

Thesis 2 (Generalized Church-Turing Thesis for Algebraic Computability) *The functions computable by finite deterministic algebraic algorithms on a many-sorted algebra A are precisely the functions **While*** computable on A.*

An account of computability on abstract structures from this algebraic point of view is given in [28].

We agree with Feferman [6] who argues that this should be termed a Church-Turing thesis for *algorithms* on abstract structures, rather than *computations*. He prefers, in this context, to reserve the term "computation" for *calculations* on concrete structures composed of finite symbolic configurations.[11]

Now (jumping to viewpoint (3)) suppose that the ***While**** computable functions are considered with the needs of computation in mind. This context of studying data types, programming and specification constructs, etc., leads to a formulation tailored to the language and use of computer scientists:

Thesis 3 (Generalized Church-Turing Thesis for Programming Languages) *Consider a deterministic programming language over an abstract data type **dt**. The functions that can be programmed in the language on an algebra A which represents an implementation of **dt**, are the same as the functions **While*** programmable on A.*

This thesis has been discussed in [29].

Finally, returning to approach (2): we note that "logical" and non-deterministic languages are suitable for specifying problems. These can be considered as languages for *specification* rather than computation. Here *projectively computable* relations (Sect. 6.2) and the use of selection functions for these, play a central role.

We define a specification language to be *adequate* for an abstract data type ***dt*** if all computations on any algebra *A* implementing ***dt*** can be specified in *A*. We then formulate a generalized Church-Turing thesis for specifiability on abstract data types:

Thesis 4 (Generalized Church-Turing Thesis for Specifiability) *Consider an adequate specification language S over an abstract data type **dt**. The relations on a many-sorted algebra A implementing **dt** that can be specified by S are precisely the projectively **While*** computable relations on A.*

8.3 Concluding Remarks

We have sketched the elements of our work over four decades on generalizing computability theory to abstract structures. A thorough exposition is to be found

[11]In Feferman's memorable slogan: "No calculation without representation."

in our survey paper [31]. In [32] we have had the opportunity to recall the diverse origins of, and influences on, our research programme.

Since [31], our research has emphasized computation on many-sorted topological partial algebras (the focus of Sect. 7 here) and its diverse applications:

- computable analysis, especially on the reals [11, 30, 34],
- classical analog systems [14, 35–37],
- analog networks of discrete and continuous processors [25, 35],
- generalized stream processing in discrete and continuous time [36, 37].

These applications bring us close to an investigation of the physical foundation of computability. In this regard, considerations of *continuity* are central (cf. the discussion in Sect. 7.1.2). This is related to the issue of stability of analog systems, and more broadly, to Hadamard's principle [12] which, as (re-)formulated by Courant and Hilbert [4, 13], states that for a scientific problem to be well posed, the solution must exist, be unique and depend continuously on the data. To this we might add: it must also be computable.

Acknowledgements We are grateful to Mark Armstrong, Sol Feferman, Diogo Poças and an anonymous referee for very helpful comments on earlier drafts of this chapter. We also thank the editors, Giovanni Sommaruga and Thomas Strahm, for the opportunity to participate in this volume, and for their helpfulness during the preparation of this chapter.

This research was supported by a grant from the Natural Sciences and Engineering Research Council (Canada).

References

1. M. Armstrong, Notions of semicomputability in topological algebras over the reals. M.Sc. Thesis, Department of Computing and Software, McMaster University, 2015. Archived at http://hdl.handle.net/11375/18334
2. L. Blum, F. Cucker, M. Shub, S. Smale, *Complexity and Real Computation* (Springer, New York, 1998)
3. V. Brattka, P. Hertling, Topological properties of real number representations. Theor. Comput. Sci. **284**(2), 241–257 (2002)
4. R. Courant, D. Hilbert, *Methods of Mathematical Physics*, vol. II (Interscience, New York, 1953). Translated and revised from the German edition [1937]
5. Y.L. Ershov, S.S. Goncharov, A.S. Nerode, J.B. Remmel (eds.), *Handbook of Recursive Mathematics* (Elsevier, Amsterdam, 1998). In 2 volumes
6. S. Feferman, Theses for computation and recursion on abstract structure, in *Turing's Revolution. The Impact of his Ideas About Computability*, ed. by G. Sommaruga, T. Strahm (Birkhäuser/Springer, Basel, 2015)
7. J.E. Fenstad, Computation theories: an axiomatic approach to recursion on general structures, in *Logic Conference, Kiel 1974*, ed. by G. Muller, A. Oberschelp, K. Potthoff (Springer, Berlin, 1975), pp. 143–168
8. J.E. Fenstad, *Recursion Theory: An Axiomatic Approach* (Springer, Berlin, 1980)
9. H. Friedman, Algebraic procedures, generalized Turing algorithms, and elementary recursion theory, in *Logic Colloquium '69*, ed. by R.O. Gandy, C.M.E. Yates (North Holland, Amsterdam, 1971), pp. 361–389

10. A. Fröhlich, J. Shepherdson, Effective procedures in field theory. Philos. Trans. R. Soc. Lond. (A) **248**, 407–432 (1956)
11. M.Q. Fu, J.I. Zucker, Models of computability for partial functions on the reals. J. Log. Algebraic Methods Program. **84**, 218–237 (2015)
12. J. Hadamard, in *Lectures on Cauchy's Problem in Linear Partial Differential Equations* (Dover, New York, 1952). Translated from the French edition [1922]
13. J. Hadamard, *La Théorie des Équations aux Dérivées Partielles* (Éditions Scientifiques, Warsaw, 1964)
14. N.D. James, J.I. Zucker, A class of contracting stream operators. Comput. J. **56**, 15–33 (2013)
15. S.C. Kleene, *Introduction to Metamathematics* (North Holland, Amsterdam, 1952)
16. G. Kreisel, J.L. Krivine, *Elements of Mathematical Logic* (North Holland, Amsterdam, 1971)
17. G. Kreisel, D. Lacombe, J. Shoenfield, Partial recursive functions and effective operations, in *Constructivity in Mathematics: Proceedings of the Colloqium in Amsterdam, 1957*, ed. by A. Heyting (North Holland, Amsterdam, 1959), pp. 290–297
18. A.I. Mal'cev, Constructive algebras I, in *The Metamathematics of Algebraic Systems. A.I. Mal'cev, Collected papers: 1936–1967* (North Holland, Amsterdam, 1971), pp. 148–212
19. W. Magnus, A. Karass, D. Solitar, *Combinatorial Group Theory* (Dover, New York, 1976)
20. J. Moldestad, V. Stoltenberg-Hansen, J.V. Tucker, Finite algorithmic procedures and computation theories. Math. Scand. **46**, 77–94 (1980)
21. Y.N. Moschovakis, Axioms for computation theories—first draft, in *Logic Colloquium '69* ed. by R.O. Gandy, C.E.M. Yates (North Holland, Amsterdam, 1971), pp. 199–255
22. M. Rabin, Computable algebra, general theory and the theory of computable fields. Trans. Am. Math. Soc. **95**, 341–360 (1960)
23. J.C. Shepherdson, Algebraic procedures, generalized Turing algorithms, and elementary recursion theory, in *Harvey Friedman's Research on the Foundations of Mathematics*, ed. by L.A. Harrington, M.D. Morley, A. Ščedrov, S.G. Simpson (North Holland, Amsterdam, 1985), pp. 285–308
24. V. Stoltenberg-Hansen, J.V. Tucker, Computable rings and fields, in *Handbook of Computability Theory*, ed. by E. Griffor (Elsevier, Amsterdam, 1999)
25. B.C. Thompson, J.V. Tucker, J.I. Zucker, Unifying computers and dynamical systems using the theory of synchronous concurrent algorithms. Appl. Math. Comput. **215**, 1386–1403 (2009)
26. G.S. Tseitin, Algebraic operators in constructive complete separable metric spaces. Dokl. Akad. Nauk SSSR **128**, 49–52 (1959). In Russian
27. G.S. Tseitin, Algebraic operators in constructive metric spaces. Tr. Mat. Inst. Steklov **67**, 295–361 (1962); In Russian. Translated in AMS Translations (2) 64:1–80. MR 27#2406
28. J.V. Tucker, Computing in algebraic systems, in *Recursion Theory, Its Generalisations and Applications*. London Mathematical Society Lecture Note Series, vol. 45, ed. by F.R. Drake, S.S. Wainer (Cambridge University Press, Cambridge, 1980), pp. 215–235
29. J.V. Tucker, J.I. Zucker, *Program Correctness Over Abstract Data Types, with Error-State Semantics*. CWI Monographs, vol. 6 (North Holland, Amsterdam, 1988)
30. J.V. Tucker, J.I. Zucker, Computation by 'while' programs on topological partial algebras. Theor. Comput. Sci. **219**, 379–420 (1999)
31. J.V. Tucker, J.I. Zucker, Computable functions and semicomputable sets on many-sorted algebras, in *Handbook of Logic in Computer Science*, vol. 5, ed. by S. Abramsky, D. Gabbay, T. Maibaum (Oxford University Press, Oxford, 2000), pp. 317–523
32. J.V. Tucker, J.I. Zucker, Origins of our theory of computation on abstract data types at the Mathematical Centre, Amsterdam, 1979–1980, in *Liber Amicorum: Jaco de Bakker*, ed. by F. de Boer, M. van der Heijden, P. Klint, J.J.M.M. Rutten (Centrum Wiskunde & Informatica, Amsterdam, 2002), pp. 197–221
33. J.V. Tucker, J.I. Zucker, Abstract versus concrete computation on metric partial algebras. ACM Trans. Comput. Log. **5**, 611–668 (2004)
34. J.V. Tucker, J.I. Zucker, Computable total functions, algebraic specifications and dynamical systems. J. Log. Algebraic Program. **62**, 71–108 (2005)

35. J.V. Tucker, J.I. Zucker, Computability of analog networks. Theor. Comput. Sci. **371**, 115–146 (2007)
36. J.V. Tucker, J.I. Zucker, Continuity of operators on continuous and discrete time streams. Theor. Comput. Sci. **412**, 3378–3403 (2011)
37. J.V. Tucker, J.I. Zucker, Computability of operators on continuous and discrete time streams. Computability **3**, 9–44 (2014)
38. A.M. Turing, On computable numbers, with an application to the Entscheidungsproblem. Proc. Lond. Math. Soc. **42**, 230–265 (1936). With correction, *ibid.*, **43**, 544–546 (1937). Reprinted in *The Undecidable*, ed. by M. Davis (Raven Press, New York, 1965)
39. K. Weihrauch, *Computable Analysis: An Introduction* (Springer, Berlin, 2000)
40. B. Xie, M.Q. Fu, J. Zucker, Characterizations of semicomputable sets of real numbers. J. Log. Algebraic Methods Program. **84**, 124–154 (2015)

Discrete Transfinite Computation

P.D. Welch

Abstract We describe various computational models based initially, but not exclusively, on that of the Turing machine, that are generalized to allow for transfinitely many computational steps. Variants of such machines are considered that have longer tapes than the standard model, or that work on ordinals rather than numbers. We outline the connections between such models and the older theories of recursion in higher types, generalized recursion theory, and recursion on ordinals such as α-recursion. We conclude that, in particular, polynomial time computation on ω-strings is well modelled by several convergent conceptions.

Keywords Turing machine • Computability • Generalized recursion theory • Constructibility

AMS Classifications: 03D10, 03D65, 03E45, 03D30

1 Introduction

There has been a resurgence of interest in models of infinitary computation in the last decade. I say resurgence because there has been for 50 years or more models of computation that deal with sets of integers, or objects of even higher type, in generalized recursion theory. Such a theory was initiated by Kleene's generalization of his equational calculus for ordinary recursion theory [16–19]. Whilst that was indeed a generalized recursion theory some commentators remarked that it was possible to view, for example, what came to be called Kleene recursion, as having a machine-like model. The difference here was that the machine would have a countable memory, a countable tape or tapes, but an ability to manipulate that countable memory in finite time, or equivalently in one step. Thus a query about a set of integers r say, say coded as an element of Cantor space, as to whether r was an element of an oracle set $A \subseteq 2^{\mathbb{N}}$, could be considered a single computational operation. Moreover r could be moved from a tape to a storage

P.D. Welch (✉)
School of Mathematics, University of Bristol, Bristol, BS8 1TW, England
e-mail: philipwelch.welch@gmail.com

G. Sommaruga, T. Strahm (eds.), *Turing's Revolution*,
DOI 10.1007/978-3-319-22156-4_6

or otherwise manipulated or altered in infinitely many places at once by one step of some procedure. This is what Addison and Spector called the "$\aleph_0$-mind" (see[30], p. 405; further Kreisel spoke of a generalized Church's thesis). We thus had essentially a generalization of the original Turing machine model. It should be noted however (in contradistinction to the transfinite models of more recent years) that the computations would all be represented by wellfounded computation trees when convergent, and would otherwise be deemed non-convergent: an infinitely long linear sequence of computations represents failure. By declaring that countably many operations on a set of integers could be done in "finite time" or all at once and so count as one step, one simply sidesteps the difficulty of thinking of infinitely many steps of a successful computation that is yet to be continued.

Thus in the past mathematicians have been more inclined to consider wellfounded computation but applied to infinite objects, rather than considering transfinite computations containing paths of transfinite *order type*. Examples of the former: Kleene recursion as HYP-recursion, or the Blum-Shub-Smale (BSS [3]) machines acting with real inputs (or indeed other mathematical objects taken from a ring with suitable functions). One could almost say that a hallmark of the generalizations of recursion theory to "higher", or "generalized" recursion theory has been that it has considered *only* wellfounded computation albeit with infinitary objects of higher types. Sacks's E-recursion [32] again considers essentially the same paradigm of building up functions applied now to sets in general: a non-convergent computation is exemplified by an infinite path in a computation tree, thus rendering the tree ill-founded.

However what if we were nevertheless to think also of transfinite stages of computation? There is a natural reaction to this: we feel that "computation" is so tied up with our notions of finitary algorithm, indeed effective algorithm, that it is too heterodox to consider transfinitely numbered steps as "computation." However why should this be? We are probably constrained to think of well defined computations as shying away from anything that smells of a *supertask*: the quandaries arising from Thomson's Lamp [36] seem to have used up a surprising lot of space and ink, for what is after all a simple definitional problem. So supertasks have been banished from computational study.

However transfinite recursion or procedures are not at all alien, if not common in the everyday experience of the mathematician—there are after all, few Π^1_1-complete sets occurring as "natural" mathematical objects—but the early canonical example arises in the Cantor-Bendixson process. Here with a simple decreasing chain of sets $X_\alpha \supseteq X_{\alpha+1}$ and intersections at limits: $X_\lambda = \bigcap_{\alpha<\lambda} X_\alpha$, one has a monotone process. Again the monotonicity phenomenon occurs centrally in Kleene's recursion theory of higher types, and we feel safer with a monotone process than a non-monotone or discontinuous one.

Notwithstanding these qualms, the current chapter reviews the recent descriptions of various machine models, including that of Turing's original machine itself, which can be given a defined behaviour at limit stages of time, enabling them to compute through recursive ordinals and beyond. This behaviour, apart from a few very elementary models, is signified by being non-monotonic, or *quasi-inductive.*

We shall see that the various models link into several areas of modern logic: besides recursion theory, set theory and the study of subsystems of second order analysis play a role. Questions arise concerning the strengths of models that operate at the level one type above that of the integers. This may be one of ordinal types: how long a well ordered sequence of steps must a machine undertake in order to deliver its output? Or it may be of possible output: if a machine produces real numbers, which ordinals can be coded as output reals? And so on and so forth.

1.1 Computation in the Limit

To start at the beginning, steps towards transfinite considerations, or at least that of considering what might have occurred on a tape at a finite stage, come immediately after considering the halting problem. The universal Turing machine can be designed to print out on an infinite output tape the code numbers e of programmes P_e that will halt on input e: thus $P_e(e)\downarrow$. This Σ^0_1 set, as is well known, is *complete*: any other Σ^0_1 is (ordinary) Turing reducible to it.

Putnam [29] (and Gold [10]) went a step further.

Definition 1 ([29]) P is a *trial and error predicate* if and only if there is a general recursive function f such that for every $x_1, \ldots, x_n$:

$$P(x_1, \ldots, x_n) \equiv \lim_{y\to\infty} f(x_1, \ldots, x_n, y) = 1$$
$$\neg P(x_1, \ldots, x_n) \equiv \lim_{y\to\infty} f(x_1, \ldots, x_n, y) = 0.$$

Running such a procedure on a Turing machine allows us to print out a Δ^0_2 set A's characteristic function on the output tape. In order to do this we are forced to allow the machine to change its mind about $n \in A$ and so repeatedly substitute a 0 for a 1 or vice versa in the n'th cell of the output tape. However, and this is the point, at most finitely many changes are to be made to that particular cell's value. It is this feature of not knowing at any finite given time whether further alterations are to made, that makes this a transition from a computable set to a non-computable one.

By a recursive division of the working area up into infinitely many infinite pieces, one can arrange for the correct computation of all $?m \in A?$ to be done on the one machine, and the correct values placed on the output tape.

However this is as far as one can go if one imposes the (very obvious, practical) rule that a cell's value can only be altered finitely often. In order to get a Σ^0_2 set's characteristic function written to the output tape, then in general one cannot guarantee that a cell's value is changed finitely often. Then immediately one is in the hazardous arena of supertasks.

Nevertheless let us play the mathematicians' game of generalizing for generalization's sake: let us by fiat declare a cell's value that has switched infinitely often $0 \to 1 \to 0$ to be 0 at "time ω". With this lim inf declaration one has, mathematically at least, written down the Σ^0_2-set on the output tape, again at time ω.

Fig. 1 A 3-tape infinite time Turing machine

			R/W							
Input:	1	1	0	1	1	0	0	0	0	⋯
Scratch:	0	1	0	0	1	1	0	0	0	⋯
Output:	1	0	0	0	1	1	0	1	0	⋯

Following this through we may contemplate continuing the computation at times $\omega+1, \omega+2, \ldots, \omega+\omega \ldots$. Let $\langle C_i(\alpha) \mid i \in \omega, \alpha \in \mathrm{On}\rangle$ denote the contents of cell C_i at time α in On. Let $l(\alpha) \in \omega$ represent the cell number being observed at time α. Similarly let $q(\alpha)$ denote the state of the machine/program at time α (Fig. 1). We merely need to specify (1) the read/write head position $l(\lambda)$ and (2) the next state $q(\lambda)$ for limit times λ, whilst $l(\alpha+1)$ and $q(\alpha+1)$ are obtained by just following the usual rules for head movement and state change according to the standard Turing transition table. (Note that the stock of programmes has not changed, hence $q(\alpha)$ names one of the finitely many states of the usual transition table; we are merely enlarging the possible behaviour.)

We intend that control of the program at a limit time λ, be placed at the beginning of the outermost loop, or subroutine call, that was called unboundedly often below λ. We thus set:

$$q(\lambda) = \mathrm{Liminf}_{\alpha<\lambda}\, q(\alpha).$$

For limit λ we set $C_i(\lambda)$ by:

$$\begin{aligned} C_i(\lambda) &= k \text{ if } \exists\alpha < \lambda \forall\beta < \lambda(\alpha < \beta \longrightarrow C_i(\beta) = k) \text{ for } k \in \{0,1\};\\ &= 0 \text{ otherwise.}\end{aligned}$$

The R/W head we place according to the above, also using a modified Liminf rule:

$$l(\lambda) = \mathrm{Liminf}^{*}\,\langle l(\alpha) \mid \alpha < \lambda\rangle$$

This is not exactly the arrangement that Hamkins and Lewis specified in [11] but it is inessentially different from it (HL specified a special *limit state* q_λ which the machine entered into automatically at limit stages, and the head was always set back to the start of the tape. They specified (which we shall keep here) that the machine be a three tape machine).

Input then can consist of a set of integers, suitably coded as an element of $2^{\mathbb{N}}$ on $\langle C_{3i}\rangle_i$ and output likewise is such an element on $\langle C_{3i+2}\rangle_i$. Thus there is little difference in a machine with an oracle $Z \subseteq \omega$ and one acting on input Z coded onto the input tape. However we immediately see the possibility of higher type computation: we may have some $\mathcal{Z} \subseteq 2^{\mathbb{N}}$ and then we add a query state which asks if, say, the scratch tape $\langle C_{3i+1}\rangle_i$'s contents is or is not an element of $\mathcal{Z}$.

We have thus completely specified the ITTM's behaviour. The scene is thus set to ask what such machines are capable of. We defer discussion of this until Sect. 2, whilst we outline the rest of this chapter here.

In one sense we have here a logician's plaything: the Turing model has been taken over and redesigned with a heavy-handed liminf rule of behaviour. This liminf operation at limit stages is almost tantamount to an infinitary logical rule, and most of the behaviour the machine exhibits is traceable to this rule. But then of course it has to be, what else is there? Nevertheless this model and those that have been studied subsequently have a number of connections or aspects with other areas of logic. Firstly, with weak subsystems of analysis: it is immediately clear that the behaviour of such machines is dependent on what ordinals are available. A machine may now halt at some transfinite stage, or may enter an infinitely repeating loop; but any theory that seeks to describe such machines fully is a theory which implies the existence of sufficiently long wellorderings along which such a machine can run (or be simulated as running). We may thus ask "What (sub-) system of analysis is needed in which to discuss such a machine"? We shall see that machine variants may require longer or shorter wellorderings, thus their theory can be discussed within different subsystems.

Secondly, we can ask how the computable functions/sets of such a model fit in with the earlier theories of generalized recursion theory of the 1960s and 1970s. For example there is naturally associated with ITTM's a so-called *Spector Class* of sets. Such classes arise canonically in the generalized recursion theories of that era through notions of definability.

Once one model has been defined it is very tempting to define variants. One such is the *Infinite Time Register Machine* (ITRM's—due to Koepke [23]) which essentially does for Shepherdson-Sturgis machines what HL does for Turing machines. Whilst at the finite level these two models are equal in power, their infinitary versions differ considerably, the ITTM's being much stronger. The ITRM model is discussed in Sect. 3.

Just as for ordinary recursion on ω the TM model with a putative tape in order type ω length is used, so when considering notions of α-recursion theory for admissible ordinals α, it is possible to think of tapes also unfettered by having finite initial segments: we may consider machines with tapes of order type α and think of computing along α with such machines. What is the relation to this kind of computation and α-recursion theory?

One can contemplate even machines with an On-length tape. It turns out (Koepke [22]) that this delivers a rather nice presentation of Gödel's constructible hierarchy. Finally discussed here is the notion of a *Blum-Shub-Smale* machine ([3]) acting transfinitely. With some continuity requirement imposed on register contents for limit times, we see that functions such as exponentiation e^x which are not BSS computable, become naturally IBBS computable. Moreover there is a nice equivalence between their decidable reals, and those produced by the *Safe Set Recursion* ("SSR") of Beckmann, Buss, and S. Friedman, which can be thought of as generalizing to transfinite sets notions of polynomial time computable functions on integers. Briefly put, a polynomial time algorithm using ω as an input string, should be halting by some time ω^n for some finite n. The IBBS computable reals are then identical to the SSR-computable reals. The background second order theory needed to run IBBS machines lies intermediate between WKL_0 and ATR_0.

The relation of ITTM's to Kleene recursion is discussed in Sect. 2.

2 What ITTM's Can Achieve?

Hamkins and Lewis in [11] explore at length the properties of ITTM's: they demonstrate the natural notion of a *universal* such machine, and hence an S^n_m-theorem and the Recursion Theorems. A number of questions immediately spring to mind:

Q. What is the halting set $H = \{e \in \omega \mid P_e(0)\downarrow\}$?

Here $\langle P_e \rangle_e$ enumerates the usual Turing machine programs/transition tables (and we use $P_e(x)\downarrow y$ to denote that the e'th program on input $x \in \mathbb{N}$ or in $2^{\mathbb{N}}$ halts with output y. (If we are unconcerned about the y we omit reference to it.) An ITTM computation such as this can now halt in ω or more many steps. But how long should we wait to see if $P_e(0)\downarrow$ or not? This is behind the following definitions.

Definition 2

(i) We write "$P_e(n)\downarrow^{\alpha} y$" if $P_e(n)\downarrow y$ in exactly α steps. We call α *clockable* if $\exists e \exists n \in \omega \exists y\, P_e(n)\downarrow^{\alpha} y$.
(ii) A real $y \in 2^{\mathbb{N}}$ is *writable* if there are $n, e \in \omega$ with $P_e(n)\downarrow y$; an ordinal β is called writable, if β has a writable code y.

We may consider a triple $s(\alpha) = \langle l(\alpha), q(\alpha), \langle C_i(\alpha)_i \rangle \rangle$ as a *snapshot* of a machine at time α, which contains all the relevant information at that moment. A *computation* is then given by a wellordered sequence of snapshots. There are two possible outcomes: there is some time α at which the computation *halts*, or else there must be some stage α_0 at which the computation enters the beginning of a loop, and from then on throughout the ordinals it must iterate through this loop. It is easy either by elementary arguments or simply by Löwenheim-Skolem, to see that such an α_0 must be a countable ordinal, and moreover that the periodicity of the cycling loop is likewise countable.

The property of being a "well-ordered sequence of snapshots in the computation $P_e(x)$" is Π^1_1 as a relation of e and x. Hence "$P_e(x)\downarrow y$" is Δ^1_2:

$\exists w$(*w codes a halting computation of* $P_e(x)$, *with y written on the output tape at the final stage*) $\Longleftrightarrow$

$\forall w$(*w codes a computation of* $P_e(x)$ *that is either halting or performs a repeating infinite loop* $\longrightarrow$ *w codes a halting computation with y on the output tape*.)

Likewise $P_e(x)\uparrow$ is also Δ^1_2. By the above discussion then it is immediate that the clockable and writable ordinals are all countable. Let $\lambda =_{\mathrm{df}} \sup\{\alpha \mid \alpha \text{ is writable}\}$; let $\gamma =_{\mathrm{df}} \sup\{\alpha \mid \alpha \text{ is clockable}\}$. Hamkins-Lewis showed that $\lambda \leq \gamma$.

Q2 Is $\lambda = \gamma$?

Definition 3

(i) $x^{\nabla} = \{e | P_e(x)\downarrow\}$ (The halting set on integers).
(ii) $X^{\blacktriangledown} = \{(e, y) | P_e^X(y)\downarrow\}$ (The halting set on reals relativised to $X \subseteq 2^{\mathbb{N}}$).

This yields the halting sets, both for computations on integers and secondly on reals where by the latter we include the instruction for the ITTM to query whether the current scratch tape's contents considered as a real, is in X.

Definition 4

(i) $R(x)$ is an ITTM-*semi-decidable* predicate if there is an index e so that:

$$\forall x (R(x) \leftrightarrow P_e(x)\downarrow 1)$$

(ii) A predicate R is ITTM-*decidable* if both R and $\neg R$ are ITTM-semi-decidable.

Q3 What are the ITTM-(semi-)decidable sets of integers, or reals? What is x^{∇}?

The last question seems somewhat impenetrable without a characterisation of the halting behaviour of ITTM's—and one version of that problem is Question 2. To analyse decidability one route might be through a version of Kleene's *Normal Form Theorem* in the context of ITTM's. However there is an obvious type difference: a successful halting computation of an ordinary Turing machine in a finite amount of time can be coded as a finite sequence of finite ordinary-TM-snapshots, and thus through the usual coding devices, by an integer. This is essentially the heart of Kleene's T-predicate argument. Thus for standard TM's Kleene demonstrated that given an index e we may effectively find an index e' so that for any n: $P_e(n)\downarrow \longrightarrow P_{e'}(n)\downarrow M$ where $M \in \mathbb{N}$ is a code for the whole computation sequence of $P_e(n)$.

However this is too simple here: $P_e(0)\downarrow$ may halt at some transfinite time $\beta \geq \omega$. Hence the halting computation is only codable by an infinite sequence of infinite ITTM-snapshots S say, of some order type τ. In order for there to be a chance of having another index e' with $P_{e'}(n)\downarrow y$ where y codes such a sequence S, one has to know at the very least that there is an e_0 so that $P_{e_0}(n)$ halts with output a code for τ. In other words that τ be *writable*. Thus we need an affirmative answer to *Q2* at the very least.

Interestingly the key to answering halting behaviour is not to aim straight for an analysis of halting *per se*, but at another phenomenon that is peculiarly significant to ITTM's. There can be computations $P_e(x)$ that whilst they have not formally halted, nevertheless from some point in time onwards, leave their output tapes alone, and just churn around for ever perhaps doodling or making entries on their scratch tape. The output has however *stabilized*. We formally define this as follows:

Definition 5

(i) Suppose for the computation $P_e(x)$ the machine does not halt then we write $P_e(x)\uparrow$; if eventually the output tape does have a stable value $y \in 2^{\mathbb{N}}$ then we write: $P_e(x)\uparrow y$ and we say that y is *eventually x-writable*.

(ii) $R(x)$ is an *eventually* ITTM-*semi-decidable* predicate if there is an index e so that:

$$\forall x(R(x) \leftrightarrow P_e(x) \uparrow 1)$$

(iii) A predicate R is *eventually* ITTM-*decidable* if both R and $\neg R$ are eventually ITTM-semi-decidable.

This proliferation of notions is not gratuitous: it turns out that answering *Q2* on clockables *vis à vis* writables, depends on successfully analysing stabilization patterns of individual cells C_i during the course of the computation $P_e(n)$. The moral is that stabilization is anterior to halting. The following lemma illustrates the point.

Theorem 1 (The $\boldsymbol{\lambda, \zeta, \Sigma}$-Theorem) (Welch cf. [38, 41])

(i) Any ITTM computation $P_e(x)$ which halts, does so by time λ^x, the latter being defined as the supremum of the x-writable ordinals.
(ii) Any computation $P_e(x)$ with eventually stable output tape, will stabilize before the time ζ^x defined as the supremum of the eventually x-writable ordinals.
(iii) Moreover ζ^x is the least ordinal so that there exists $\Sigma^x > \zeta^x$ with the property that

$$L_{\zeta^x}[x] \prec_{\Sigma_2} L_{\Sigma^x}[x];$$

(iv) Then λ^x is also characterised as the least ordinal satisfying:

$$L_{\lambda^x}[x] \prec_{\Sigma_1} L_{\zeta^x}[x].$$

If we unpack the contents here, answers to our questions are given by (iii) and (iv). Let us take $x = \varnothing$ so that we may consider the unrelativised case. Our machine-theoretic structure and operations are highly absolute and it is clear that running the machine inside the constructible hierarchy of L_α's yields the same snapshot sequence as considering running the machine in V. If $P_e(n)\downarrow$ then this is a Σ_1-statement (in the language of set theory). As halting is merely a very special case of stabilization, then we have that

$$P_e(n)\downarrow \leftrightarrow (P_e(n)\downarrow)^{L_\zeta} \leftrightarrow (P_e(n)\downarrow)^{L_\lambda}$$

(the latter because $L_\lambda \prec_{\Sigma_1} L_\zeta$). Hence the computation must halt before λ. Hence the answer to *Q2* is affirmative: every halting time (of an integer computation) is a writable ordinal. One quickly sees that a set of integers is ITTM-decidable if and only if it is an element of L_λ. It is ITTM-semi-decidable if and only if it is $\Sigma_1(L_\lambda)$.

Since the limit rules for ITTM's are intrinsically of a Σ_2-nature, with hindsight it is perhaps not surprising that this would feature in the (ζ, Σ) pair arising as they do: after all the snapshot of the universal ITTM at time ζ is going to be coded into the Σ_2-Theory of this L_ζ. The universality of the machine is then apparent in the fact that by stage ζ it will have "constructed" all the constructible sets in L_ζ.

As a corollary one obtains then:

Theorem 2 (Normal Form Theorem) *(Welch)*

(a) For any ITTM computable function φ_e we can effectively find another ITTM computable function $\varphi_{e'}$ so that on any input x from $2^{\mathbb{N}}$ if $\varphi_e(x)\downarrow$ then $\varphi_{e'}(x)\downarrow y \in 2^{\mathbb{N}}$, where y codes a wellordered computation sequence for $\varphi_e(x)$.
(b) There is a universal predicate $\mathfrak{T}_1$ *which satisfies $\forall e \forall x$:*

$$P_e(x)\downarrow z \quad \leftrightarrow \quad \exists y \in 2^{\mathbb{N}}[\mathfrak{T}_1(e,x,y) \wedge Last(y) = z].$$

Moreover as a corollary (to Theorem 1):

Corollary 1

(i) $x^{\nabla} \equiv_1 \Sigma_1\text{-Th}(\langle L_{\lambda^x}[x], \in, x\rangle)$—the latter the Σ_1-theory of the structure.
*(ii) Let $x^{\infty} =_{\mathrm{df}} \{e \mid \exists y P_e(x) \uparrow y\}$ be the set of x-*stable indices, *of those program numbers whose output tapes eventually stabilize. Then*

$$x^{\infty} \equiv_1 \Sigma_2\text{-Th}(\langle L_{\zeta^x}[x], \in, x\rangle).$$

The conclusions are that the Σ_1-Theory of L_λ is recursively isomorphic to the ITTM-jump 0^{∇}. One should compare this with Kleene's $\mathcal{O}$ being recursively isomorphic to the hyperjump, or again the Σ_1-Theory of $L_{\omega_1^{ck}}$. The second part of the corollary gives the analogous results for the index set of the eventually stable programs: here we characterise 0^{∞} as the Σ_2-Theory of L_ζ. The relativisations to inputs x are immediate.

One should remark that extensions of Kleene's $\mathcal{O}$ from the ? case to the ITTM case are straightforward: we can define $\mathcal{O}^+$ by adding in to $\mathcal{O}$ those indices of Turing programs that now halt at some transfinite time. After all, we are keeping the programs the same for both classes of machines, so we can keep the same formalism and definitions (literally) but just widen the class of what we consider computations. Similarly we can expand $\mathcal{O}^+$ to $\mathcal{O}^{\infty}$ by adding in those indices of eventually stabilising programs. This is done in detail in [20]. We thus have:

$$\frac{\mathcal{O}}{L_{\omega_1^{ck}}} \approx \frac{\mathcal{O}^+}{L_\lambda} \approx \frac{\mathcal{O}^{\infty}}{L_\zeta}.$$

2.1 Comparisons with Kleene Recursion

We have alluded to Kleene recursion in the introduction. His theory of recursion in higher types [16–19] was an equational calculus, a generalization of that for the Gödel-Herbrand generalized recursive functions. In this theory numbers were objects of type 0, whilst a function $f : \mathbb{N}^m \longrightarrow \mathbb{N}$ is an object of type 1; and

$F : \mathbb{N}^l \times (2^{\mathbb{N}})^m \longrightarrow \mathbb{N}$ one of type 2 etc. The *e'th procedure* (whether thought of as the e'th program of the kind of machine as outlined in the introduction, or else as e'th equation system in his calculus) then allowed a computation with inputs $\vec{n}, \vec{x}$ with oracle $\mathcal{I}$ to be presented in the form $\{e\}(\vec{n}, \vec{x}, \mathcal{I})$. The oracle $\mathcal{I}$ was usually taken to include an oracle for existential quantification $\mathcal{E}$ where, for $x \in 2^{\mathbb{N}}$:

$$\mathcal{E}(x) = \begin{array}{l} 0 \text{ if } \exists n x(n) = 0 \\ \\ 1 \text{ otherwise.} \end{array}$$

The reason for this was, although for any oracle $\mathcal{I}$ the class of relations semi-decidable in $\mathcal{I}$ was closed under $\forall^{\mathbb{N}}$ quantification, when semi-decidable additionally in $\mathcal{E}$ it becomes closed under $\exists^{\mathbb{N}}$ quantification. The Kleene semi-decidable sets then would include the arithmetic sets in $\mathbb{N} \times 2^{\mathbb{N}}$ (or further products thereof). (Ensuring computations be relative to $\mathcal{E}$ also guarantees that we have the *Ordinal Comparison Theorem.*)

The decidable relations turn out to be the hyperarithmetic ones, and the semi-decidable are those Kleene-reducible to WO, the latter being a complete Π^1_1 set of reals. Thus:

Theorem 3 (Kleene) *The hyperarithmetic relations $R(\vec{n}, \vec{x}) \subseteq \mathbb{N}^k \times (\mathbb{N}^{\mathbb{N}})^l$ for any $k, l \in \mathbb{N}$, are precisely those computable in $\mathcal{E}$.*
The Π^1_1 relations are precisely those semi-computable in $\mathcal{E}$.

Then a reducibility ordering comes from:

Definition 6 (Kleene Reducibility) Let $A, B \subseteq \mathbb{R}$; we say that A is *Kleene-semi-computable in B* iff there is an index e and $y \in \mathbb{R}$ so that

$$\forall x \in \mathbb{R}(x \in A \longleftrightarrow \{e\}(x, y, B, \mathcal{E}) \downarrow 1)).$$

A is *Kleene computable in B*, written, $A \leq_K B$, iff both A and its complement are Kleene-semi-computable in B.

The presence of the real y deserves some explanation. We want to think of the degree structure as being between sets of reals; the set y throws in a countable amount of information to the computation, and we are thus thinking of two sets of reals $A =_K B$ as being of the same complexity up to this countable amount of data. It implies that each Kleene degree contains continuum many sets of reals, but moreover is closed under continuous pre-images—it thus forms also a union of *Wadge degrees.*

We thus shall have that besides $\varnothing, \mathbb{R}$ the bottommost Kleene degree contains precisely all the Borel sets, whilst the degree of WO contains all co-analytic sets. As one sees the notion is very tied up with hyperarithmeticity.

If we have a transitive reducibility notion $\leq$ on sets of integers x say, together with a concomitant *jump operator* $x \longrightarrow x'$ then an ordinal assignment $x \longrightarrow \tau^x \in On$ is said to be a *Spector criterion* if we have:

$$x \leq y \longrightarrow (x' \leq y \longleftrightarrow \tau^x < \tau^y). \qquad (*)$$

As an example if we take here *hyperdegree* $x \leq_h y$ (abbreviating "x is hyperarithmetic in y") and the *hyperjump* operation, $x \longrightarrow x^h$ where x^h is a complete $\Pi_1^{1,x}$ set of integers, then the assignment $x \longrightarrow \omega_{1\,\mathrm{ck}}^x$ (where the latter is the least ordinal not (ordinary) Turing recursive in x) satisfies the Spector Criterion $(*)$ above. For sets of reals B we may extend this notation and let $\omega_{1\,\mathrm{ck}}^{B,x}$ be the ordinal height α of the least model of KP set theory (so the least admissible set) of the form $L_\alpha[x, B] \models \mathrm{KP}$.

With this we may express $A \leq_K B$ as follows:

Lemma 1 *$A \leq_K B$ iff there are Σ_1-formulae in $\mathcal{L}_{\in,\dot{X}}$ $\varphi_1(\dot{X}, v_0, v_1), \varphi_2(\dot{X}, v_0, v_1)$, and there is $y \in \mathbb{R}$, so that*

$$\begin{aligned} \forall x \in \mathbb{R}(x \in A &\Longleftrightarrow L_{\omega_1^{B,y,x}}[B, y, x] \models \varphi_1[B, y, x] \\ &\Longleftrightarrow L_{\omega_1^{B,y,x}}[B, y, x] \models \neg\varphi_2[B, y, x]). \end{aligned}$$

Back to ITTM-semidecidability:

The notion of semi-decidability comes in two forms.

Definition 7

(i) A set of integers x is *semi-decidable* in a set y if and only if:

$$\exists e \forall n \in x \left[P_e^y(n) \downarrow 1 \longleftrightarrow n \in x \ \right]$$

(ii) A set of integers x is *decidable* in a set y if and only if both x and its complement is semi-decidable in y. We write $x \leq_\infty y$ for the reducibility ordering.

(iii) A set of integers x is eventually-(semi)-decidable in a set y if and only if the above holds with $\uparrow$ replacing $\downarrow$. For this reducibility ordering we write $x \leq^\infty y$.

We then get the analogue of the Spector criterion using x^∇ as the jump operator:

Lemma 2

(i) The assignment $x \rightarrowtail \lambda^x$ satisfies the Spector Criterion:

$$x \leq_\infty y \longrightarrow (x^\nabla \leq_\infty y \leftrightarrow \lambda^x < \lambda^y).$$

(ii) Similarly for the assignment $x \rightarrowtail \zeta^x$:

$$x \leq^\infty y \longrightarrow (x^\infty \leq^\infty y \leftrightarrow \zeta^x < \zeta^y)$$

One can treat the above as confirmation that the ITTM degrees and jump operation are more akin to hyperarithmetic degrees and the hyperjump, than to the (standard) Turing degrees and Turing jump. Indeed they are intermediate between hyperdegrees and Δ^1_2-degrees.

To see this, we define a notion of degree using definability and Turing-invariant functions on reals (by the latter we mean a function $f : \mathbb{R} \longrightarrow \omega_1$ such that $x \leq_T y \longrightarrow f(x) \leq f(y)$). Now assume that f is Σ_1-definable over $(\mathrm{HC}, \in)$ without parameters, by a formula in $\mathcal{L}_{\dot{\in}}$.

Definition 8 Let f be as described; let Φ be a class of formulae of $\mathcal{L}_{\dot{\in}}$. Then $\Gamma = \Gamma_{f,\Phi}$ is the pointclass of sets of reals A so that $A \in \Gamma$ if and only if there is $\varphi \in \Phi$ with:

$$\forall x \in \mathbb{R}(x \in A \leftrightarrow L_{f(x)}[x] \models \varphi[x]).$$

With the function $f(x) = \omega^x_{1\,\mathrm{ck}}$ and Φ as the class of Σ_1-formulae we have that $\Gamma_{f,\Phi}$ coincides with the Π^1_1-sets of reals (by the Spector-Gandy Theorem). Replacing f with the function $g(x) = \lambda^x$ then yields the (lightface) ITTM-semi-decidable sets. Lemma 1 is then the relativisation of Kleene recursion which yields the relation $A \leq_K B$.

We now make the obvious definition:

Definition 9

(i) A set of reals A is *semi-decidable* in a set of reals B if and only if:

$$\exists e \forall x \in 2^{\mathbb{N}} \left[P^B_e(x) \downarrow 1 \leftrightarrow x \in A\right]$$

(ii) A set of reals A is *decidable* in a set of reals B if and only if both A and its complement is semi-decidable in B.

(iii) If in the above we replace $\downarrow$ everywhere by $\uparrow$ then we obtain the notion in (i) of A is *eventually decidable in* B and in (ii) of A is *eventually semi-decidable in* B.

Then the following reducibility generalizes that of Kleene recursion.

Definition 10

(i) $A \leq_\infty B$ iff for some $e \in \omega$, for some $y \in 2^{\mathbb{N}}$: A is decidable in (y, B).

(ii) $A \leq^\infty B$ iff for some $e \in \omega$, for some $y \in 2^{\mathbb{N}}$: A is eventually decidable in (y, B).

Again a real parameter has been included here in order to have degrees closed under continuous pre-images. We should expect that these reducibilities are dependent on the ambient set theory, just as they are for Kleene degrees: under $V = L$ there are many incomparable degrees below that of the complete semi-decidable degree, and under sufficient determinacy there will be no intermediate degrees between the latter and **0**, and overall the degrees will be wellordered. Now we get the promised analogy lifting Lemma 1, again generalizing in two ways depending on the reducibility.

Lemma 3

(i) $A \leq_\infty B$ *iff there are* Σ_1*-formulae in* $\mathcal{L}_{\in,\dot{X}}$ $\varphi_1(\dot{X}, v_0, v_1), \varphi_2(\dot{X}, v_0, v_1)$*, and* $y \in \mathbb{R}$*, so that*

$$\forall x \in \mathbb{R}(x \in A \iff L_{\zeta^{B,y,x}}[B, y, x] \models \varphi_1[B, y, x] \iff L_{\zeta^{B,y,x}}[B, y, x] \models \neg\varphi_2[B, y, x]).$$

(ii) $A \leq^\infty B$ *iff there are* Σ_2*-formulae in* $\mathcal{L}_{\in,\dot{X}}$ $\varphi_1(\dot{X}, v_0, v_1), \varphi_2(\dot{X}, v_0, v_1)$*, and* $y \in \mathbb{R}$*, so that*

$$\forall x \in \mathbb{R}(x \in A \iff L_{\zeta^{B,y,x}}[B, y, x] \models \varphi_1[B, y, x] \iff L_{\zeta^{B,y,x}}[B, y, x] \models \neg\varphi_2[B, y, x]).$$

We have not formally defined all the terms here: $\lambda^{B,y,x}$ is the supremum of the ordinals written by Turing programs acting transfinitely with oracles for B, y. The ordinal $\zeta^{B,y,x}$ is the least that is not ITTM-(B, x, y)-eventually-semi-decidable. There is a corresponding λ-ζ-Σ-theorem and thus we have also that this ζ is least such that $L_{\zeta^{B,y,x}}[B, y, x]$ has a proper Σ_2-elementary end-extension in the $L[B, y, x]$ hierarchy.

2.2 Degree Theory and Complexity of ITTM Computations

Corollary 1 shows that the ITTM-jump of a set of integers x is essentially a mastercode, or a Σ_1-truth set if you will, namely that of $L_{\lambda^x}[x]$. The analogy here then is with $\mathcal{O}^x$, the hyperjump of x, which is a complete $\Pi^{1,x}_1$ set of integers, as being also recursively isomorphic to Σ_1-(Th$(L_{\omega^x_{1,\text{ck}}}[x])$. This again indicates that the degree analogy here should be pursued with hyperdegrees rather than Turing degrees. It is possible to iterate the jump hierarchy through the $=_\infty$-degrees, and one finds that, *inside L*, the first ζ-iterations form a linearly ordered hierarchy with least upper bounds at limit stages. We emphasise this as being inside L since one can show that there is no least upper bound to $\{0^{\nabla_n} \mid n < \omega\}$, but rather continuum many minimal upper bounds (see [37]). We don't itemize these results here but refer the reader instead to [38].

A more general but basic open question is:

Q If $D = \{d_n : n < \omega\}$ *is a countable set of* $=_\infty$*-degrees, does D have a minimal upper bound?*

The background to this question is varied: for hyperdegrees this is also an open question. Under Projective Determinacy a positive answer is known for Δ^1_{2n}-degrees, but for Δ^1_{2n+1}-degrees this is open, even under PD. Minimal infinite time ∞-degrees can be shown to exist by similar methods, using perfect set forcing, to those of Sacks for minimal hyperdegrees (again see [37]).

One can also ask at this point about the nature of Post's problem for semi-decidable sets of integers. By the hyperdegree analogy one does not expect there to be incomparable such sets below 0^∇ and indeed this turns out to be the case [12].

2.3 Truth and Arithmetical Quasi-Inductive Sets

It is possible to relate ITTM's closely to an earlier notion due to Burgess [4] of *arithmetical quasi-inductive definitions.* We first make a general definition:

Definition 11 Let $\Phi : \mathcal{P}(\mathbb{N}) \to \mathcal{P}(\mathbb{N})$ be a Γ-operator, that is "$n \in \Phi(X)$" is a Γ-relation. We define the Γ- *quasi-inductive operator* using iterates of Φ as:
$\Phi_0(X) = X; \quad \Phi_{\alpha+1}(X) = \Phi(\Phi_\alpha(X));$
$\Phi_\lambda(X) = \liminf_{\alpha\to\lambda} \Phi_\alpha(X) =_{df} \bigcup_{\alpha<\lambda} \bigcap_{\lambda>\beta>\alpha} \Phi_\beta(X).$
We set the *stability set* to be $\Phi_{\mathrm{On}}(X)$.

By the nature of the lim inf operation at limits, it is easy to see that the operation of an ITTM is an example of a recursive quasi-inductive operator on $\mathbb{N}$. Recall that a set of integers B say is *inductive* if it is (1–1) reducible to the least fixed point of a monotone Π^1_1-operator. Burgess defined such a B to be *arithmetically quasi-inductive* if it was (1–1) reducible to the *stability set* $\Phi_{\mathrm{On}}(\varnothing)$.

In order to prove that an AQI halts, or reaches a stability point, one needs to know that one has sufficiently long wellorderings, and a certain amount of second order number theory is needed to prove that such ordinals exist. For the case of the ITTM's we know which ordinals we need: Σ^x for a computation involving integers and the input real x. We then adopt this idea of a "repeat pair" of ordinals for a quasi-inductive operator Φ: the least pair $(\zeta, \Sigma) = (\zeta(\Phi, x), \Sigma(\Phi, x))$ with $\Phi_\zeta(x) = \Phi_\Sigma(x) = \Phi_{\mathrm{On}}(x)$.

Definition 12 AQI is the sentence: "For every arithmetic operator Φ, for every $x \subseteq \mathbb{N}$, there is a wellordering W with a repeat pair $(\zeta(\Phi, x), \Sigma(\Phi, x))$ in Field(W)". If an arithmetic operator Φ acting on x has a repeat pair, we say that Φ *converges* (with input x).

We may simulate an AQI with starting set $x \subseteq \mathbb{N}$ as an ITTM with input x. Since we know how long such a machine takes to halt or loop, this gives the length of ordering needed to determine the extent of the AQI. Given the characterisation from the (relativized) λ-ζ-Σ-Theorem one arrives at the fact that

Theorem 4 *The theories* Π^1_3-CA_0, Δ^1_3-CA_0+AQI, *and* Δ^1_3-CA_0 *are in strictly descending order of strength, meaning that each theory proves the existence of a* β*-model of the next.*

What was engaging Burgess was an analysis of a *theory of truth* due to Herzberger [15]. The latter had defined a *Revision Sequence* which was essentially a quasi-inductive operator, just a bit beyond the arithmetic as follows.
$H_0 = \emptyset$:
$H_{\alpha+1} = \{\ulcorner\sigma\urcorner : \langle \mathbb{N}, +, \times, \cdots, H_\alpha\rangle \models \sigma\}$; with H_α interpreting T;
$H_\lambda = \bigcup_{\alpha<\lambda} \bigcap_{\lambda>\beta>\alpha} H_\beta.$

Burgess then defined the AQI sets as above and calculated that the ordinals (ζ, Σ) formed exactly the repeat pair needed for AQI's or for the Herzberger revision sequence. This was much earlier than the invention of ITTM's and was unknown to workers in the latter area around 2000, until Benedikt Löwe pointed out [27] the similarity between the Herzberger revision sequence formalism and that of the machines. It can be easily seen that any Herzberger sequence with starting distribution of truth values x say, can be mimicked on an ITTM with input x. Thus this is one way of seeing that Herzberger sequences must have a stability pair lexicographically no later than (ζ, Σ). Burgess had shown that H-sequences then loop at no earlier pair of points. More recently Field [7] has used a revision theoretic definition with a Π^1_1-quasi-inductive operator to define a variant theory of truth. For all three formalisms, Fields, Burgess's AQI, and ITTM's, although differing considerably in theory, the operators are all essentially equivalent as is shown in [40], since they produce recursively isomorphic stable sets. The moral to be drawn from this is that in essence the strength of the liminf rule is at play here, and seems to swamp all else.

3 Variant ITTM Models

Several questions readily occur once one has formulated the ITTM model. Were any features chosen crucial to the resulting class of computable functions? Do variant machines produce different classes? Is it necessary to have three tapes in the machine? The answer for the latter question is both yes and no. First the affirmative part: it was shown in [14] the class of functions $f : \mathbb{N} \longrightarrow \mathbb{N}$ remains the same if 3 tapes are replaced by 1, but not the class of functions $f : 2^{\mathbb{N}} \longrightarrow 2^{\mathbb{N}}$. The difficulty is somewhat arcane: one may simulate a 3-tape machine on a 1-tape machine, but to finally produce the output on the single tape and halt, some device is needed to tell the machine when to finish compacting the result down on the single tape, and they show that this cannot be coded on a 1-tape machine. On the other hand [39] shows that if one adopts an alphabet of three symbols this can be done and the class of functions $f : 2^{\mathbb{N}} \longrightarrow 2^{\mathbb{N}}$ is then the same. One may also consider a B for "Blank" as the third symbol, and change the liminf rule so that if cell C_i has varied cofinally in a limit ordinal λ, then $C_i(\lambda)$ is set to be blank (thus nodding towards ambiguity of the cell value). With this alphabet and liminf rule a 1-tape machine computes the same classes as a 3-tape machine, and these are both the same as computed by the original ITTM.

What of the liminf rule itself? We have just mentioned a variant in the last paragraph. Our original liminf rule is essentially of a Σ_2 nature: a value of 1 is in a cell $C_i(\mu)$ at limit time μ if there is an $\alpha < \mu$ such that for all $\beta \in (\alpha, \beta)$ $C_i(\beta) = 1$. Running a machine inside L_μ one sees that the snapshot $s(\mu)$ is a predicate that is Σ_2-definable over L_μ. It was observed in [38] that the liminf rule is *complete* for all other rules Σ_2-definable over limit levels L_μ in that for any other such rule the stability set obtained for the universal machine (on 0 input) with such a rule is (1–1) Σ_2-definable over L_ζ and thus is (1–1) in the Σ_2-truth set for L_ζ. However the latter is recursively isomorphic to the stability set for the universal ITTM by Corollary 1 and hence the standard stability set subsumes that of another machine with a different Σ_2-rule. Given the Σ_2 nature of the limit rule, with hindsight one sees that it is obvious that with (ζ', Σ') defined to be the lexicographically least pair with $L_{\zeta'} \prec_{\Sigma_2} L_{\Sigma'}$, then we must have that the universal ITTM enters a loop at ζ'. That it cannot enter earlier of course is the λ-ζ-Σ-Theorem, but a vivid way to see that this is the case is afforded by the construction in [8] which demonstrated that there was a non-halting ITTM program producing on its output tape continually sets of integers that coded levels L_α of the constructible hierarchy for ever larger α below Σ; at stage Σ it would perforce produce the code for L_ζ and then forever cycle round this loop producing codes for levels $\alpha \in [\zeta, \Sigma)$.

More complex rules lead to more complex machines. These were dubbed "hypermachines" in [9], where a machine was defined with a Σ_3-limit rule, and this was shown to be able to compute codes for L_α for $\alpha < \Sigma(3)$, where now $\zeta(3) < \Sigma(3)$ was the lexicographically least pair with $L_{\zeta(3)} \prec_{\Sigma_3} L_{\Sigma(3)}$. The stability set was now that from the snapshot at stage $\zeta(3)$ and was (1–1) to the Σ_3-truth set for this level of L. Inductively then one defines $\Sigma_4, \Sigma_5, \ldots, \Sigma_n, \ldots$ limit rules with the analogous properties. I think it has to be said though that the definitions become increasingly complex and even for $n = 3$, mirror more the structure of L in these regions with its own "stable ordinals" rather than anything machine-inspired. With these constructions one can then "compute" any real that is in L_τ where $\tau = \sup_n \zeta(n)$.

3.1 Longer Tapes

Generalizations of the ITTM machine are possible in different directions. One can consider machines with tapes not of cells of order type ω but of longer types. Some modifications are needed: what do we do if the program asks the R/W head to move one step leftwards when hovering over a cell C_λ for λ a limit ordinal? There are some inessentially different choices to be made which we do not catalogue here, but assume some fixed choices have been made.

We consider first the extreme possibility that the tape is of length On, that is of the class of all ordinals. We now have the possibility that arbitrary sets may be computed by such machines. Independently Dawson and Koepke came up with this concept. There are some caveats: how do we know that we can "code" sets by transfinite strings of 0, 1's at all? Dawson [6] formulated an *Axiom of Computability* that said every set could appear coded on the output tape of such a machine at some stage whilst it was running; thus for any set z there would be a program number e with P_e (not necessarily halting) with a code for z appearing on its output tape. He then argued that the class of such sets is a model of ZFC, and by studying the two dimensional grid of snapshots produced a Löwenheim-Skolem type argument to justify that the Axiom of Computability implied the Generalized Continuum Hypothesis. That the class of computable sets satisfied AC falls out of the assumption that sets can be coded by strings and such can be ordered. Since this machine's operations are again very absolute, it may be run inside L, thus demonstrating that "computable sets" are nothing other than the constructible sets. Koepke in [21] and later with Koerwien in [22] considered instead halting computations starting with an On-length tape marked with finitely many 1's in certain ordinal positions $(n, \xi_1, \dots, \xi_n)$, and asked for a computation as to whether $(\varphi_n(\xi_1, \dots \xi_{n-1}))^{L_{\xi_n}}$ was true. Thus the machine was capable of computing a truth predicate for L. This leads to:

Theorem 5 (Koepke [21]) *A set* $x \subseteq \mathrm{On}$ *is* On-ITTM-*computable from a finite set of ordinal parameters if and only if it is a member of the constructible hierarchy.*

One might well ask whether the computational approach to L might lead to some new proofs, or at least new information, on some of the deeper fine structural and combinatorial properties of L. However this hope turned out to be seemingly thwarted by the Σ_2-nature of the limit rule. Fine structural arguments are very sensitive to definability issues, and in constructions such as that for Jensen's $\Box$ principle, say, we need to know when or how ordinals are singularised for any n including $n = 1$ and the limit rule works against this. Moreover alternatives such as the *Silver Machine* model which was specifically designed to by-pass Jensen's fine structural analysis of L, make heavy use of a *Finiteness Property* that everything appearing at a successor stage can be defined from the previous stages and a finite set of parameters; just does not seem to work for On-ITTM's.

However this does bring to the fore the question of shortening the tapes to some admissible ordinal length $\alpha > \omega$ say, and asking what are the relations between α-ITTM's and the α-recursion theory developed in the late 1960s and early 70s. The definitions of that theory included that a set $A \subseteq \alpha$ which is $\Sigma_1(L_\alpha)$ was called *α-recursively enumerable* (α-r.e.). It was *α-recursive* if both it and its complement is α-r.e. and thus is $\Delta_1(L_\alpha)$. A notion of *relative α-recursion* was defined but then noticed to be intransitive; a stronger notion was defined and denoted by $A \leq_\alpha B$.

Koepke and Seyfferth in [24] define *A is computable in B* to mean that the characteristic function of A can be computed by a machine in α many stages from an oracle for B. This is exactly the relation that $A \in \Delta_1(L_\alpha[B])$. This has the advantage that the notion of α-computability and the associated α-computable enumerability (α-c.e.) tie up exactly with the notions of α-recursiveness and α-r.e. They then reprove the Sacks-Simpson theorem for solving Post's problem: namely that there can be two α-c.e. sets neither of which are mutually computable in their sense from the other.

However the relation "*is computable in*" again suffers from being an intransitive one. Dawson defines the notion of *α-sequential computation* that requires the output to the α-length tape be written in sequence without revisions. This gives him a transitive notion of relative computability: a set is α-computable if and only if it is α-recursive, and it is α-computably enumerable if and only if it is both α-r.e. and *regular*. Since Sacks had shown [31] that any α-degree of α.-r.e. sets contains a regular set, he then has that the structure of the α-degrees of the α-r.e. sets in the classical, former, sense, is isomorphic to that of the α-degrees of the α-c.e. sets. This implies that theorems of classical α-recursion theory about α-r.e. sets whose proofs rely on, or use regular α-r.e. sets will carry over to his theory. This includes the Sacks-Simpson result alluded to. The Shore Splitting Theorem [34] which states that any regular α-r.e. set A may be split into two disjoint α-r.e. sets B_0, B_1 with $A \not\leq_\alpha B_i$, is less amenable to this kind of argument but with some work the Shore Density theorem [35] that between any two α-r.e. sets $A <_\alpha B$ there lies a third α-r.e. C: $A <_\alpha C <_\alpha B$ can be achieved. As Sacks states in his book, the latter proof seems more bound up with the finer structure of the constructible sets than the other α-recursion theory proofs. Dawson generalizes this by lifting his notion of α-computation to that of a $\mathbb{B}$-α-computation where now $\underline{\mathbb{B}} =_{\mathrm{df}} \langle J_\alpha^{\mathbb{B}}, \in, \mathbb{B}\rangle$ is an admissible, *acceptable*, and *sound* structure for a $\mathbb{B} \subseteq \alpha$. These assumptions make $J_\alpha^{\mathbb{B}}$ sufficiently L-like to rework the Shore argument to obtain:

Theorem 6 (Dawson—The α-c.e. Density Theorem) *Let $\underline{\mathbb{B}}$ be as above. Let A, B be two $\mathbb{B}$-α-c.e. sets, with $A <_{\mathbb{B},\alpha} B$. Then there is C also $\mathbb{B}$-α-c.e., with $A <_{\mathbb{B},\alpha} C <_{\mathbb{B},\alpha} B$.*

4 Other Transfinite Machines

Once the step has been taken to investigate ITTM's, one starts looking at other machine models and sending them into the transfinite. We look here at *Infinite Time Register Machines* (ITRM's) both with integer and ordinal registers, and lastly comment on *Infinite Time Blum-Shub-Smale Machines* (IBSSM's).

4.1 *Infinite Time Register Machines (ITRM's)*

A (standard) *register machine* as devised by Shepherdson and Sturgis [33], or Minsky [28], consists of finite number of natural number registers R_i for $i < N$, running under a program consisting of a finite list of instructions $\vec{I} = I_0, \ldots, I_q$. The latter consist of zeroising, transferring of register contents one to another, or conditional jump to an instruction number in the program, when comparing two registers. At time α we shall list the N-vector of register contents as $\vec{R}(\alpha)$. The next instruction the machine is about to perform we shall denote by $I(\alpha)$. We adopt a liminf rule again. Thus the next instruction to be performed at limit stage λ, is $I(\lambda) =_{\mathrm{df}} \liminf_{\alpha\to\lambda} I(\alpha)$. As discussed before for ITTM's, this can be seen to place control at the beginning of the outermost loop, or subroutine, entered cofinally often before stage λ. We shall use a lim inf* rule for register contents: if a register's contents edges up to infinity at time λ it is reset to 0:

$$R_i(\lambda) =_{\mathrm{df}} \liminf_{\alpha\to\lambda} R_i(\alpha) \text{ if this is finite; otherwise we set } R_i(\lambda) = 0.$$

Although perhaps not apparent at this point, it is this "resetting to zero" that gives the ITRM its surprising strength: specifying that the machine, or program, crash with no output if a register becomes unbounded results in a substantially smaller class of computable functions. A function $F : \mathbb{N}^N \to \mathbb{N}$ is then *ITRM-computable* if there is an ITRM program P with $P(\vec{k}){\downarrow}F(\vec{k})$ for every $\vec{k} \in \mathbb{N}^N$. In order to accommodate computation from a set of integers $Z \subseteq \mathbb{N}$ say, we add an oracle query instruction $?k \in Z?$ and receive as $0/1$ the answer to a register as a result.

These machines were defined by Koepke and investigated by him and co-workers in [5, 23]. A *clockable ordinal* has the same meaning here as for ITTM's, except that here these ordinals for man initial segment of On. Defining a *computable ordinal* as one which has real code whose characteristic function is ITRM-computable, they show that the clockables ordinals coincide with the computable ordinals. To analyse what they are capable of, first note as a crude upper bound that they could be easily simulated on an ITTM. However ITRM's can detect whether an oracle set $Z \subseteq \mathbb{N}$ codes a wellfounded relation: a backtracking algorithm that searches for leftmost paths can be programmed. Thus Π^1_1-sets are ITRM-decidable.

It also turns out that, in contradistinction to the finite case, the strength of the infinite version of register machines diverges from that of the Turing machine, but moreover there is no universal ITRM. We outline the arguments for this.

Definition 13 (*N*-Register Halting Set)

$$H_N =_{\mathrm{df}} \{\langle e, r_0, \ldots, r_{N-1}\rangle \mid P_e(r_0, \ldots, r_{N-1}) \downarrow\}.$$

(There is an obvious generalization H^Z_n for machines with oracle Z.)

Koepke and Miller show that if there is some instruction I' and register contents vector $\overrightarrow{R_i}$ such that the snapshot $(I', \overrightarrow{R_i})$ reoccurs in the course of computation at least ω^ω times in order type, then the computation is in a loop and will go on for ever.

Theorem 7 (Koepke-Miller [23]) *For any N the N-halting problem: "$\langle e, \vec{r}\rangle \in H_N$" is decidable by an ITRM. Similarly for any oracle Z, the* (N, Z)-halting problem *"$\langle e, \vec{r}\rangle \in H^Z_N$" is decidable by a Z-ITRM with an oracle for Z.*

The number of registers has to be increased to calculate H_N for increasing N. The corollary to this is that there can be no one single universal ITRM. We can get an exact description of the strength of ITRM's by assessing bounds on the ordinals needed to see that a machine either halts or is looping. It is discussed in [26] and shown there that if one has an ITRM with a single register then it has either halted or is in an infinite loop by the second admissible ordinal ω^{ck}_2. One cannot replace this with $\omega_1 = \omega^{\mathrm{ck}}_1$: if $\mathrm{Liminf}_{\beta\to\omega_1} R_0(\beta) = p < \omega$ then a Π_2-reflection argument shows that the same instruction number is used, and the value in the register is this p, on a set of ordinals closed and unbounded in ω_1. By the Koepke-Miller criterion mentioned above this would indeed mean that the computation was looping. However it can be the case that $\mathrm{Liminf}_{\beta\to\omega_1} R_0(\beta) = \omega$ and then this would have to be reset to 0: $R_0(\omega_1) = 0$. Then again $\mathrm{Liminf}_{\beta\to\omega_1+\omega_1} R_0(\beta)$ may also be ω, but the instruction number could now differ. However by ω^{ck}_2 the criterion will have already applied and the computation if still running will be looping. One then shows by induction that each extra register added to the architecture requires a further admissible ordinal in run time to guarantee looping behaviour. One then thus arrives at the property that any ordinal below $\omega_\omega^{\mathrm{ck}}$—the first limit of admissibles, is clockable by such an ITRM, and thence that the halting sets H_n can be computed by a large enough device. We can state this more formally:

Thus the assertion that these machines either halt or exhibit looping behaviour turns out to be equivalent to a well known subsystem of second order number theory, namely, Π^1_1-CA_0. Let ITRM_N be the assertion: "The *N-register halting set* H_N exists." Further, let ITRM be the similar relativized statement that "For any $Z \subseteq \omega$, for any $N < \omega$ the *N-register halting set* H^Z_N exists." Then more precisely:

Theorem 8 (Koepke-Welch [26])

(i) Π^1_1-$\mathsf{CA}_0 \vdash \mathsf{ITRM}$. *In particular:*
$\mathsf{KP}+$ *"there exist* $N+1$ *admissible ordinals* $> \omega$*"* $\vdash \mathsf{ITRM}_N$.
(ii) $\mathsf{ATR}_0 + \mathsf{ITRM} \vdash \Pi^1_1$-$\mathsf{CA}_0$.
In particular there is a fixed $k < \omega$ *so that for any* $N < \omega$

$$\mathsf{ATR}_0 + \mathsf{ITRM}_{N\cdot k} \vdash \text{“}\,\mathrm{HJ}(N, \varnothing) \text{ exists.”}$$

An analysis of Post's problem in this context is effected in [13].

4.2 Ordinal Register Machines (ORM's)

We mention finally here the notion studied by Koepke and Siders of *Ordinal Register Machines* (ORM's [25]): essentially these are the devices above but extended to have ordinal valued registers. Platek (in private correspondence) indicated that he had originally considered his equational calculus on recursive ordinals as being implementable on some kind of ordinal register machine. Siders also had been thinking of such machines and in a series of papers with Koepke considered the unbounded ordinal model. The resetting Liminf* rule is abandoned, and natural Liminf's are taken. Now ordinal arithmetic can be performed. Remarkably given the paucity of resources apparently available one has the similar theorem to that of the On-ITTM:

Theorem 9 (Koepke-Siders [25]) *A set* $x \subseteq \mathrm{On}$ *is* ORM*-computable from a finite set of ordinals parameters if and only if it is a member of the constructible hierarchy.*

They implement an algorithm that computes the truth predicate $T \subseteq \mathrm{On}$ for L and which is ORM-computable on a 12 register machine (even remarking that this can be reduced to 4!). From T a class of sets S can be computed which is a model of their theory SO, which is indeed the constructible hierarchy.

5 Infinite Time Blum-Shub-Smale Machines (IBSSM's)

Lastly we consider the possible transfinite versions of the Blum-Shub-Smale machines. These can be viewed as having registers $R_1, \ldots, R_N$ containing now Euclidean reals $r_1, \ldots, r_n \in \mathbb{R}$. There is a finite program or flow-chart with instructions divided into function nodes or conditional branching nodes. We shall assume that function nodes have the possibility of applying a rational function computation of the registers (we test each time that we are not dividing by zero). So far this accords with the finite BSS version. We now make the, rather stringent, condition that at a limit stage λ if any register R_i does not converge to a limit in the usual sense, then the whole computation is deemed to have crashed and so be undefined. The value of $R_\iota(\lambda)$ is then set to be the ordinary limit of the contents of $R_i(\alpha)$ as $\alpha \to \lambda$. It has been noted that a BSS machine cannot calculate the functions e^x, $\sin x$ etc., but an IBSSM can, indeed in ω many steps (by simply calculating increasingly long initial segments of the appropriate power series).

Koepke and Seyfferth [24] have investigated such machines with continuous limits. To simulate other sorts of machines on an IBSSM requires some ingenuity: a register that is perhaps simulating a register of one of the ITRM's discussed earlier, may have some contents x, that tends to infinity and be then reset. Here then it is better to calculate with $\frac{1}{x}$ in order to ensure a continuous limit of 0. Else if the register is simulating the contents of the scratch tape of an ITTM, then perhaps at successor stages continual division by 2 ensures again a continuous limit at the next limit ordinal of time. They show that a machine with n nodes in its flow diagram can halt on rational number input at ordinal times without any limit below ω^{n+1}. Thus any such machine will halt, crash or be looping by time ω^ω.

The question is naturally what is the computational power of such machines? Clearly, by absoluteness considerations, on rational input such a machine can be run inside the constructible hierarchy, and indeed from what they showed on ordinal lengths of computations, inside L_{ω^ω}. They then naturally ask whether any real in L_{ω^ω} can be produced by an IBSSM machine?

We can answer this affirmatively below. However at the same time we combine this with yet another characterisation. It is possible to give another characterisation of the sets in L_{ω^ω} by using the notions of *Safe Recursive Set Functions* (SRSF) of Beckmann et al. [1]. They are generalizing the notion of safe recursion of Bellantoni and Cook [2] used to define polynomial time computations. Here variables are divided into two types *safe* and *normal*. In the notation $f(\vec{a}/\vec{b})$ recursion is only allowed on the safe variables in $\vec{b}$. This allows for the definition by recursion of addition and multiplication but crucially not exponentiation. One of the aims of Beckmann et al. [1] is to have a notion of set recursion that corresponds to "polynomial time". On input an ω-string in $2^{\mathbb{N}}$ for example, one wants a computation that halts in polynomial time from ω—the length of the input. Hence the calculation should halt by some ω^n for an $n < \omega$. They have:

Theorem 10 ([1]) *Let f be any SRSF. Then there is a ordinal polynomial q_f in variables $\vec{\alpha}$ so that*

$$rk(f(\vec{a}/\vec{b})) \leq \max_i rk(a_i) + q_f(r\vec{k}(a)).$$

Thus the typing of the variables ensures that the ranks of sets computed as outputs from an application of an SRSFunction are polynomially bounded in the ranks of the input. Using an adaptation of Arai, such functions on finite strings correspond to polynomial time functions in the ordinary sense. For ω-strings we have that such computations halt by a time polynomial in ω. As mentioned by Schindler, it is natural to define "polynomial time" for ITTM's to be those calculations that halt by stage ω^ω, and a polynomial time ITTM function to be one that, for some $N < \omega$, terminates on all inputs by time ω^N. We thus have:

Theorem 11 *The following classes of functions of the form $F : (2^{\mathbb{N}})^k \to 2^{\mathbb{N}}$ are extensionally equivalent:*

(I) *Those functions computed by a continuous* IBSSM *machine;*
(II) *Those functions that are polynomial time* ITTM*;*
(III) *Those functions that are safe recursive set functions.*

Proof We take $k = 1$. We just sketch the ideas and the reader may fill in the details. By Koepke-Seyfferth for any IBSSM computable function there is $N < \omega$ so that the function is computable in less than ω^N steps. We may thus consider that computation to be performed inside $L_{\omega^N}[x]$ and so potentially simulable in polynomial time (in ω^M steps, for some M) by an ITTM. However this can be realised: a code for any $L_\alpha[x]$ for $\alpha \leq \omega^N$, $x \in 2^{\mathbb{N}}$, and its theory, may be computable by an ITTM (uniformly in the input x) by time ω^{N+3} by the argument of Lemma 2 of Friedman and Welch [8] (Friedman-Welch). Since we have the theory, we have the digits of the final halting IBSSM-output (or otherwise the fact that it is looping or has crashed respectively, since these are also part of the set theoretical truths of $L_{\omega^N}[x]$). Thus (II) $\supseteq$ (I). If F is in the class (II), then for some $N < \omega$, $F(x)$ is computable within $L_{\omega^N}[x]$ and by setting up the definition of the ITTM program P computing F we may define some α such that the output of that program P on x (i.e. $F(x)$) is the α'th element of $2^{\mathbb{N}}$ in $L_{\omega^N}[x]$ uniformly in x. However the set $L_{\omega^N}[x]$ is SRSF-recursive from $\omega \cup \{x\}$ (again uniformly in x) as is a code for α. This yields the conclusion that we may find uniformly the output of $P(x)$ using the code for α, again as the output of an SRSF-recursive-in-x function. This renders (II) $\subseteq$ (III).

Finally if F is in (III), (and we shall assume that the variable x is in a safe variable place—but actually the case where there are normal and safe variables is handled no differently here) then there is (cf. [1], 3.5) a finite N and a Σ_1-formula $\varphi(v_0, v_1)$ so that $F(x) = z$ iff $L_{\omega^N}[x] \models \varphi[x, z]$ (using here that $\mathrm{TC}(x) = \omega$ and thus $\mathrm{rk}(x) = \omega$). Indeed we may assume that z is named by the canonical Σ_1-Skolem function h for, say, $L_{\omega^N+\omega}[x]$ as $h(i, n)$ for some $n < \omega$. Putting this together we have some

Σ_1 $\psi(v_0)$ (in $\mathcal{L}_{\dot{x},\dot{\in}}$) so that $F(x)(k) = z(k) = 1$ iff $L_{\omega^N+\omega}[x] \models \psi[k]$. In short to be able to determine such $F(x)$ by an IBSSM it suffices to be able to compute the Σ_1-truth sets for $L_\alpha[x]$ for all $\alpha < \omega^\omega$ by IBSSM's. There are a variety of ways one could do this, but it is well known that calculating the α'th iterates of the Turing jump relativised to x for $\alpha < \omega^\omega$ would suffice. To simplify notation we shall let x also denote the set of integers in the infinite fractional expansion of the real x. So fix a $k < \omega$, to see that we may calculate $x^{(\beta)'}$ for $\beta < \omega^k$. One first constructs a counter to be used in general iterative processes, using registers $C_0, \ldots C_{k-1}$ say, whose contents represent the integer coefficients in the Cantor normal form of $\beta < \omega^k$ where we are at the β'th stage in the process. (The counter of course must conform to the requirement that registers are continuous at limits $\lambda \leq \omega^k$. This can be devised using reciprocals and repeated division by 2 rather than incrementation by 1 each time.) We assume this has been done so that in particular that $C_0 = C_1 = \cdots = C_{k-1} = 0$ occurs first at stage ω^k. We then code the characteristic function of $\{m \in \omega \mid m \in W_m^{x^{(\beta)'}}\}$ as 1/0's in the digits at the s'th-places after the decimal point of R_1 where s is of the form $p_{k+m}.p_0^{n_0+1}.\cdots.p_{k-1}^{n_{k-1}+1}$ where $p_0 = 2, p_1 = 3$, etc., enumerates the primes, and n_j the exponent of ω^j in the Cantor Normal form of ω^β. For limit stages $\lambda < \omega^k$, continuity of the register contents automatically ensures that this real in R_1 also codes the disjoint union of the $x^{(\beta)'}$ for $\beta < \lambda$, and at stage ω^k we have the whole sequence of jumps encoded as required. Q.E.D.

6 Conclusions

The avenues of generalization of the Turing machine model into the transfinite which we have surveyed, give rise to differing perspectives and a wealth of connections. Higher type recursion theory, to which the models mostly nearly approximate, to a lesser or greater extent, was a product of Kleene's generalization of the notion of an equational calculus approach to recursive functions. Here discussed are machines more on the Turing side of the balance. Some of the other generalizations of recursion theory, say to meta-recursion theory, as advocated by Kreisel and elucidated by Sacks and his school, and which later became ordinal α-recursion theory, we have not really discussed here in great detail, but again their motivations came from the recursion theoretic-side, rather than any "computational-model-theoretic" direction. The models discussed in this chapter thus fill a gap in our thinking.

Referring to the last section, we find that, rather like a Church's thesis, we have here an effective system for handling ω-strings in polynomial time, as formalized by the SRSF's, and a natural corresponding computational model of ITTM's working with calculations halting by time earlier than ω^ω. The model of computation with the continuous limit IBSSM's then also computes the same functions. Note that assertions such as that "*every continuous IBSSM halts, loops, or becomes discontinuous*" when formalized in second order arithmetic, are intermediate between ACA_0

and ATR_0. There is much to be said for the IBSSM model over its finite version: we have remarked that the infinite version calculates power series functions, such as sin, e^x. With a little work one sees also that if any differentiable function $f : \mathbb{R} \longrightarrow \mathbb{R}$ is IBSSM computable, then so is its derivative f'.

On the other hand the class of sets that ITTM's compute form a Spector class, and so we can bring to bear general results about such classes on the ITTM semi-decidable, and eventually semi-decidable classes; their strength we saw was very strong: between Π^1_2-CA_0 and Π^1_3-CA_0. Finally the On-tape version of the ITTM, gives us a new presentation of the constructible hierarchy as laid out by an ordinary Turing program progressing throughout On time.

References

1. A. Beckmann, S. Buss, S.-D. Friedman, *Safe Recursive Set Functions* (Centre de Recerca Matematica Document Series, Barcelona, 2012)
2. S. Bellantoni, S. Cook, A new recursion-theoretic characterization of the poly-time functions. Comput. Complex. **2**, 97–110 (1992)
3. L. Blum, M. Shub, S. Smale, On a theory of computation and complexity over the real numbers. Not. Am. Math. Soc. **21**(1), 1–46 (1989)
4. J.P. Burgess, The truth is never simple. J. Symb. Log. **51**(3), 663–681 (1986)
5. M. Carl, T. Fischbach, P. Koepke, R. Miller, M. Nasfi, G. Weckbecker, The basic theory of infinite time register machines. Arch. Math. Log. **49**(2), 249–273 (2010)
6. B. Dawson, Ordinal time Turing computation. Ph.D. thesis, Bristol (2009)
7. H. Field, A revenge-immune solution to the semantic paradoxes. J. Philos. Log. **32**(3), 139–177 (2003)
8. S.-D. Friedman, P.D. Welch, Two observations concerning infinite time Turing machines, in *BIWOC 2007 Report*, ed. by I. Dimitriou (Hausdorff Centre for Mathematics, Bonn, 2007), pp. 44–47. Also at http://www.logic.univie.ac.at/sdf/papers/joint.philip.ps
9. S.-D. Friedman, P.D. Welch, Hypermachines. J. Symb. Log. **76**(2), 620–636 (2011)
10. E. Gold, Limiting recursion. J. Symb. Log. **30**(1), 28–48 (1965)
11. J.D. Hamkins, A. Lewis, Infinite time Turing machines. J. Symb. Log. **65**(2), 567–604 (2000)
12. J.D. Hamkins, A. Lewis, Post's problem for supertasks has both positive and negative solutions. Arch. Math. Log. **41**, 507–523 (2002)
13. J.D. Hamkins, R. Miller, Post's problem for ordinal register machines: an explicit approach. Ann. Pure Appl. Log. **160**(3), 302–309 (2009)
14. J.D. Hamkins, D. Seabold, Infinite time Turing machines with only one tape. Math. Log. Q. **47**(2), 271–287 (2001)
15. H.G. Herzberger, Notes on naive semantics. J. Philos. Log. **11**, 61–102 (1982)
16. S.C. Kleene, Recursive quantifiers and functionals of finite type I. Trans. Am. Math. Soc. **91**, 1–52 (1959)
17. S.C. Kleene, Turing-machine computable functionals of finite type I, in *Proceedings 1960 Conference on Logic, Methodology and Philosophy of Science* (Stanford University Press, 1962), pp. 38–45
18. S.C. Kleene, Turing-machine computable functionals of finite type II. Proc. Lond. Math. Soc. **12**, 245–258 (1962)
19. S.C. Kleene, Recursive quantifiers and functionals of finite type II. Trans. Am. Math. Soc. **108**, 106–142 (1963)
20. A. Klev, *Magister Thesis* (ILLC, Amsterdam, 2007)
21. P. Koepke, Turing computation on ordinals. Bull. Symb. Log. **11**, 377–397 (2005)

22. P. Koepke, M. Koerwien, Ordinal computations. Math. Struct. Comput. Sci. **16**(5), 867–884 (2006)
23. P. Koepke, R. Miller, An enhanced theory of infinite time register machines, in *Logic and the Theory of Algorithms*, ed. by A. Beckmann et al. Springer Lecture Notes Computer Science, vol. 5028 (Springer, Swansea, 2008), pp. 306–315
24. P. Koepke, B. Seyfferth, Ordinal machines and admissible recursion theory. Ann. Pure Appl. Log. **160**(3), 310–318 (2009)
25. P. Koepke, R. Siders, Computing the recursive truth predicate on ordinal register machines, in *Logical Approaches to Computational Barriers*, ed. by A. Beckmann et al. Computer Science Report Series (Swansea, 2006), p. 21
26. P. Koepke, P.D. Welch, A generalised dynamical system, infinite time register machines, and Π^1_1-CA_0, in *Proceedings of CiE 2011, Sofia*, ed. by B. Löwe, D. Normann, I. Soskov, A. Soskova (2011)
27. B. Löwe, Revision sequences and computers with an infinite amount of time. J. Log. Comput. **11**, 25–40 (2001)
28. M. Minsky, *Computation: Finite and Infinite Machines* (Prentice-Hall, Upper Saddle River, 1967)
29. H. Putnam, Trial and error predicates and the solution to a problem of Mostowski. J. Symb. Log. **30**, 49–57 (1965)
30. H. Rogers, *Recursive Function Theory*. Higher Mathematics (McGraw, New York, 1967)
31. G.E. Sacks, Post's problem, admissible ordinals and regularity. Trans. Am. Math. Soc. **124**, 1–23 (1966)
32. G.E. Sacks, *Higher Recursion Theory*. Perspectives in Mathematical Logic (Springer, New York, 1990)
33. J. Shepherdson, H. Sturgis, Computability of recursive functionals. J. Assoc. Comput. Mach. **10**, 217–255 (1963)
34. R.A. Shore, Splitting an α recursively enumerable set. Trans. Am. Math. Soc. **204**, 65–78 (1975)
35. R.A. Shore, The recursively enumerable α-degrees are dense. Ann. Math. Log. **9**, 123–155 (1976)
36. J. Thomson, Tasks and supertasks. Analysis **15**(1), 1–13 (1954/1955)
37. P.D. Welch, Minimality arguments in the infinite time Turing degrees, in *Sets and Proofs: Proc. Logic Colloquium 1997, Leeds*, ed. by S.B.Cooper, J.K.Truss. London Mathematical Society Lecture Notes in Mathematics, vol. 258 (Cambridge University Press, Cambridge, 1999)
38. P.D. Welch, Eventually infinite time Turing degrees: infinite time decidable reals. J. Symb. Log. **65**(3), 1193–1203 (2000)
39. P.D. Welch, Post's and other problems in higher type supertasks, in *Classical and New Paradigms of Computation and Their Complexity Hierarchies. Papers of the Conference Foundations of the Formal Sciences III*, ed. by B. Löwe, B. Piwinger, T. Räsch. Trends in Logic, vol. 23 (Kluwer, Dordrecht, 2004), pp. 223–237
40. P.D. Welch, Ultimate truth *vis à vis* stable truth. Rev. Symb. Log. **1**(1), 126–142 (2008)
41. P.D. Welch, Characteristics of discrete transfinite Turing machine models: halting times, stabilization times, and normal form theorems. Theor. Comput. Sci. **410**, 426–442 (2009)

Semantics-to-Syntax Analyses of Algorithms

Yuri Gurevich

> *The real question at issue is "What are the possible processes which can be carried out in computing a number?"*
>
> Turing

> *Give me a fulcrum, and I shall move the world.*
>
> Archimedes

Abstract Alan Turing pioneered semantics-to-syntax analysis of algorithms. It is a kind of analysis where you start with a large semantically defined species of algorithms, and you finish up with a syntactic artifact, typically a computation model, that characterizes the species. The task of analyzing a large species of algorithms seems daunting if not impossible. As in quicksand, one needs a rescue point, a fulcrum. In computation analysis, a fulcrum is a particular viewpoint on computation that clarifies and simplifies things to the point that analysis become possible. We review from that point of view Turing's analysis of human-executable computation, Kolmogorov's analysis of sequential bit-level computation, Gandy's analysis of a species of machine computation, and our own analysis of sequential computation.

Keywords Abstract state machines • Algorithms • Analysis of algorithms • Church's thesis • Computability • Concept of algorithms • Definition of algorithms • Human computers • Semantics-to-syntax analysis • Sequential algorithm • Turing's thesis

AMS Classification: 68W40

Y. Gurevich (✉)
Microsoft Research, Redmond, WA, USA
e-mail: gurevich@microsoft.com

G. Sommaruga, T. Strahm (eds.), *Turing's Revolution*,
DOI 10.1007/978-3-319-22156-4_7

1 Introduction

This article is a much revised and extended version of our Foundational Analyses of Computation [16].

1.1 Terminology

1.1.1 Species of Algorithms

For the sake of brevity we introduce the term *species of algorithms* to mean a class of algorithms given by semantical constraints. We are primarily interested in large species like sequential algorithms or analog algorithms.

Q[1]: Contrary to biological species, yours are not necessarily disjoint. In fact, one of your species may include another as a subspecies.

A: This is true. For example, the species of sequential-time algorithms, that execute step after step, includes the species of sequential algorithms, with steps of bounded complexity.

Q: The semantic-constraint requirement seems vague.

A: It is vague. It's purpose is just to distinguish the analyses of algorithms that we focus upon here from other analyses of algorithms in the literature.

1.1.2 Analyses of Algorithms

We are interested in the semantics-to-syntax analyses of algorithms like that of Turing. You study a species of algorithms. The original definition of the species may have been vague, and you try to explicate it which may narrow the species in the process. And you finish up with a syntactic artifact, like a particular kind of machines that execute all and only the algorithms of the (possibly narrowed) species in consideration.

Q: Did Turing's analysis cover all algorithms or only those of a particular species.

A: There are limitations on algorithms covered by Turing's analysis, and we address some of them in Sect. 2.

Q: In computer science, they teach the analysis of algorithms. I guess that isn't semantics-to-syntax analysis.

A: The analysis of algorithms in such a course is normally the analysis of known algorithms, say known sorting algorithms or known string algorithms. A significant part of that is resource analysis: estimate how much time or space or some other resource is used by an algorithm. It all is important and useful, but it isn't a semantics-to-syntax analysis.

[1] Q is our inquisitive friend Quisani, and A is the author.

1.1.3 Algorithms and Computations

By default, in this paper, computations are algorithmic. This is the traditional meaning of the term computation in logic and computer science. However, a wider meaning of the term is not uncommon these days. For an interesting and somewhat extreme example see [27] where computations are viewed as "processes generating knowledge."

The concepts of algorithms and computations are closely related. Whatever algorithms are syntactically, semantically they specify computations. For our purposes, the analysis of algorithms and the analysis of computation are one and the same.

Q: One algorithm may produce many different computations.

A: That is why we defined species as collections of algorithms rather than computations.

1.1.4 Algorithms and Computable Functions

In mathematical logic, theory of algorithms is primarily the theory of recursive functions. But there may be much more to an algorithm than its input-output behavior. In general algorithms perform tasks, and computing functions is a rather special class of tasks. Note in this connection that, for some useful algorithms, non-termination is a blessing, rather than a curse. Consider for example an algorithm that opens and closes the gates of a railroad crossing.

1.1.5 Sequential Algorithms

Q: You made a distinction between sequential and sequential-time computations. Give me an example of a sequential-time algorithm that isn't sequential.

A: Consider the problem of evaluating a finite logic circuit in the form of a tree. Each leaf is assigned 0 or 1, and each internal node is endowed with a Boolean operation to be applied to the Boolean values of the children nodes. Define the height of a leaf to be 0, and the height of an internal node to be 1 plus the maximum of the heights of its children. At step 1, the nodes of height 1 fire (i.e. apply their Boolean operations). At step 2, the nodes of height 2 fire. And so on, until the top node fires and produces the final Boolean value.

Q: So the nodes of any given height fire in parallel. Presumably sequential algorithms don't do parallel operations.

A: Well, bounded parallelism is permissible in sequential algorithms. For example, a typical Turing machine can do up to three operations in parallel: change the control state, modify the active-cell symbol, and move the tape.

Q: It seems that sequential algorithms are sequential-time algorithms with bounded parallelism.

A: No, things are a bit more complicated. A sequential-time algorithm may be interactive, even intra-step interactive, while it is usually assumed that a sequential algorithm does not interact with its environment, certainly not while performing a step.

Q: I find the term sequential algorithm not very cogent.

A: The term is not cogent but traditional. Elsewhere we used alternative terms: small step algorithms [2], classical algorithms [8], classical sequential algorithms [15]. The term sequential time was coined in [14].

Q: You mentioned that the steps of a sequential algorithm are of bounded complexity. What do you mean?

A: We will confront this question in Sect. 5.

Q: I presume a sequential algorithm can be nondeterministic.

A: Many authors agree with you, but we think that true sequential algorithms are deterministic, and so did Turing: "When such a machine reaches one of these ambiguous configurations, it cannot go on until some arbitrary choice has been made by an external operator." Nondeterminism involves intra-step interaction [14, Sect. 9], because an external intervention is needed to resolve nondeterminism; otherwise the algorithm hangs helplessly. In this connection, recall Yogi Berra's famous nondeterministic algorithm: "When you come to a fork in the road, take it!"

Q: Then why do many authors allow sequential algorithms to be nondeterministic.

A: Because, for many purposes, it is convenient to hide the nondeterminism resolving mechanism.

Q: I guess you'll be talking plenty about interaction in the rest of the paper.

A: No, not in this paper. But we analyzed interactive algorithms elsewhere [3, 4].

1.2 What's in the Paper?

We review selected semantics-to-syntax analyses of species of algorithms in the literature. In each case, we try to identify the species and explicate the fulcrum that made the analysis possible.

Q: The fulcrum? The literal meaning of the word is the support point about which a lever turns, as in the Archimedes's "Give me a fulcrum, and I shall move the world." But that isn't what you mean.

A: We use the term as a metaphor. Suppose that you study a large species of algorithms, and you ask yourself whether there are syntactic means to describe the species. The problem seems overwhelming, impossible to solve. There are so many vastly different algorithms there. You don't even know where to start. Yet you persist and continue to think about the problem. Now imagine that one day an idea occurs to you to look at the problem differently, from another angle so to speak. And suddenly the problem looks more feasible though not necessarily easy. It is that idea that we metaphorically call a fulcrum.

Q: Your metaphorical use of the term seems by no means restricted to the semantics-to-syntax analysis of algorithms.

A: That's right. Here's an example from another domain. In the 1960s people tried various ways to organize large quantities of data. There are so many different kinds of data, and they seem to require different presentations. But eventually it occurred to Edgar Frank Codd to view a database as a relational structure of mathematical logic [7]. Relational database theory was born. The rest is history.

In Sect. 2, we review Alan Turing's celebrated analysis of computation in his 1936 paper [23].

Q: Was it the first semantics-to-syntax analysis?

A: No, Alonzo Church's analysis [6] preceded that of Turing. There was also analysis of recursion that culminated with Kurt Gödel's definition of recursive numerical[2] functions [17]. These classical analyses are richly covered in the logic literature, and Turing's analysis is the deepest and by far the most convincing.

In Sect. 3, we discuss Andrey Kolmogorov's analysis of computation of the 1950s [18, 19].

Q: We spoke about Kolmogorov machines earlier [12]. They compute the same numerical functions as Turing machines do, and there had been other computation models of that kind introduced roughly at the same time or earlier. In particular, Emil Post introduced his machines already in 1936 [21]. Kolmogorov's compatriot Andrey Markov worked on his normal algorithms in the early 1950s [20]. Why do you single out Kolmogorov's analysis?

A: As far as we know, among the analyses of algorithms of that post-Turing period, only Kolmogorov's analysis seems to be a true semantics-to-syntax analysis. By the way, even though Kolmogorov machines do not compute more numerical functions than Turing machines do, they implement more algorithms.

Q: That raises the question when two algorithms are the same.

A: We addressed the question in [5].

In Sect. 4, we review Robin Gandy's analysis of machine computations [9]. In Sect. 5, we review our own analysis of sequential algorithms.

Q: You review your own analysis? Why?

A: Well, most of the semantics-to-syntax analyses of species of algorithms in the literature are devoted to sequential algorithms, and we think, rightly or wrongly, that our analysis of sequential algorithms is definitive. Also we use this opportunity to spell out, for the first time, our fulcrum.

Finally, in Sect. 6, we discuss limitations of semantics-to-syntax analyses.

---

[2] Here and below a numerical function is a function $f(x_1, \dots, x_j)$, possibly partial, of finite arity j, where the arguments x_i range over natural numbers, and the values of f—when defined—are natural numbers.

2 Turing

Alan Turing analyzed computation in his 1936 paper "On Computable Numbers, with an Application to the Entscheidungsproblem" [23]. All unattributed quotations in this section are from that paper.

The Entscheidungsproblem is the problem of determining whether a given first-order formula is valid. The validity relation on first-order formulas can be naturally represented as a real number, and the Entscheidungsproblem becomes whether this particular real number is computable. "Although the subject of this paper is ostensibly the computable numbers, it is almost equally easy to define and investigate computable functions of an integral variable or a real or computable variable, computable predicates, and so forth. The fundamental problems involved are, however, the same in each case, and I have chosen the computable numbers for explicit treatment as involving the least cumbrous technique."

Q: Hmm, input strings of a Turing machine are finite which does not suffice to represent real numbers.

A: Turing himself did not insist that input strings are finite.

2.1 Turing's Species of Algorithms

Turing's intention might have been to consider all algorithms. But algorithms of the time were sequential, and the computers were humans.[3] So Turing analyzed sequential algorithms performed by idealized human computers. In the process he explicated some inherent constraints of the species and imposed—explicitly or implicitly—some additional constraints. Here are the more prominent constraints of Turing's analysis.

Digital Computation is digital (or symbolic, symbol-pushing).
"Computing is normally done by writing certain symbols on paper."

Q: Is the digital constraint really a constraint?

A: These days we are so accustomed to digital computations that the digital constraint may not look like a constraint. But it is. Non-digital computations have been performed by humans from ancient times. Think of ruler-and-compass computations or of Euclid's algorithm for lengths [15, Sect. 3].

[3]"Numerical calculation in 1936 was carried out by human beings; they used mechanical aids for performing standard arithmetical operations, but these aids were not programmable" (Gandy [10, p. 12]).

Sequential Time Computation splits into a sequence of steps.
"Let us imagine the operations performed by the computer to be split up into [a sequence of] 'simple operations'."

Elementary Steps The simple operations mentioned above "are so elementary that it is not easy to imagine them further divided."

Q: The elementary steps are elementary indeed but, as you mentioned, they involve parallel actions: changing the control state, modifying the active-cell, moving the tape. There are multi-tape Turing machines where more actions are performed in parallel.

A: This is true, but the parallelism remains bounded.

2.1.1 Interaction with the Environment

Q: Earlier you mentioned an important constraint on sequential algorithms which, I guess, applies to Turing's analysis: the computation does not interact with the environment. Of course the environment supplies inputs and presumably consumes the outputs but the computation itself is self-contained. It is determined by the algorithm and the initial state. No oracle is consulted, and nobody interferes with the computation.

A: Actually[4] Turing introduced nondeterministic machines already in [23] and oracle machines in [24].

Q: I didn't know that. Come to think of it, oracle machines make good sense in the context of human computing. The human computer may consult various tables, may ask an assistant to perform an auxiliary computation, etc. But I don't see how nondeterministic algorithms come up in the analysis of human computing.

A: It wasn't the analysis of human computing that brought Turing to nondeterminism. "For some purposes we might use machines (choice machines or *c*-machines) whose motion is only partially determined by the configuration .... When such a machine reaches one of these ambiguous configurations, it cannot go on until some arbitrary choice has been made by an external operator. This would be the case if we were using machines to deal with axiomatic systems."

Q: I see, a nondeterministic machine can deal gracefully with axiomatic systems; it can guess and then verify a proof. You argued already that nondeterministic algorithms are really interactive. But can one really use a choice machine for meaningful interaction?

A: Well, a choice machine may be programmed to play chess. The moves of the other player can be entered as choices "made by an external operator."

[4]This was pointed out to us by the anonymous referee.

2.2 Turing's Fulcrum

How can one analyze the great and diverse variety of computations performed by human computers? Amazingly Turing found a way to do that. We believe that his fulcrum was as follows. *Ignore what a human computer has in mind and concentrate on what the computer does and what the observable behavior of the computer is.* In other words, Turing treated the idealized human computer as an operating system of sorts.

One may argue that Turing did not ignore the computer's mind. He spoke about the state of mind of the human computer explicitly and repeatedly. Here is an example. "The behaviour of the computer at any moment is determined by the symbols which he is observing, and his 'state of mind' at that moment." But Turing postulated that "the number of states of mind which need be taken into account is finite." The computer just remembers the current state of mind, and even that is not necessary: "we avoid introducing the 'state of mind' by considering a more physical and definite counterpart of it. It is always possible for the computer to break off from his work, to go away and forget all about it, and later to come back and go on with it. If he does this he must leave a note of instructions (written in some standard form) explaining how the work is to be continued. This note is the counterpart of the 'state of mind'."

2.3 On Turing's Results and Argumentation

Turing introduced abstract computing machines that became known as Turing machines, and he constructed a universal Turing machine. A real number is Turing computable "if its decimal can be written down by a [Turing] machine." His thesis was that Turing computable numbers "include all numbers which could naturally be regarded as computable." He used the thesis to prove the undecidability of the Entscheidungsproblem. To convince the reader of his thesis, Turing used three arguments [23, Sect. 9].

Reasonableness He gave examples of large classes of real numbers which are [Turing] computable.

Robustness He sketched an alternative definition of computability "in case the new definition has a greater intuitive appeal" and proved it equivalent to his original definition. The alternative definition is based on provability in a finitely axiomatizable fragment of the first-order theory of arithmetic. The robustness argument was strengthened in the appendix where, after learning about Church's definition of computability [6], he proved the equivalence of their definitions.

Appeal to Intuition He analyzed computation appealing to intuition directly.

The first two arguments are important but insufficient. There are other reasonable and robust classes of computable real numbers, e.g. the class of primitive recursive real numbers. The direct appeal to intuition is crucial.

2.4 *Two Critical Quotes*

Q: I came across a surprising remark of Gödel that Turing's argument "is supposed to show that mental procedures cannot go beyond mechanical procedures" [11]. It is hard for me to believe that this really was Turing's goal. Anyway, Gödel continues thus. "What Turing disregards completely is the fact that mind, in its use, is not static, but constantly developing, i.e., that we understand abstract terms more and more precisely as we go on using them, and that more and more abstract terms enter the sphere of our understanding. There may exist systematic methods of actualizing this development, which could form part of the procedure" [11].
Do you understand that? Apparently Gödel thought that gifted mathematicians may eventually find a sophisticated decision procedure for the Entscheidungsproblem that is not mechanical. But if gifted mathematicians are able to reliably execute the procedure, they should be able to figure out how to program it, and then the procedure is mechanical.

A: Maybe Gödel was just pointing out that, in solving instances of Entscheidungsproblem, human creativity would outperform any mechanical procedure. Turing would surely agree with that.

Q: Let me change the topic. Here is another interesting quote. "For the actual development of the (abstract) theory of computation, where one must build up a stock of particular functions and establish various closure conditions, both Church's and Turing's definitions are equally awkward and unwieldy. In this respect, general recursiveness is superior" (Sol Feferman, [10, p. 6]). Do you buy that?

A: Indeed, the recursive approach has been dominant in mathematical logic. It is different though in computer science where Turing's approach dominates. Turing's machine model enabled computational complexity theory and even influenced the early design of digital computers. Church's λ-calculus has been influential in programming language theory.

3 Kolmogorov

Andrey Kolmogorov's analysis is reflected in a 1953 talk to the Moscow Mathematical Society [18] and in a paper [19] with his student Vladimir Uspensky.[5] Kolmogorov's approach isn't as known as it deserves to be. In this connection, here are some relevant references: [2, 12, 25, 26].

3.1 Kolmogorov's Species of Algorithms

Like Turing, Kolmogorov might have intended to analyze all algorithms. The algorithms of his time still were sequential. In the 1953 talk, Kolmogorov stipulated that every algorithmic process satisfies the following constraints.

Sequentiality An algorithmic process splits into steps whose complexity is bounded in advance.

Elementary Steps Each step consists of a direct and unmediated transformation of the current state S to the next state S^*.

Locality Each state S has an active part of bounded size. The bound does not depend on the state or the input size, only on the algorithm itself. The direct and unmediated transformation of S to S^* is based only on the information about the active part of S and applies only to the active part.

Implicitly Kolmogorov presumes also that the algorithm does not interact with its environment, so that a computation is a sequence $S_0, S_1, S_2, \dots$ of states, possibly infinite, where every $S_{n+1} = S_n^*$

Q: The second stipulation does not seem convincing to me. For example, a sequential algorithm may multiply and divide integers in one step. Such transformations do not look direct and immediate in some absolute sense.

A: Kolmogorov restricts attention to sequential algorithms working on the lowest level of abstraction, on the level of single bits.

3.2 Kolmogorov's Fulcrum

Kolmogorov's ideas gave rise to a new computation-machine model different from Turing's model [19]. Instead of a linear tape, a Kolmogorov (or Kolmogorov-Uspensky) machine has a graph of bounded degree (so that there is a bound on

[5] Uspensky told us that the summary [18] of the 1953 talk was written by him after several unsuccessful attempts to make Kolmogorov to write a summary.

the number of edges attached to any vertex), with a fixed number of types of vertices and a fixed number of types of edges. We speculated in [12] that "the thesis of Kolmogorov and Uspensky is that every computation, performing only one restricted local action at a time, can be viewed as (not only being simulated by, but actually being) the computation of an appropriate KU machine." Uspensky agreed [25, p. 396].

We do not know much about the analysis that led Kolmogorov and Uspensky from the stipulations above to their machine model. "As Kolmogorov believed," wrote Uspensky [25, p. 395], "each state of every algorithmic process ... is an entity of the following structure. This entity consists of elements and connections; the total number of them is finite. Each connection has a fixed number of elements connected. Each element belongs to some type; each connection also belongs to some type. For every given algorithm the total number of element types and the total number of connection types are bounded." In that approach, the number of non-isomorphic active zones is finite (because of a bound on the size of the active zones), so that the state transition can be described by a finite program.

Leonid Levin told us that Kolmogorov thought of computation as *a physical process developing in space and time* (Levin, private communication, 2003). That seems to be Kolmogorov's fulcrum. In particular, the edges of the state graph of a Kolmogorov machine reflect physical closeness of computation elements. One difficulty with this approach is that there may be no finite bound on the dimensionality of the computation space [14, footnote 1].

Q: I would think that Kolmogorov's analysis lent support to the Church-Turing thesis.

A: It did, to the extent that it was independent from Turing's analysis. We discuss the issue in greater detail in [8, Sect. 1.2].

Q: You mentioned that Turing's machine model enabled computational complexity theory. Was the Kolmogorov-Uspensky machine model useful beyond confirming the Church-Turing thesis?

A: Very much so; please see [2, Sect. 3].

4 Gandy

Gandy analyzed computation in his 1980 paper "Church's Thesis and Principles for Mechanisms" [9]. In this section, by default, quotations are from that paper.

4.1 Gandy's Species of Algorithms

The computers of Gandy's time were machines, or "mechanical devices", rather than humans, and that is Gandy's departure point.

> Turing's analysis of computation by a human being does not apply directly to mechanical devices ... Our chief purpose is to analyze mechanical processes and so to provide arguments for ...
> **Thesis M.** *What can be calculated by a machine is computable.*

Since mechanical devices can perform parallel actions, Thesis M "must take parallel working into account." But the species of all mechanical devices is too hard to analyze, and Gandy proceeds to narrow it to a species of mechanical devices that he is going to analyze.

> (1) In the first place I exclude from consideration devices which are *essentially* analogue machines. ... I shall distinguish between "mechanical devices" and "physical devices" and consider only the former. The only physical presuppositions made about mechanical devices ... are that there is a lower bound on the linear dimensions of every atomic part of the device and that there is an upper bound (the velocity of light) on the speed of propagation of changes.
> (2) Secondly we suppose that the progress of calculation by a mechanical device may be described in discrete terms, so that the devices considered are, in a loose sense, digital computers.
> (3) Lastly we suppose that the device is deterministic; that is, the subsequent behaviour of the device is uniquely determined once a complete description of its initial state is given.
> After these clarifications we can summarize our argument for a more definite version of Thesis M in the following way.
> **Thesis P.** *A discrete deterministic mechanical device satisfies principles I–IV below.*

Gandy's Principle I asserts in particular that, for any mechanical device, the states can be described by hereditarily finite sets[6] and there is a transition function F such that, if x describes an initial state, then $Fx, F(Fx), \ldots$ describe the subsequent states. Gandy wants "the form of description to be sufficiently abstract to apply uniformly to mechanical, electrical or merely notional devices," so the term mechanical device is treated liberally.

Principles II are III are technical restrictions on the state descriptions and the transition function respectively. Principle IV generalizes Kolmogorov's locality constraint to parallel computations.

> We now come to the most important of our principles. In Turing's analysis the requirement that the action depend only on a bounded portion of the record was based on a human limitation. We replace this by a physical limitation [Principle IV] which we call the *principle of local causation.* Its justification lies in the finite velocity of propagation of effects and signals: contemporary physics rejects the possibility of instantaneous action at a distance.

A preliminary version of Principle IV gives a good idea about the intentions behind the principle.

> **Principle IV** (Preliminary version). The next state, Fx, of a machine can be reassembled from its restrictions to overlapping "regions" s and these restrictions are locally caused. That is, for each region s of Fx there is a "causal neighborhood" $t \subseteq \mathrm{TC}(x)$ of bounded size such that $Fx \restriction s$ [the restriction of Fx to s] depends only on $x \restriction t$ [the restriction of x to t].

[6] A set x is *hereditarily finite* if its transitive closure $\mathrm{TC}(x)$ is finite. Here $\mathrm{TC}(x)$ is the least set t such that $x \in t$ and such that $z \in y \in t$ implies $z \in t$.

4.2 Gandy's Fulcrum

It seems to us that Gandy's fulcrum is his Principle I. It translates the bewildering world of mechanical devices into the familiar set-theoretic framework.

4.3 Comments

Gandy pioneered the axiomatic approach in the area of semantics-to-syntax analyses of algorithms. He put forward the ambitious Thesis M asserting that a numerical function can be calculated by a machine only if it is Turing computable. In our view, Gandy's decision to narrow the thesis is perfectly justified; see Sect. 6 in this connection.

But his narrowing of Thesis M is rather severe. A computing machine may have various features that Turing's analysis rules out, e.g. asynchronous parallel actions, analog computations. Gandy allows only synchronous parallelism, that is sequential-time parallelism. His species is a subspecies of the species of synchronous parallel algorithms (which was analyzed later, already in the new century, in [1]).

One reason that Gandy's species is a subspecies of synchronous parallel algorithm is Principle I. Hereditarily finite sets are finite. Taking into account that Gandy's machines do not interact with the environment, this excludes many useful algorithms. For example it excludes a simple algorithm that consumes a stream of numbers keeping track of the maximal number seen so far.

Q: Isn't this algorithm inherently interactive?

A: In a sense yes. But it is a common abstraction in programming to pretend that the whole input stream is given in the beginning. The abstraction is realizable in the Turing machine model.

Also, the principle of local causality (Principle IV) does not apply to all synchronous parallel algorithms. Gandy himself mentions one counterexample, namely Markov's normal algorithms [20]. The principle fails in the circuit model of parallel computation, the oldest model of parallel computation in computer theory. The reason is that the model allows gates to have unbounded fan-in. We illustrate this on the example of a first-order formula $\forall x R(x)$ where $R(x)$ is atomic. The formula gives rise to a collection of circuits C_n of depth 1. Circuit C_n has n input gates, and any unary relation R on $\{1, \dots, n\}$ provides an input for C_n. Circuit C_n computes the truth value of the formula $\forall x R(x)$ in one step, and the value depends on the whole input.

Our additional critical remarks of Gandy's analysis are found in [15, Sect. 4]. (Wilfried Sieg adopted Gandy's approach and simplified Gandy's axioms, see [22] and references there, but—as far as our critique is concerned—the improvements do not make much difference.)

Q: If Turing thought of synchronous parallelism, he could have claimed that, without loss of generality, the parallel actions performed during one step can be executed sequentially, one after another.

A: Gandy complains that the claim seems obvious to people. One should be careful though about executing parallel actions sequentially. Consider for example a parallel assignment $x, y := y, x$. If you start with $x := y$, you'd better save the value of x so that $y := x$ can be performed as intended. In any case, the claim can be proved in every model of synchronous parallelism in the literature, including the most general model of [1].

5 Sequential Algorithms

5.1 Motivation

By the 1980s, there were plenty of computers and software. A problem arose how to specify software. The most popular theoretical approaches to this problem were declarative. And indeed, declarative specifications (or specs) tend to be of higher abstraction level and easier to understand than executable specs. But executable specs have their own advantages. You can "play" with them: run them, test, debug.

Q: If your spec is declarative then, in principle, you can verify it mathematically.

A: That is true, and sometimes you have to verify your spec mathematically; there are better and better tools to do that. In practice though, mathematical verification is out of the question in an overwhelming majority of cases, and the possibility to test specs is indispensable, especially because software evolves. In most cases, it is virtually impossible to keep a declarative spec in sync with the implementation. In the case of an executable spec, you can test whether the implementation conforms to the spec (or, if the spec was reverse-engineered from an implementation, whether the spec is consistent with the implementation).

A question arises whether an executable spec has to be low-level and detailed? This leads to a foundational problem whether any algorithm can be specified, in an executable way, on its intrinsic level of abstraction.

Q: A natural-language spec would not do as it is not executable.

A: Besides, such a spec may (and almost invariably does) introduce ambiguities and misunderstanding.

Q: You can program the algorithm in a conventional programming language but this will surely introduce lower-level details.

A: Indeed, even higher-level programming languages tend to introduce details that shouldn't be in the spec.

Turing and Kolmogorov machines are executable but low-level. Consider for example two distinct versions of Euclid's algorithm for the greatest common divisor of two natural numbers: the ancient version where you advance by means of differences, and a modern (and higher-level) version where you advance by means of divisions. The chances are that, in the Turing machine implementation, the distinction disappears.

Can one generalize Turing and Kolmogorov machines in order to solve the foundational problem in question? The answer turns out to be positive, at least for sequential algorithms [14], synchronous parallel algorithms [1], and interactive algorithms [3, 4]. We discuss here only the first of these.

5.2 The Species

Let's restrict attention to the species of sequential algorithms but without any restriction on the abstraction level. It could be the Gauss Elimination Procedure for example. Informally, paraphrasing the first stipulation in Sect. 3, an algorithm is sequential if it computes in steps whose complexity is bounded across all computations of the algorithm. In the rest of this section, algorithms are by default sequential.

We use the axiomatic method to explicate the species. The first axiom is rather obvious.

Axiom 1 (Sequential Time) *Any algorithm A is associated with a nonempty collection $\mathcal{S}(A)$ of states, a sub-collection $\mathcal{I}(A) \subseteq \mathcal{S}(A)$ of initial states and a (possibly partial) state transition map $\tau_A : \mathcal{S}(A) \longrightarrow \mathcal{S}(A)$.*

Definition 1 Two algorithms are *behaviorally equivalent* if they have the same states, the same initial states and the same transition function.

Q: I guess the computation of a sequential algorithm A is determined by the initial state.

A: Actually we can afford to be more general and allow the environment to intervene between the steps of the algorithm A, provided that every such intervention results in a legitimate state of A, so that A can continue to run. Thus, in general, the steps of A are interleaved with those of the environment, and the computation of A depends on the environment. But, for a given environment, replacing A with a behaviorally equivalent algorithm results in exactly the same computation.

Recall that a first-order structure X is a nonempty set (the base set of X) with relations and operations; the vocabulary of X consists of the names of those relations and operations. For example, if the vocabulary of X consists of one binary relation then X is a directed graph.

Axiom 2 (Abstract State) *The states of an algorithm A can be faithfully represented by first-order structures of the same finite vocabulary, which we call the vocabulary of A, in such a way that*

- τ_A *does not change the base set of a state,*
- *collections* $\mathcal{S}(A)$ *and* $\mathcal{I}(A)$ *are closed under isomorphisms, and*
- *any isomorphism from a state X to a state Y is also an isomorphism from* $\tau_A(X)$ *to* $\tau_A(Y)$.

Q: You claim that first-order structures are sufficiently general to faithfully represent the states of any algorithm?

A: We do, and we have been making that claim since the 1980s. The collective experience of computer science seem to corroborate the claim.

Q: The states of real-world algorithms seems to be finite. Do you need infinite structures?

A: The states of real-world algorithms often are infinite. For example, consider the C programming language. A programming language can be viewed as an algorithm that runs a given program on the given data. In C, data structures like multi-sets or trees need to be programmed but arrays are readily available; they pre-exist in the initial state of C. So the states of C are infinite. And of course the states of the Gauss Elimination Procedure are infinite.

Q: OK, I have another question about the axiom. That base-set preservation sounds restrictive. A graph algorithm may extend the graph with new nodes.

A: And where will the algorithm take those nodes? From some reserve? Make (a possibly abstracted version of) that reserve a part of your initial state.

Q: Now, why should the collection of states be closed under isomorphisms, and why should the state transition respect isomorphisms?

A: Because an algorithm works at a particular level of abstraction, and lower-level details are abstracted away. Consider a graph algorithm for example. In an implementation, nodes may be integer numbers, but the algorithm can't examine whether a node is even or odd because implementation details are irrelevant. If the algorithm does take advantage of the integer representation of nodes then its vocabulary should reflect the relevant part of arithmetic.

According to the informal definition of sequential algorithms, there is a bound on the complexity of the steps of the algorithm. But how to measure step complexity? The abstract-state axiom allows us to address the problem. The next state $\tau_A(X)$ of an algorithm A depends only on the current state X of A. The executor does not need to remember any history (even the current position in the program); all of that is reflected in the state. If the executor is human and writes something on scratch paper, that paper should be a part of the computation state.

In order to change the given state X into $\tau_A(X)$, the algorithm A explores a portion of X and then performs the necessary changes of the values of the predicates and operations of X. We argue in [14] that an abstracted version of Kolmogorov's locality constraint is valid on any level of abstraction. The "active zone" of state X is bounded. The change from X to $\tau_A(X)$, let us call it $\Delta_A(X)$, depends only on the results of the exploration of the active zone. Formally, $\Delta_A(X)$ can be defined as a collection of assignments $F(\bar{a}) := b$ where F is a vocabulary function.

Q: What about vocabulary relations? Are they necessarily static?

A: We view relations as Boolean-valued functions, so the vocabulary relations may be updatable as well.

Q: How does the algorithm know what to explore and what to change?

A: That information is supplied by the program, and it is applicable to all the states. In the light of the abstract-state axiom, it should be given symbolically, in terms of the vocabulary of A.

Axiom 3 (Bounded Exploration) *There exists a finite set T of terms (or expressions) in the vocabulary of algorithm A such that $\Delta_A(X) = \Delta_A(Y)$ whenever states X, Y of A coincide over T.*

Now we are ready to define (sequential) algorithms.

Definition 2 A (sequential) algorithm is any entity that satisfies the sequential-time, abstract-state and bounded-exploration axioms.

Abstract state machines (ASMs) were defined in [13]. Here we restrict attention to sequential ASMs which are undeniably algorithms.

Theorem 1 ([14]) *For every algorithm A, there exists a sequential ASM that is behaviorally equivalent to A.*

5.3 *The Fulcrum*

Every sequential algorithm has its native level of abstraction. On that level, the states can be faithfully represented by first-order structures of a fixed vocabulary in such a way that state transitions can be expressed naturally in the language of the fixed vocabulary.

Q[7]: I can't ask Turing, Kolmogorov or Gandy how they arrived to their fulcrums, but I can ask you that question.

A: In 1982 we moved abruptly from logic to computer science. We tried to understand what was that emerging science about. The notion of algorithm seemed central. Compilers, programming languages, operating systems are all algorithms. As far as sequential algorithms are concerned, the sequential-time axiom was obvious.

Q: How do you view a programming language as an algorithm?

A: It runs a given program on given data.

[7] This discussion is provoked by the anonymous referee who thought that the two sentences above are insufficient for this subsection.

Q: What about all those levels of abstraction?

A: It had been well understood by 1982 that a real-world computation process can be viewed at different levels of abstraction. The computer was typically electronic, but you could abstract from electronics and view the computation on the level of single bits. More abstractly, you could view the computation on the assembly-language level, on the level of the virtual machine for the programming language of the program, or on the level of the programming language itself. But there is no reason to stop there. The abstraction level of the algorithm that the program executes is typically higher yet.
Besides, we became interested in software specifications and took the position that software specs are high-level algorithms. The existing analyses of algorithms did not deal with the intrinsic levels of abstraction of algorithms.

Q: Is the intrinsic abstraction level of any algorithm well determined?

A: This is one of the critical questions. It was our impression that the abstraction level of an algorithm is determined by that of its states and by the operations that the algorithm executes in one step. That led us eventually to the abstract-state axiom and to abstract state machines originally called evolving algebras [13].

Q: Was your logic experience of any use?

A: Yes, it was particularly useful to know how ubiquitous first-order structures were. Any static mathematical object that we knew could be faithfully represented by a first-order structure, and the states of an algorithm are static mathematical objects.

Q: When did you realize that sequential algorithms could be axiomatized?

A: This took years. We mentioned above that, in a semantics-to-syntax analysis, a fulcrum makes the analysis more feasible but not necessarily easy. The sequential-time and abstract-state axioms are implicit in the definition of abstract state machines [13]. We had not thought of them as axioms yet. Our thesis, restricted to sequential algorithms, was that sequential abstract state machines are able to simulate arbitrary sequential algorithms "in a direct and essentially coding-free way" [13]. The thesis was successfully tested in numerous applications of abstract state machines, and our confidence in the thesis as well as the desire to derive it from "first principles" grew. The problem was solved in [14].

6 Final Remarks

In [15] we argue that "the notion of algorithm cannot be rigorously defined in full generality, at least for the time being." The reason is that the notion is expanding. In addition to sequential algorithms, in use from antiquity, we have now parallel algorithms, interactive algorithms, distributed algorithms, real-time algorithms, analog algorithms, hybrid algorithms, quantum algorithms, etc. New kinds of algorithms may be introduced and most probably will be. Will the notion of algorithms ever crystallize to support rigorous definitions? We doubt that.

"However the problem of rigorous definition of algorithms is not hopeless. Not at all. Large and important strata of algorithms have crystallized and became amenable to rigorous definitions" [15]. In Sect. 5, we explained the axiomatic definition of sequential algorithms. That axiomatic definition was extended to synchronous parallel algorithms in [1] and to interactive sequential algorithms in [3, 4].

The axiomatic definition of sequential algorithms was also used in to derive Church's thesis from the three axioms plus an additional Arithmetical State axiom which asserts that only basic arithmetical operations are available initially [8].

Q: I wonder whether there is any difference between the species of all algorithms and that of machine algorithms.

A: This is a good point, though there may be algorithms executed by nature that machines can't do. In any case, our argument that the species of all algorithms can't be formalized applies to the species of machine algorithms. The latter species also evolves and may never crystallize.

Acknowledgements Many thanks to Andreas Blass, Bob Soare, Oron Shagrir and the anonymous referee for useful comments.

References

1. A. Blass, Y. Gurevich, Abstract state machines capture parallel algorithms. ACM Trans. Comput. Log. **4**(4), 578–651 (2003). Correction and extension, same journal **9**, 3 (2008), article 19
2. A. Blass, Y. Gurevich, Algorithms: a quest for absolute definitions, in *Current Trends in Theoretical Computer Science*, ed. by G. Paun et al. (World Scientific, 2004), pp. 283–311; in *Church's Thesis After 70 Years*, ed. by A. Olszewski (Ontos Verlag, Frankfurt, 2006), pp. 24–57
3. A. Blass, Y. Gurevich, Ordinary interactive small-step algorithms. ACM Trans. Comput. Log. **7**(2), 363–419 (2006) (Part I), plus 8:3 (2007), articles 15 and 16 (Parts II and III)
4. A. Blass, Y. Gurevich, D. Rosenzweig, B. Rossman, Interactive small-step algorithms. Log. Methods Comput. Sci. **3**(4), 1–29, 1–35 (2007), papers 3 and 4 (Part I and Part II)
5. A. Blass, N. Dershowitz, Y. Gurevich, When are two algorithms the same? Bull. Symb. Log. **15**(2), 145–168 (2009)
6. A. Church, An unsolvable problem of elementary number theory. Am. J. Math. **58**, 345–363 (1936)
7. E.F. Codd, Relational model of data for large shared data banks. Commun. ACM **13**(6), 377–387 (1970)
8. N. Dershowitz, Y. Gurevich, A natural axiomatization of computability and proof of Church's thesis. Bull. Symb. Log. **14**(3), 299–350 (2008)
9. R.O. Gandy, Church's thesis and principles for mechanisms, in *The Kleene Symposium*, ed. by J. Barwise et al. (North-Holland, Amsterdam, 1980), pp. 123–148
10. R.O. Gandy, C.E.M. (Mike) Yates (eds.), *Collected Works of A.M. Turing: Mathematical Logic* (Elsevier, Amsterdam, 2001)
11. K. Gödel, A philosophical error in Turing's work, in *Kurt Gödel: Collected Works*, vol. II, ed. by S. Feferman et al. (Oxford University Press, Oxford, 1990), p. 306

12. Y. Gurevich, On Kolmogorov machines and related issues. Bull. Eur. Assoc. Theor. Comput. Sci. **35**, 71–82 (1988)
13. Y. Gurevich, Evolving algebra 1993: Lipari guide, in *Specification and Validation Methods*, ed. by E. Börger, (Oxford University Press, Oxford, 1995), pp. 9–36
14. Y. Gurevich, Sequential abstract state machines capture sequential algorithms. ACM Trans. Comput. Log. **1**(2), 77–111 (2000)
15. Y. Gurevich, What is an algorithm? in *SOFSEM 2012: Theory and Practice of Computer Science*. Springer LNCS, vol. 7147, ed. by M. Bielikova et al. (Springer, Berlin, 2012). A slight revision will appear in Proc. of the 2011 Studia Logica conference on Church's Thesis: Logic, Mind and Nature
16. Y. Gurevich, Foundational analyses of computation, in *How the World Computes. Turing Centennial Conference*. Springer LNCS, vol. 7318, ed. by S.B. Cooper et al. (Springer, Berlin, 2012), pp. 264–275
17. S.C. Kleene, *Introduction to Metamathematics* (D. Van Nostrand, Princeton, 1952)
18. A.N. Kolmogorov, On the concept of algorithm. Usp. Mat. Nauk **8**(4), 175–176 (1953). Russian
19. A.N. Kolmogorov, V.A. Uspensky, On the definition of algorithm. Usp. Mat. Nauk **13**(4), 3–28 (1958). Russian. English translation in AMS Translations **29**, 217–245 (1963)
20. A.A. Markov, *Theory of Algorithms*. Transactions of the Steklov Institute of Mathematics, vol. 42 (1954). Russian. English translation by the Israel Program for Scientific Translations, 1962; also by Kluwer, 2010
21. E.L. Post, Finite combinatorial processes—formulation I. J. Symb. Log. **1**, 103–105 (1936)
22. W. Sieg, On computability, in *Handbook of the Philosophy of Mathematics*, ed. by A. Irvine (Elsevier, Amsterdam, 2009), pp. 535–630
23. A.M. Turing, On computable numbers, with an application to the Entscheidungsproblem. Proc. Lond. Math. Soc. Ser. 2 **42**, 230–265 (1936/1937)
24. A.M. Turing, Systems of logic based on ordinals. Proc. Lond. Math. Soc. Ser. 2 **45**, 161–228 (1939)
25. V.A. Uspensky, Kolmogorov and mathematical logic. J. Symb. Log. **57**(2), 385–412 (1992)
26. V.A. Uspensky, A.L. Semenov, *Theory of Algorithms: Main Discoveries and Applications* (Nauka, Moscow, 1987) in Russian; (Kluwer, Dordrecht, 2010) in English
27. J. Wiedermann, J. van Leeuwen, Rethinking computation, in *Proceedings of 6th AISB Symposium on Computing and Philosophy*. Society for the Study of Artificial Intelligence and the Simulation of Behaviour, ed. by M. Bishop, Y.J. Erden (Exeter, London, 2013), pp. 6–10

The Information Content of Typical Reals

George Barmpalias and Andy Lewis-Pye

Abstract The degrees of unsolvability provide a way to study the continuum in algorithmic terms. Measure and category, on the other hand, provide notions of size for subsets of the continuum, giving rise to corresponding notions of "typicality" for real numbers. We give an overview of the order-theoretic properties of the degrees of typical reals, presenting old and recent results, and pointing to a number of open problems for future research on this topic.

Keywords Category • Genericity • Measure • Randomness • Turing degrees

AMS Classifications: 03D28

1 Introduction

The study of the structure of the degrees of unsolvability dates back to Kleene and Post [10]. The same is true of the application of Baire category methods in this study, while the application of probabilistic techniques in relative computation dates back to de Leeuw et al. [6]. In this article we give an overview of the state of the art in degree theory in terms of category and measure. Recent interest in this topic has been motivated by questions in algorithmic randomness, but there is an

G. Barmpalias (✉)
State Key Lab of Computer Science, Institute of Software, Chinese Academy of Sciences, 100190 Beijing, China

School of Mathematics, Statistics and Operations Research, Victoria University, Wellington, New Zealand
e-mail: barmpalias@gmail.com
http://www.barmpalias.net

A. Lewis-Pye
Department of Mathematics, Columbia House, London School of Economics, Houghton Street, London WC2A 2AE, UK
e-mail: andy@aemlewis.co.uk
http://aemlewis.co.uk

G. Sommaruga, T. Strahm (eds.), *Turing's Revolution*,
DOI 10.1007/978-3-319-22156-4_8

essential distinction to be drawn between much of the research that takes place in algorithmic randomness and our interests here: while in algorithmic randomness one is concerned with understanding the properties of random reals quite generally, here we are interested specifically in the properties of the *degrees* of random reals.

Formalising a notion of typicality essentially amounts to defining a notion of size, and then defining the typical objects to be those which belong to all large sets from some restricted (normally countable) class. It is then interesting to note that, beyond cardinality, there are two basic notions of size for sets of reals. One can think in terms of measure, the large sets being those of measure 1, or in terms of category, the large sets being those which are comeager. In [17] Kunen provided, in fact, a way of formalising the question as to whether these are the only "reasonable" notions of size for sets of reals.

Of course one cannot expect a real not to belong to any set of measure 0, or not to belong to any meager set, and so to formalise a level of typicality one restricts attention to sets of reals which are definable in some specific sense—one might consider all arithmetically definable sets of reals which are of measure 0, for example, giving a corresponding class of "typical reals" which do not belong to any such set. Working both in terms of category and measure, we are then interested in establishing the order-theoretic properties of the "typical" degrees, i.e. those containing typical reals. Despite the large advances in degree theory over the last 60 years, our knowledge on this issue is rather limited. On the other hand, some progress has been made recently, which also points to concrete directions and methodologies for future research. We present old and new results on this topic, and ask a number of questions whose solutions will help obtain a better understanding of the properties of typical degrees.

In order to increase readability we have presented a number of results in tables, rather than a list of theorem displays. For the same reason, a number of citations of results are suppressed. For the remainder of this introductory section we discuss some of the background concepts which we shall require from computability theory. For a thorough background in classical computability theory we refer to Odifreddi [25], Downey and Hirshfeldt [5], and Nies [23], while [1] is a thorough historical survey on the topic.

1.1 The Algorithmic View of the Continuum

While it is certainly sometimes of interest, in the computability theoretic context, to consider the reals imbued with their full structure as the complete ordered field, most of the time it is common practice to identify reals with their infinite binary expansions. In this context a real is seen principally as coding certain information—as an infinite binary database. One might consider, for example, the problem of deciding which Diophantine equations with integer coefficients have at least one (integer) solution. Each Diophantine equation can be given a natural number code (in some algorithmic fashion which may or may not involve redundancies of various kinds) and, given any such coding, some reals then specify exactly which equations have solutions. Given access to the bits of the real, we may be able to derive the correct answer just by checking if a particular bit of the real is a 0 or a 1. So reals are identified with subsets of the natural numbers, which in turn can be viewed as solutions to problems, and in this context we are often not overly concerned with the properties of the real as an element of the standard algebraic structure, but rather in the nature of the information that it encodes and perhaps the features of the encoding itself.

The Turing reducibility is a preorder that compares reals according to their information content and was introduced (implicitly) by Turing in [32] and was explicitly studied by Kleene and Post in [10]. The formal definition[1] makes use of *oracle* Turing machines, which are Turing machines with an extra oracle tape upon which the (potentially non-computable) binary expansion of a real may be written. We say that A is Turing reducible to B (in symbols, $A \leq_T B$) if there is an oracle Turing machine which calculates the bits of A when the binary expansion of B is written on the oracle tape. One can think of this as meaning quite simply that one can move in an algorithmic fashion from the binary expansion of B to the binary expansion of A. So the reducibility formalises the notion that the information in A is encoded in (and hence, can be algorithmically retrieved from) B.

When $A \leq_T B$ and $B \leq_T A$, we regard A and B as having the same information content. Hence the preorder $\leq_T$ induces an equivalence relation on the continuum which identifies reals with the same information content. These equivalence classes are called Turing degrees, and we consider the natural ordering on these degrees inherited by the Turing reducibility. Considering reals as solutions to problems (or rather problem sets indexed by natural numbers), the degrees of unsolvability can then be viewed as ordering problems or computational tasks according to their level of difficulty. We often denote by $\mathbf{a}, \mathbf{b}, \mathbf{c}$ the degrees of the problems A, B, C respectively. Hence if $\mathbf{a} \leq \mathbf{b}$ then a solution to (any problem in) $\mathbf{b}$ is enough to provide a solution to (any problem in) $\mathbf{a}$.

1.2 Properties of Degrees

The Turing degree structure provides a way to formally address and study the classification of problems according to their difficulty. A large part of this study has focused on establishing basic properties of the partial order, beginning already in [10]. It was observed there, for example, that the degree structure forms an upper semi-lattice with a least element and the countable predecessor property. Early research then focused on exhibiting degrees with certain structural properties, for instance minimal pairs and minimal degrees. We shall focus here on natural properties which are first order definable in the language for the structure, i.e. the language of partial orders.

[1] Turing's original definition did not involve an extra tape. However it is equivalent to the standard formulation that we give here.

Constructing a degree with certain properties reduces to the construction of a representative real in the degree, with analogous properties. In this context, the connection to Baire category, in particular, becomes immediately apparent. Normally the construction of a real whose degree satisfies a given set of properties, proceeds by first breaking down those properties into a countable list of sub-requirements. If the real A to be constructed must be non-computable, for example, then one might break this down into a list of requirements $\mathcal{P}_i : A \neq \phi_i$, where ϕ_i is the ith partial computable function. Very often (at least in many simple cases) this list of requirements can be satisfied by proceeding according to a *finite extension argument*: one defines A as the infinite sequence extending a sequence of finite strings $\{\tau_s\}_{s\in\omega}$, which are defined in stages. At stage $s+1$, given τ_s, we define τ_{s+1} to be a finite extension of τ_s, chosen in such a way that τ_{s+1} being an initial segment of A ensures satisfaction of $\mathcal{P}_s$. When the proof is of this form we are invariably able to conclude that, in fact, the set of reals satisfying the required properties is not only non-empty but is comeager. Given this immediate connection, it is perhaps not surprising that most of the early results describing the first order properties of degrees of typical reals relate to category rather than measure. Generally it is also fair to say that the proofs required for the measure theoretic results tend to be more complicated, and so one might expect that they would appear later in the subject's development.

1.3 A History of Measure and Category Arguments in the Turing Degrees

Measure and Baire category arguments in degree theory are as old as the subject itself. For example, Kleene and Post [10] used arguments that resemble the Baire category theorem construction in order to build Turing degrees with certain basic properties. Moreover de Leeuw et al. [6] used a so-called "majority vote argument" in order to show that if a subset of ω can be enumerated relative to every set in a class of positive measure then it has an unrelativised computable enumeration. Spector [30] used a measure theoretic argument in order to produce incomparable hyperdegrees. Myhill [22], on the other hand, advocated for and demonstrated the use of Baire category methods in degree theory. Sacks' monograph [27] included a chapter on "Measure-theoretic, category and descriptive set-theoretic arguments", where he shows that the minimal degrees have measure 0.

A highly influential yet unpublished manuscript by Martin (1967, Measure, category, and degrees of unsolvability, Unpublished manuscript) showed that more advanced degree-theoretic results are possible using these classical methods. By that time degree theory was evolving into a highly sophisticated subject and the point of this paper was largely that category and measure can be used in order to obtain advanced results, which go well beyond the basic methods of Kleene and Post [10]. Of the two results in Martin (1967, Measure, category, and degrees of unsolvability,

Unpublished manuscript) the first was that the Turing upward closure of a meager set of degrees that is downward closed amongst the non-zero degrees, but which does not contain **0**, is meager (see [25, Sect. V.3] for a concise proof of this). Given that the minimal degrees form a meager class, an immediate corollary of this was the fact that there are non-zero degrees that do not bound minimal degrees. The second result was that the measure of the hyperimmune degrees is 1. Martin's paper was the main inspiration for much of the work that followed in this topic, including [8, 26, 33].

Martin's early work seemed to provide some hope that measure and category arguments could provide a simple alternative to conventional degree-theoretic constructions which are often very complex. This school of thought received a serious blow, however, with [26]. Paris answered positively a question of Martin which asked if the analogue of his category result in Martin (1967, Measure, category, and degrees of unsolvability, Unpublished manuscript) holds for measure: are the degrees that do not bound minimal degrees of measure 1? Paris' proof was considerably more involved than the measure construction in Martin (1967, Measure, category, and degrees of unsolvability, Unpublished manuscript) and seemed to require sophisticated new ideas. The proposal of category methods as a simple alternative to "traditional" degree theory had a similar fate. Yates [33] started working on a new approach to degree theory that was based on category arguments and was even writing a book on this topic. Unfortunately the merits of his approach were not appreciated at the time (largely due to the heavy notation that he used) and he gave up research on the subject altogether.

Yates' work in [33] deserves a few more words, however, especially since it anticipated much of the work in [8]. Inspired by Martin (1967, Measure, category, and degrees of unsolvability, Unpublished manuscript), Yates started a systematic study of degrees in the light of category methods. A key feature in this work was an explicit interest in the level of effectivity possible in the various category constructions and the translation of this level of effectivity into category concepts (like "$\mathbf{0}'$-comeager" etc.). Using his own notation and terminology, he studied the level of genericity that is sufficient in order to guarantee that a set belongs to certain degree-theoretic comeager classes, thus essentially defining various classes of genericity already in 1974. He analysed Martin's proof that the Turing upper closure of a meager class which is downward closed amongst the non-zero degrees but which does not contain **0** is meager, for example (see [33, Sect. 5]), and concluded that no 2-generic degree bounds a minimal degree. Moreover, he conjectured (see [33, Sect. 6]) that there is a 1-generic that bounds a minimal degree. These concerns occurred later in a more appealing form in Jockusch [8], where simpler terminology was used and the hierarchy of n-genericity was explicitly defined and studied.

With Jockusch [8], the heavy notation of Yates was dropped and a clear and systematic calibration of effective comeager classes (mainly the hierarchy of n-generic sets) and their Turing degrees was carried out. A number of interesting results were presented along with a long list of questions that set a new direction for future research. The latter was followed up by Kumabe [12–16] (as well as other authors, e.g. [3]) who answered a considerable number of these questions.

The developments in the measure approach to degree theory were similar but considerably slower, at least in the beginning. Kurtz's thesis [18] is probably the first systematic study of the Turing degrees of the members of effectively large classes of reals, in the sense of measure. Moreover the general methodology and the types of questions that Kurtz considers are entirely analogous to the ones proposed in [8] for the category approach (e.g. studying the degrees of the n-random reals as opposed to the n-generic reals, minimality, computable enumerability and so on). Kučera [11] focused on the degrees of 1-random reals. Kautz [9] continued in the direction of Kurtz [18] but it was not until the last 10 years (and in particular with the writing of Downey and Hirshfeldt [5, Chap. 8]) that the study of the degrees of n-random reals became well known and this topic became a focused research area.

1.4 Overview

Building on the long history of category and measure argument for the study of the degrees of unsolvability, this paper can be seen as an explicit proposal for a systematic analysis of the *order theoretically* definable properties satisfied by the typical Turing degree, where typicality is gauged in terms of category or measure. The main part of this article is organised in three sections. In Sect. 2 we give precise definitions of typicality in terms of measure and category, and how one can calibrate typicality by refining these definitions in terms of the classical definability hierarchies. We also discuss for which properties of degrees it makes sense to ask our main question (i.e. whether they are typical) and what kind of answers can be expected. We focus on a list of rather basic properties, introduced in Table 1. These properties have received special attention in the study of degrees over the years. In Sect. 3 we describe all known results concerning which of these properties are typical. We also consider the question as to which properties are inherited by all (non-zero) predecessors of sufficiently typical degrees.

Table 1 Properties of a degree $\mathbf{a}$

Join	$\forall \mathbf{x} < \mathbf{a}\, \exists \mathbf{y} < \mathbf{a}\, (\mathbf{x} \neq \mathbf{0} \rightarrow (\mathbf{x} \vee \mathbf{y} = \mathbf{a}))$
Meet	$\forall \mathbf{x} < \mathbf{a}\, \exists \mathbf{y} < \mathbf{a}\, ((\mathbf{y} \neq \mathbf{0})\, \&\, \mathbf{x} \wedge \mathbf{y} = \mathbf{0})$
Cupping	$\forall \mathbf{y} > \mathbf{a}\, \exists \mathbf{x} < \mathbf{y}\, (\mathbf{a} \vee \mathbf{x} = \mathbf{y})$
Complementation	$\forall \mathbf{x} < \mathbf{a}\, \exists \mathbf{y} < \mathbf{a}\, (\mathbf{x} \neq \mathbf{0} \rightarrow (\mathbf{x} \wedge \mathbf{y} = \mathbf{0}\, \&\, \mathbf{x} \vee \mathbf{y} = \mathbf{a}))$
Top of a diamond	$\exists \mathbf{x}\, \exists \mathbf{y}\, (\mathbf{x} \neq \mathbf{0}\, \&\, \mathbf{x} \vee \mathbf{y} = \mathbf{a}\, \&\, \mathbf{x} \wedge \mathbf{y} = \mathbf{0})$
Being a minimal degree	$\mathbf{a} > \mathbf{0}\, \&\, (\mathbf{0}, \mathbf{a}) = \emptyset$
Bounding a minimal degree	$\exists \mathbf{x} \leq \mathbf{a}\, [\mathbf{x} > \mathbf{0}\, \&\, (\mathbf{0}, \mathbf{x}) = \emptyset]$
Being a minimal cover	$\exists \mathbf{x} < \mathbf{a}\, [(\mathbf{x}, \mathbf{a}) = \emptyset]$
Having a minimal cover	$\exists \mathbf{y} > \mathbf{a}\, [(\mathbf{a}, \mathbf{y}) = \emptyset]$
Being a strong minimal cover	$\exists \mathbf{x} < \mathbf{a}\, [(\mathbf{0}, \mathbf{a}) = (\mathbf{0}, \mathbf{x}]]$
Having a strong minimal cover	$\exists \mathbf{y} > \mathbf{a}\, [(\mathbf{0}, \mathbf{a}] = (\mathbf{0}, \mathbf{y})]$

In Sect. 4 we discuss the similarities and differences between the two faces of typicality that we have considered. After some superficial remarks, we present the recent results of Shore on the theories of the lower cones of typical degrees. Both of these results are motivated by the fact that, fixing a notion of typicality and given two typical degrees, the first order theories of the corresponding lower cones (in the language of partial orders and with the inherited ordering) are equal. The first result determines the level of typicality that is required for this fact to hold (namely, arithmetical genericity or randomness). The second result shows that the two different kinds of typicality give rise to different corresponding theories.

2 Typical Degrees and Calibration of Typicality

2.1 Large Sets and Typical Reals

Given a property which refers to a real (or a degree), measure and category arguments may aim to show that there exist reals that satisfy this property by demonstrating that the property is typical. In other words "most" reals satisfy this property. This approach is based on:

(a) a formalisation of the notion of "large";
(b) a restriction of the sets of reals that we consider to a countable class;
(c) the definition "typical real" as a real which occurs in every large set in this restricted class.

Of course, in terms of category "large" means comeager, while in terms of measure, "large" means "of measure 1". Restricting attention to those reals which belong to every member of a countable class of large sets, still leaves us with a large class:

(1) the intersection of countably many comeager sets is comeager;
(2) the intersection of countably many measure 1 sets has measure 1.

Those reals which are typical, we call *generic* for the category case, and *random* if working with measure—both of these notions clearly depending on the countable class specified in (b). The "default" for the generic case is to consider the countable collection of sets which are definable in first order arithmetic, giving "arithmetical genericity" (or "arithmetical randomness" accordingly). It is quite common when discussing genericity to suppress the prefix "arithmetical", so that by "generic" is often meant arithmetical genericity. Alternative choices are possible, resulting in stronger or weaker genericity and randomness notions. For example, we may consider a randomness notion that is defined with respect to the hyperarithmetical sets—for more information on such strong notions of randomness we refer to Nies [23, Chap. 9].

In the computability context though, it is normally appropriate to consider a finer hierarchy and much weaker randomness and genericity notions. To this end, we define a set of reals $\mathcal{C}$ to be *effectively* null, if there is an algorithm which given any rational $\epsilon > 0$ as input, enumerates an open covering of $\mathcal{C}$ of measure $< \epsilon$—to enumerate an open set means to enumerate $W \subset 2^{<\omega}$ which specifies the set, in the sense that its elements are precisely those infinite strings extending members of W. This definition relativises in the obvious way, so that $\mathcal{C}$ is effectively null relative to the real A, if the same condition holds when the algorithm has access to an oracle for A. Then we define a real to be A-random if it does not belong to any set which is effectively null relative to A. A simple but crucial observation is that *any* null set is effectively null relative to *some* A, so that once we have observed that a set is of measure 1, there will always be some level of randomness which suffices to ensure membership of the set. Finally, the hierarchy we normally work with is defined as follows: 1-random means $\emptyset$-random, 2-random means $\emptyset'$-random and, more generally, n-random means $\emptyset^{(n-1)}$-random, so that arithmetical randomness is equivalent to being n-random for all $n \in \omega$.

The generic case works similarly. The definition we give here is not the original, but is equivalent to it, and is more commonly used. Given $W \subseteq 2^{<\omega}$, we say that A *forces* W if there exists $\sigma \subset A$ (denoting that σ is an initial segment of A) such that either:

(1) $\sigma \in W$, or;
(2) σ has no extension in W.

So A forces W if it has some initial segment which is already enough to decide whether A has an initial segment in W. The hierarchy is then defined as follows: a real is 1-generic if it forces every W which is c.e. (computably enumerable), and more generally, is n-generic if it forces every W which is c.e. relative to $\emptyset^{(n-1)}$. Analogously to the random case, if a set of reals $\mathcal{C}$ is meager then there exists some A such that being 1-generic relative to A ensures failure to belong to $\mathcal{C}$. It is also easily seen that arithmetical genericity is equivalent to being n-generic for all $n \in \omega$.

Finally, the following notions allow us to consider an even finer hierarchy. We say a real is weakly n-random if it does not belong to any Π^0_n null set. We say that a real is weakly n-generic if it forces every W which is c.e. relative to $\emptyset^{(n-1)}$ and dense (i.e. any finite string has an extension in W).

Thus the (non-collapsing) hierarchies of genericity and randomness provide a calibration of typicality on the continuum, and have become a subject in their own right. Of course, we are interested in answering the following question:

$$\text{Which definable properties are typical of a degree of unsolvability?} \tag{1}$$

Before we embark to the pursuit of this quest, let us examine the answers that are possible. It is reasonable to expect that some structural properties hold for all generic reals while they fail to hold for all random reals? Indeed, as we noted above, while genericity and randomness might be said to formalise the same intuitive notion of typicality, on a technical level they are very different notions. In fact, as we remark in Sect. 4, the two classes and their Turing degrees form disjoint classes.

2.2 Properties of Degrees and Definability

Is it reasonable to expect that for every property $\mathcal{P}$ which is definable in some sense, either $\mathcal{P}$ is met by all typical reals or else that $\neg\mathcal{P}$ is met by all typical reals? We have already discussed the fact that there are many levels of typicality that one may consider, for both randomness and genericity. Let $\mathcal{T}$ be a certain level of genericity or randomness. In the case that $\mathcal{P}$ is met by all reals in $\mathcal{T}$ or $\neg\mathcal{P}$ is met or by all reals in $\mathcal{T}$ we say that $\mathcal{P}$ is decided by $\mathcal{T}$. Hence we are essentially asking whether every definable property $\mathcal{P}$ is decided by some typicality level $\mathcal{T}$. At this point let us consider the cases of randomness and genericity separately. For the case of randomness, Kolmogorov's 0-1 law states that any (Lebesgue) measurable tailset[2] is either of measure 0 or 1. We may identify a set of Turing degrees with the set of reals contained in those degrees (i.e. with the union). In this sense, every set of Turing degrees is a tailset. Hence any measurable set of Turing degrees must either be of measure 0 or 1. So if we restrict question (1) to arithmetically definable properties of degrees $\mathcal{P}$ (i.e. sets of degrees definable in the structure, and for which the union is arithmetically definable as a set of reals), then the satisfying class is a set of measure 0 or 1, so either all arithmetically random degrees $\mathbf{x}$ satisfy $\neg\mathcal{P}$ or all arithmetically random degrees satisfy $\mathcal{P}$ (respectively).

More generally we could restrict question (1) to Borel properties $\mathcal{P}$, and the same considerations would hold. Ultimately, we would like to consider all properties $\mathcal{P}$ which are definable in the structure of the Turing degrees, in the first order language of partial orders. As was demonstrated in [2, Sect. 3], however, whether all definable sets of degrees are measurable is independent from *ZFC*. Hence in this more general form, question (1) may not always admit a clear answer. In our discussions on randomness, most of the properties $\mathcal{P}$ that we consider are arithmetically definable. In the cases where $\mathcal{P}$ is not evidently arithmetically definable (see the cupping property in Table 1) we show that (arithmetical) randomness suffices in order to decide this property.

In the case of genericity we can consider the topological 0-1 law which says that tailsets satisfying the property of Baire[3] are either meager or comeager. Since all Borel sets of reals have the property of Baire, if we restrict question (1) to Borel properties $\mathcal{P}$ we can expect a definite answer. In other words, in this case there is a level of genericity that decides $\mathcal{P}$. In the case where $\mathcal{P}$ is restricted to the arithmetically definable properties, we can expect that n-genericity for some $n \in \omega$ decides $\mathcal{P}$. We may also want to consider the more general case where $\mathcal{P}$ is definable in the structure of the degrees (again, in the first-order language of partial orders). Unfortunately, it was observed in [2, Sect. 3] that in this case we cannot expect a definite answer to question (1). Indeed, it is independent of *ZFC* whether

---

[2]A set of reals C is a tailset if for every real A and every $\sigma \in 2^{<\omega}$, $\sigma * A \in C$ iff $A \in C$ (where $*$ denotes concatenation).

[3]A set of reals has the property of Baire if its symmetric difference from some open set is meager.

Table 2 Validity of the properties for a typical degree

	Generic			Random		
Properties	Validity	Level	Fails	Validity	Level	Fails
Join	✓	1	w1	✓	2	1
Meet	✓	2	w1	?	?	w1
Cupping	✓	w2	1	✗	2	w2
Complementation	✓	2	w1	?	?	1
Top of a diamond	✓	1	w1	✓	w2	w1
Being a minimal degree	✗	1	w1	✗	1	w1
Bounding a minimal degree	✗	2	1	✗	2	w2
Being a minimal cover	✓	2	w1	?	?	1
Having a minimal cover	✓	0	–	✓	0	–
Being a strong minimal cover	✗	1	w1	✗	1	w1
Having a strong minimal cover	✗	w2	1	✓	2	w2

all definable sets of degrees are either meager or comeager. On the other hand, it is well known that under the axiom of determinacy *AD* every set of reals has the property of Baire. Hence in $ZF + AD$ every property $\mathcal{P}$ is decided by some level of genericity.

In conclusion, our project consists of considering various structural properties $\mathcal{P}$ of the Turing degrees and establishing a level of randomness or genericity that decides $\mathcal{P}$. If $\mathcal{P}$ is Borel then we can expect this task to have a clear solution. Otherwise it is possible that none of the standard levels of randomness or genericity decides $\mathcal{P}$. The properties that we consider are in a certain sense simple, although not always (obviously) Borel. Moreover, according to the known results, they all have a typicality level that decides them, and this is often much lower than the level that is guaranteed by their complexity. This latter observation is evident from an inspection of Tables 1 and 2.

2.3 *Very Basic Questions Remain Open*

Our search for the properties of the typical degree starts by considering a small collection of properties which can be considered as "natural" in the sense that they are encountered in most considerations in classical degree theory. Let us start by noting that there are very simple properties, like density, for which it is unknown whether they are typical. This may come as a surprise to the reader, given that the study of the degrees of unsolvability dates back to Kleene and Post [10].

Question 1 *Are the random degrees dense?*

Formally, is it true that given any two sufficiently random degrees $\mathbf{a} < \mathbf{b}$, the interval $(\mathbf{a}, \mathbf{b})$ is nonempty? A variation of this question can be stated in terms of a property of a single degree: is it true that for any sufficiently random degree $\mathbf{b}$ and any $\mathbf{a} < \mathbf{b}$ the interval $(\mathbf{a}, \mathbf{b})$ is nonempty? In other words (see the relevant entry in Table 1) we ask whether every random degree fails to be a minimal cover. This property can be expressed with 5 alternating quantifiers in arithmetic, so we can expect that either every 5-random real satisfies it or every 5-random real fails to satisfy it. In particular, arithmetical randomness decides this property. Ultimately, this question can be expressed without reference to random degrees:

Question 2 *What is the measure of minimal covers?*

In Sect. 3 we are going to see that in terms of genericity, typical degrees are minimal covers. On the other hand it is a well known fact that no typical degree is minimal, both in terms of measure and category. Other basic properties that we consider are displayed in Table 1, along with their formal definitions. All of these properties have straightforward first order definitions in arithmetic, except for the cupping property and the property of having a (strong) minimal cover. However, as we discuss in the following, these latter properties are decided by arithmetical randomness (indeed, 2-randomness). In Sect. 3 we give a full account of the known status of the properties in Table 1 (i.e. whether they are typical). In general, more is known for genericity than randomness.

3 Properties of the Typical Degrees and Their Predecessors

The properties of degrees that we are going to be examining are displayed in Table 1. From an algebraic point of view, they are quite basic properties that give essential information about a given structure that one might be interested (in our case, the degrees of unsolvability). Moreover, such algebraic statements can serve as building blocks in definability investigations in degree structures (e.g. see [28]), a theme that has been especially popular in the last 50 years of investigations in to the Turing degrees. For further motivation with regard to the role of the properties of Table 1 in the interplay between definability and algebraic structure in the Turing degrees, we refer to Lewis [21].

3.1 Some Properties of the Typical Degrees

We summarise the status of the properties of Table 1 with respect to the typical degrees in the validity columns of Table 2. Especially in the case of randomness, there are some notable gaps in our knowledge, including the property of being a minimal cover that was discussed in Sect. 2.3. These gaps are indicated with a question mark in the corresponding entries of the validity columns of Table 2. There are many more open problems here, however, than just those indicated by question marks. In fact, there are open questions associated with almost every row of the table. This is because, as discussed previously, our project does not end upon deciding whether a property is typical. We are also interested in determining the exact levels of the typicality hierarchies which suffice to decide each property. This information is displayed in the columns "Level" and "Fails" of Table 2. Here "2" means 2-genericity or 2-randomness, and similarly for "1". Also "w2" denotes weak 2-randomness or weak 2-genericity, and similarly for "w1". The level of the typicality hierarchy that is indicated under column "Level" is the lowest level of the hierarchy which is known to decide the corresponding property. Similarly, the level of the typicality hierarchy that is indicated under column "Fails" is the highest level where reals have been found that give opposite answers to the validity of the corresponding property—or simply the highest level at which it is known that the property can fail, in the case that we do not know whether typical degrees satisfy the property. An optimal result has been achieved in the cases where the two levels are consecutive. All of the other cases can be seen as open problems.

As an example, let us examine the join property. By Barmpalias et al. [2] all 2-random degrees satisfy the join property. On the other hand, in [20] it was shown that all low 1-random degrees fail to satisfy the join property. We do not know the answer with respect to the weakly 2-random degrees. In terms of genericity, it was shown in [8] that all 2-generic degrees satisfy the join property. This was extended in [2] (via a different argument) to all 1-generic degrees. On the other hand, by Kurtz [18, 19] every hyperimmune degree[4] contains a weakly 1-generic set. Hence the join property fails for some weakly 1-generic degrees.

Another example is the property of being the join of a minimal pair of degrees. We express this fact by saying that the degree is the "top of a diamond". It is a rather straightforward observation that every 1-generic degree is the top of a diamond. Such basic facts about the generic degrees were established in [8] (an analogous observation, following from van Lambalgen's Theorem, the fact that bases for 1-randomness are Δ^0_2 and the fact that weakly 2-random degrees form a minimal pair with $\mathbf{0}'$, is that every weakly 2-random degree is the top of a diamond). On the other hand, using the fact from [18, 19] that every hyperimmune degree contains a set which is weakly 1-generic (and a set which is weakly 1-random), it follows that there are degrees of weakly 1-generic sets (and degrees of weakly 1-random sets) which are not the top of a diamond.

Many of the results in Table 2 regarding genericity were obtained in [12, 14–16], solving a number of questions in [8]. For example, it was shown that every 2-generic degree is a minimal cover. We do not know if there exists a 1-generic degree or a weakly 2-generic degree which is not a minimal cover. Kumabe also showed that every 2-generic degree satisfies the complementation property. It is not known if this can be extended to 1-generic degrees, and the complementation property for random reals is an open problem.

[4] A degree is called hyperimmune if it computes a function which is not dominated by any computable function.

A general theme in the study of typical reals from an algorithmic point of view, is that information introduces order and hence makes reals special and less typical. In other words, a degree that has high information content (e.g. it can compute the halting problem) fails to be typical. More precisely, 1-generic reals are incomplete (i.e. fail to compute the halting problem) and weakly 2-random reals are incomplete. There are a number of results that support this intuition, both in terms of genericity and in terms of randomness. Another more sophisticated example from [31] is that weakly 2-random reals cannot compute a complete extension of Peano arithmetic. Despite these facts, it turns out that there is no bound on the information that joins of typical reals can have. Since we have not found this basic fact in the literature, we present it here.

Theorem 3.1 *Let $\mathcal{V}$ be a null or a meager set of degrees, and let* $\mathbf{d}$ *be a degree. Then there exist degrees* $\mathbf{x}, \mathbf{y}$ *which are not in $\mathcal{V}$ and* $\mathbf{d} < \mathbf{x} \vee \mathbf{y}$.

Proof Let D, X, Y denote representatives of the degrees $\mathbf{d}, \mathbf{x}, \mathbf{y}$ respectively. Since there are no maximal degrees, it suffices to show that $\mathbf{d} \leq \mathbf{x} \vee \mathbf{y}$. Let V be the union of the degrees in $\mathcal{V}$. Suppose that $\mathcal{V}$ is null, so that V is a null set of reals. Then there exists a closed set P of positive measure that is disjoint from V. It is a basic fact from measure theory that P contains a real Z which has an indifferent set of digits (s_i) with respect to P, in the sense that any modification in the bits of Z on positions s_i results in a real which is in P. Algorithmic refinements of this fact were studied in [7]. Without loss of generality let us assume that (s_i) is increasing. For each $t \in \omega$ which is not a term of (s_i) define $X(t) = Y(t) = Z(t)$. Moreover for each i let $X(s_i) = D(i)$ and $Y(s_i) = 1 - D(i)$. Then clearly $X \oplus Y$ can compute D.

The case when $\mathcal{V}$ is meager is similar, based on the fact that every comeager set of reals contains a real which is indifferent in P with respect to a sequence of positions. Algorithmic versions of this fact were studied in [4]. □

3.2 Properties of Typical Degrees are Inherited by the Non-zero Degrees They Compute

Is it reasonable to expect that the properties of typical degrees also hold for the non-zero degrees that they bound? This is equivalent to the expectation that the upward closure of any "small" class of degrees that does not contain $\mathbf{0}$ is "small". This is not true in general, although the upper cone of degrees above any given non-zerodegree is both meager and null. Martin's category theorem (from Martin, 1967, Measure, category, and degrees of unsolvability, Unpublished manuscript, see [8]) says that if $\mathcal{C}$ is a meager downward closed set of degrees then the upward closure of $\mathcal{C} - \{\mathbf{0}\}$ is meager. On the other hand if a meager class $\mathcal{C}$ is not downward closed then the upward closure of $\mathcal{C} - \{\mathbf{0}\}$ can be comeager (see [8]). An application of Martin's category theorem was that the set of degrees that bound minimal degrees is meager. An analogous result in terms of measure was shown in [26]. In particular, it was shown that if $\mathcal{C}$ is the null class of minimal degrees, then the upward closure of $\mathcal{C}$ is also null. However there exist downward closed null classes $\mathcal{C}$ of degrees such that $\mathcal{C} - \{\mathbf{0}\}$ has measure 1. For example, let $\mathcal{C}$ consist of the degrees bounded by 1-generics. Then $\mathcal{C}$ is null and by Kurtz [18] its upward closure has measure 1. Hence the exact analogue of Martin's theorem in terms of measure is not true. We are yet to find a counterexample, however, which is "naturally" definable (as a subset of the Turing degrees, rather than as a set of reals definable in arithmetic), and we do not know if there is some analogue of Martin's category theorem in terms of measure. Let us express this problem in terms of the following, somewhat vague, question:

Question 3 *Which null classes of degrees have null upward closure?*

Our discussion shows that whether the non-zeropredecessors of a typical degree **a** inherit a property of **a** depends on the type of the property in question. However many of the basic degree theoretic properties that are known to hold for generic degrees are also known to hold for the nonzero degrees that are bounded by generic degrees. A number of results in [8] follow this heuristic principle. For example, the cupping and join properties are satisfied by all non-zero degrees bounded by any 2-generic degree. Curiously enough, the same phenomenon is common in the random degrees. For example, the join property is shared by all non-zero degrees that are bounded by a 2-random degree, and the cupping property fails for all such degrees. This observation also relates to many of the arguments that are used to obtain these results. In [2] we presented a methodology for examining whether a property is typical of a random degree, which extends to a methodology for examining whether the property is typical of a nonzero degree which is bounded by a random degree. Usually, the latter argument tends to be more involved than the former, but both rest on the same ideas. This methodology rests on the work of other authors, for example [26] where it was shown that sufficiently random degrees do not bound minimal degrees. However it is refined considerably, which allows to obtain more precise classifications like the fact that 2-random degrees do not bound minimal degrees. Such results are often shown to be tight by providing counter-examples for lower levels of typicality. For example, it was shown that there are weakly 2-random degrees that bound minimal degrees.

In Table 3 we display the properties that are known to hold for all non-zero degrees that are bounded by typical degrees. The reader may observe that most of the properties of Table 2 that hold for typical degrees also hold for the non-zero degrees that are bounded by typical degrees (in fact we do not have natural counter-examples). For example, 2-random degrees all have strong minimal covers, as do all non-zero degrees that are bounded by 2-randoms. The fact that all non-zero degrees bounded by 2-randoms satisfy join, however, implies that there are no strong minimal covers below any 2-random degree. Weak 2-randoms do sometimes bound strong minimal covers, since they sometimes bound minimal degrees. These results are from [2]. In the same paper it was shown that every degree that is bounded by a 2-random degree is the top of a diamond. This result can be seen as a weak version

Table 3 Validity of the properties, for a nonzerodegree which is bounded above by a typical degree

	Bounded by a generic		Bounded by a random	
Properties	Validity	Level	Validity	Level
Join	✓	2	✓	2
Meet	?	?	?	?
Cupping	✓	2	✗	2
Complementation	?	?	?	?
Top of a diamond	?	?	✓	2
Being a minimal degree	✗	2	✗	2
Bounding a minimal degree	✗	2	✗	2
Being a minimal cover	?	?	?	?
Being a strong minimal cover	✗	2	✗	2
Having a strong minimal cover	✗	2	✓	2

of the complementation property. The latter as well as the meet property are open problems, as we indicate in Tables 2 and 3.

In terms of genericity, we do not know whether complementation is satisfied by all non-zero degrees bounded by 2-generics, or even whether such degrees will always satisfy the meet property (although the latter may not be a difficult problem). We do not know whether every non-zero degree that is bounded by a 2-generic degree is a minimal cover.

4 Genericity and Randomness

A discussion comparing randomness and genericity can be found in [5, Sect. 8.20]. As reals, generics and randoms are certainly very different. For example, their Turing degrees form disjoint classes. In fact, no 1-random real is computable in a 1-generic. Furthermore, every generic degree forms a minimal pair with every sufficiently random degree, and this already happens on the level of 2-generics and 2-randoms (see [24]). A result which reveals additional relations between the two notions, was proved in [2]. It was shown that every nonzerodegree that is bounded by a 2-random degree is the join of a minimal pair of 1-generic degrees.

An inspection of Table 2 shows that all of the properties that we have considered are decided by level 2 of the genericity hierarchy. Moreover, some of them are even decided earlier, for example the property of having a strong minimal cover which does not hold for any weakly 2-generic degree (see [2, Sect. 8.1]) but there are some 1-generic degrees which satisfy it (by Kumabe [16]). For the case of random degrees, there are some unknown cases, but most of the properties that we consider are also decided on the second level of the hierarchy of randomness.

The following question therefore comes into focus. Is it possible that there is a finite level H_n of the hierarchy which is sufficient to decide all sentences φ for the lower cone, in the sense that $\forall \mathbf{x} \in H_n,\ \mathcal{D}(\leq \mathbf{x}) \models \varphi$ or $\forall \mathbf{x} \in H_n,\ \mathcal{D}(\leq \mathbf{x}) \models \neg\varphi$? This question was raised in an early draft of Barmpalias et al. [2, Sect. 12] and for the case of generic degrees it was independently raised by Jockusch much earlier (personal communication with Richard Shore). It was recently answered in the negative.

Theorem 4.1 ([29]) *There are sentences φ_n such that, for $n > 2$, $\mathcal{D}(\leq \mathbf{x}) \models \varphi_n$ for every $(n+1)$-generic or $(n+1)$-random $\mathbf{x}$ but such that $\mathcal{D}(\leq \mathbf{x}) \not\models \varphi_n$ for some n-generics and n-randoms.*

We note that the sentences φ_n are the same for the generic and the random case. Moreover, they are obtained via a process of interpreting arithmetic inside the corresponding degree structures. As a result of this, they are not considered "natural" and one may consider the task of discovering familiar properties which separate at least the first few levels of the randomness and genericity hierarchies.

Another issue that was raised in [2, Sect. 12] was whether the theory below an arithmetically generic degree is the same as the theory below an arithmetically random degree. This question makes sense, since given any two arithmetically generic degrees $\mathbf{x}, \mathbf{y}$, the theories of the structures $\mathcal{D}(\leq \mathbf{x})$ and $\mathcal{D}(\leq \mathbf{y})$ are equal. Similarly the theories of the lower cone for any two arithmetically random degrees are equal. An inspection of Table 2 shows that the only properties there where the generic and random degrees differ, are the cupping property and the property of having a strong minimal cover. These, however, are not properties pertaining to the lower cone, so they do not provide an answer to our question. An answer was given in [29], again via the methodology of coding:

Theorem 4.2 ([29]) *There is a sentence φ such that $\mathcal{D}(\leq \mathbf{x}) \models \varphi$ for every 3-random degree $\mathbf{x}$ and $\mathcal{D}(\leq \mathbf{x}) \models \neg\varphi$ for every 3-generic degree $\mathbf{x}$.*

We note that, as with the case of Theorem 4.1, the question remains as to whether one can find "natural" examples of sentences φ which separate the theories of the lower cones below arithmetically random and arithmetically generic degrees. Table 2 points to some candidates, for example the complementation and meet properties, and the property of being a minimal cover.

Acknowledgements Barmpalias was supported by the 1000 Talents Program for Young Scholars from the Chinese Government, and the Chinese Academy of Sciences (CAS) President's International Fellowship Initiative No. 2010Y2GB03. Additional support was received by the CAS and the Institute of Software of the CAS. Partial support was also received from a Marsden grant of New Zealand and the China Basic Research Program (973) grant No. 2014CB340302. Lewis-Pye was 559 formally named Lewis, and was supported by a Royal Society University Research Fellowship.

References

1. K. Ambos-Spies, P.A. Fejer, Degrees of unsolvability, in *Handbook of the History of Logic*, vol. 9. Computational Logic, ed. D.M. Gabbay, J.H. Siekmann, J. Woods (Elsevier, Amsterdam, 2014)
2. G. Barmpalias, A.R. Day, A.E.M. Lewis, The typical Turing degree. Proc. Lond. Math. Soc. **109**(1), 1–39 (2014)
3. C.-T. Chong, R.G. Downey, On degrees bounding minimal degrees. Ann. Pure Appl. Log. **48**, 215–225 (1990)
4. A.R. Day, Indifferent sets and genericity. J. Symb. Log. **78**, 113–138 (2013)
5. R. Downey, D. Hirshfeldt, *Algorithmic Randomness and Complexity* (Springer, Heidelberg, 2010)
6. K. de Leeuw, E.F. Moore, C.E. Shannon, N. Shapiro, Computability by probabilistic machines. in *Automata Studies*, ed. by C.E. Shannon, J. McCarthy (Princeton University Press, Princeton, 1955), pp. 183–212
7. S. Figueira, J.S. Miller, A. Nies, Indifferent sets. J. Log. Comput. **19**(2), 425–443 (2009)
8. C. Jockusch Jr., Degrees of generic sets, in *Recursion Theory: Its Generalizations and Applications, Proceedings of Logic Colloquium '79*, Leeds, August 1979, ed. by F.R. Drake, S.S. Wainer , (Cambridge University Press, Cambridge, 1980), pp. 110–139
9. S.M. Kautz, Degrees of random sets. Ph.D. Dissertation, Cornell University, 1991
10. S.C. Kleene, E. Post, The upper semi-lattice of degrees of recursive unsolvability. Ann. Math. (2) **59**, 379–407 (1954)
11. A. Kučera, Measure, Π^0_1-classes and complete extensions of PA, in *Recursion Theory Week (Oberwolfach, 1984)*. Lecture Notes in Mathematics, vol. 1141 (Springer, Berlin, 1985), pp. 245–259
12. M. Kumabe, A 1-generic degree which bounds a minimal degree. J. Symb. Log. **55**, 733–743 (1990)
13. M. Kumabe, Relative recursive enumerability of generic degrees. J. Symb. Log. **56**(3), 1075–1084 (1991)
14. M. Kumabe, Every n-generic degree is a minimal cover of an n-generic degree. J. Symb. Log. **58**(1), 219–231 (1993)
15. M. Kumabe, Generic degrees are complemented. Ann. Pure Appl. Log. **59**(3), 257–272 (1993)
16. M. Kumabe, A 1-generic degree with a strong minimal cover. J. Symb. Log. **65**(3), 1395–1442 (2000)
17. K. Kunen, Random and Cohen reals, in *Handbook of Set-Theoretic Topology*, ed. by K. Kunen, J. Vaughan (North Holland, Amsterdam, 1984), pp. 887–911
18. S. Kurtz, Randomness and genericity in the degrees of unsolvability. Ph.D. Dissertation, University of Illinois, Urbana, 1981
19. S. Kurtz, Notions of weak genericity. J. Symb. Log. **48**, 764–770 (1983)
20. A.E.M. Lewis, A note on the join property. Proc. Am. Math. Soc. **140**, 707–714 (2012)
21. A.E.M. Lewis, Properties of the jump classes. J. Log. Comput. **22**, 845–855 (2012)
22. J. Myhill, Category methods in recursion theory. Pac. J. Math. **11**, 1479–1486 (1961)
23. A. Nies, *Computability and Randomness* (Oxford University Press, Oxford, 2009)
24. A. Nies, F. Stephan, S.A. Terwijn, Randomness, relativization and Turing degrees. J. Symb. Log. **70**(2), 515–535 (2005)
25. P.G. Odifreddi, *Classical Recursion Theory*, vol. I (North-Holland Publishing Co., Amsterdam, 1989)
26. J. Paris, Measure and minimal degrees. Ann. Math. Log. **11**, 203–216 (1977)
27. G.E. Sacks, *Degrees of Unsolvability*. Annals of Mathematical Studies, vol. 55, 2nd edn. (Princeton University Press, Princeton, 1966)
28. R. Shore, Natural definability in degree structures, in *Computability Theory and Its Applications: Current Trends and Open Problems*, ed. by P. Cholak, S. Lempp, M. Lerman, R. Shore (American Mathematical Society, Providence, 2000)

29. R. Shore, The Turing degrees below generics and randoms. J. Symb. Log. **79**, 171–178 (2014)
30. C. Spector, Measure-theoretic construction of incomparable hyperdegrees. J. Symb. Log. **23**, 280–288 (1958)
31. F. Stephan, Martin-löf random and PA-complete sets, in *Logic Colloquium '02*. Lecture Notes in Logic, vol. 27 (Association for Symbolic Logic, La Jolla, 2006), pp. 342–348
32. A.M. Turing, Systems of logic based on ordinals. Proc. Lond. Math. Soc. **45**, 161–228 (1939)
33. C.E.M. Yates, Banach-Mazur games, comeager sets and degrees of unsolvability. Math. Proc. Camb. Philos. Soc. **79**, 195–220 (1976)

Proof Theoretic Analysis by Iterated Reflection

L.D. Beklemishev

Abstract Progressions of iterated reflection principles can be used as a tool for the ordinal analysis of formal systems. Moreover, they provide a uniform definition of a proof-theoretic ordinal for any arithmetical complexity Π^0_n. We discuss various notions of proof-theoretic ordinals and compare the information obtained by means of the reflection principles with the results obtained by the more usual proof-theoretic techniques. In some cases we obtain sharper results, e.g., we define proof-theoretic ordinals relevant to logical complexity Π^0_1.

We provide a more general version of the fine structure relationships for iterated reflection principles (due to Ulf Schmerl). This allows us, in a uniform manner, to analyze main fragments of arithmetic axiomatized by restricted forms of induction, including $I\Sigma_n$, $I\Sigma_n^-$, $I\Pi_n^-$ and their combinations.

We also obtain new conservation results relating the hierarchies of uniform and local reflection principles. In particular, we show that (for a sufficiently broad class of theories T) the uniform Σ_1-reflection principle for T is Σ_2-conservative over the corresponding local reflection principle. This bears some corollaries on the hierarchies of restricted induction schemata in arithmetic and provides a key tool for our generalization of Schmerl's theorem.

Keywords Ordinal analysis • Reflection principles • Turing progressions • Partial conservativity • Parameter-free induction

AMS Classifications: 03F15, 03F30, 03F40, 03F45

This article is a reprint of [6] with a new sect. 1 "Preliminary Notes" added.

L.D. Beklemishev (✉)
Steklov Mathematical Institute, Gubkina 8, 119991 Moscow, Russia
e-mail: bekl@mi.ras.ru

G. Sommaruga, T. Strahm (eds.), *Turing's Revolution*,
DOI 10.1007/978-3-319-22156-4_9

1 Preliminary Notes

A.M. Turing's 1939 paper *System of logics based on ordinals*, based on his Princeton Ph.D. Thesis, is one of the longest but perhaps lesser-known among his various contributions. It is mainly cited because, almost as a side remark, he gave in it a definition of computability relative to an oracle or an *oracle machine*, a technical notion that proved to be fundamentally important in the subsequent development of the theory of recursive functions and degrees of unsolvability.[1] However, the main topic of that paper, its motivations and, mostly negative, results belong to a different part of logic, proof theory. They are in many respects remarkable, but at the same time sufficiently complicated to be easily explained to a non-specialist.

A detailed survey of Turing's pioneering paper, which includes a very useful restatement of Turing's results in the modern language, has been provided by Feferman [11]. Therefore, here my comments are restricted to necessary minimum intended to explain the relationships between Turing's work and my paper reprinted in this volume.

The subject of ordinal logics or, as S. Feferman later called them, *transfinite recursive progressions of axiomatic systems,* emerges as inevitably as any good mathematical notion. Practically anyone thinking about the meaning of Gödel's incompleteness theorems sooner or later comes to the question: "If, as Gödel claims, the consistency of a (consistent) formal system S is unprovable in S itself,

$$S \nvdash \mathrm{Con}(S),$$

what if we adjoin the formal statement of consistency $\mathrm{Con}(S)$ to S as a new axiom?" Of course, the resulting system $S_1 = S + \mathrm{Con}(S)$, if consistent, satisfies the assumptions of Gödel's theorem and is still incomplete. Then one obtains $S_1 \nvdash \mathrm{Con}(S_1)$ and the process can be continued generating a sequence of systems $S_1, S_2, \ldots, S_\omega, S_{\omega+1}, \ldots$, where we add consistency assertions at successor stages and take unions of theories at limit stages. Thereby one arrives at a general idea of transfinite progressions.

One would ask basic questions such as how long such sequences can be, whether one can achieve completeness at the limit of such a process, whether one can meaningfully classify logical sentences by stages of this process. Here is how Turing formulates the latter idea [37, p. 200]:

> We might also expect to obtain an interesting classification of number-theoretic theorems according to "depth". A theorem which required an ordinal α to prove it would be deeper than one which could be proved by the use of an ordinal β less than α. However, this presupposes more than is justified.

Of course, the "depth" Turing mentions here must not be equated with the depth of mathematical proofs as understood in the ordinary discourse. Even elementary proofs formalizable in some basic system S can be deeper than trivial proofs

[1] See also Rathjen [30] for some uses of oracles in proof theory.

requiring additional assumptions such as the consistency of S. The last sentence in the above quotation refers to some of Turing's results that only come later in his paper. As he realized, to put the idea of ordinal logics on a rigorous mathematical basis is not so simple.

Since the systems of a progression have to be effectively given—otherwise, we would in general have problems even formulating statements such as $\mathrm{Con}(S)$—we have to deal with computable descriptions of ordinals rather than with the ordinals in the usual set-theoretic sense. Moreover, these descriptions have to be formalized in S. Since one and the same ordinal may have different computable descriptions, this raises the question of invariance of ordinal logics, that is, whether the systems S_a and S_b have to be equivalent if a and b denote the same ordinal.

The answers Turing obtained are rather disappointing. He shows that, under quite general conditions, the invariance of ordinal logics is incompatible with their expected properties such as monotonicity and completeness. By the *completeness* of an ordinal logic Turing understands completeness for Π^0_2-sentences: for each true Π^0_2-sentence π, there is an ordinal notation a such that $S_a \vdash \pi$.

Throughout his paper Turing emphasized the role of arithmetical Π^0_2-sentences which he dubbed somewhat unfortunately "number-theoretic theorems". He realized that many interesting statements about numbers (and programs), notably including the Riemann hypothesis, can be reformulated as Π^0_2-sentences. (It has later been shown by G. Kreisel that Riemann hypothesis is equivalent to a Π^0_1-sentence.)

Having failed to establish the invariance of ordinal logics, Turing succeeds in proving a partial completeness result. He shows that a progression based on the iteration of the local reflection schema for Π^0_2-sentences (which he denotes Λ_P) is complete for Π^0_1-sentences. Later, contrary to Turing's expectations, Feferman showed that Λ_P is not complete for Π^0_2-sentences, whereas Turing's proof, in fact, works even for the locally weaker progression based on the iteration of ordinary consistency. However, Turing makes a pessimistic remark about his own result [37, p. 213]:

> This completeness theorem as usual is of no value. Although it shows, for instance, that it is possible to prove Fermat's last theorem with Λ_P (if it is true) yet the truth of the theorem would really be assumed by taking a certain formula as an ordinal formula.[2]

In other words, we would only know that a number a such that S_a proves Fermat's last theorem denotes an ordinal if we already knew that this theorem were true. Thus, ordinal logics do not really allow to "overcome" the incompleteness phenomenon, but only shift the problem to the one of recognizing artificial ordinal descriptions.

A major contribution to the study of recursive progressions was the already quoted work of Feferman [9] who gave them a thorough and technically clean treatment based on the use of Kleene's system of ordinal notation $\mathcal{O}$. He showed, among other things, that the progression based on the iteration of the unrestricted uniform reflection principle (as opposed to the local reflection principle considered

[2] These are his notations for recursive well-orderings.

by Turing) is complete for arbitrary arithmetical sentences. Moreover, there exists a path within $\mathcal{O}$ such that the progression is complete along this particular path. However, as in Turing's case, these results rely on the use of artificial ordinal notations and essentially reconfirm the main conclusion of Turing's work.

In view of the failure of the original purpose of ordinal logics (to overcome Gödelian incompleteness), one is justified to ask if there were any other uses for them. Of the ideas generated by Turing's paper, perhaps, the study of the so-called *autonomous progressions* had the greatest impact on the development of proof theory. These are sequences of systems S_a restricted to those notations a that can be recognized as denoting an ordinal in one of the previously constructed systems S_b, for some $b < a$. The idea of autonomous progressions has been proposed by Kreisel [16, 17] as a means of explication of "informally defined concepts of proof" such as Hilbert's *finitistic* reasoning. Kreisel's analysis essentially identified finitism with an autonomous progression of length ϵ_0 based on the iteration of uniform Σ_1-reflection principle over the primitive recursive arithmetic (disregarding many details). Later, Feferman [10] worked out a similar proposal for the analysis of the important notion of *predicativity* in the context of theories of ramified analysis. He (and independently Schütte [33, 34]) associated the ordinal Γ_0 as a bound to predicative methods. This work has become an important milestone in the development of proof theory which experienced a tremendous growth in the years that followed (see [29] for a survey).

The current paper goes into a different direction that can be seen as the development of Turing's ideas concerning the classification of arithmetical sentences according to "depth". Apparently, such a classification is not possible if the progression at hand is not invariant, however Turing remarks [37, p. 209]:

> We can still give a certain meaning to the classification into depths with highly restricted kinds of ordinals. Suppose that we take a particular ordinal logic Λ and a particular ordinal formula Ψ representing the ordinal α say (preferably a large one), and that we restrict ourselves to ordinal formulae of the form $\mathrm{Inf}(\Psi, a)$.[3] We then have a classification into depths, but the extents of all the logics which we so obtain are contained in the extent of a single logic.

This passage describes fairly well what is going on in [6]: We sacrifice the completeness of ordinal logics in favor of invariance. The price that we pay is that we are only able to classify arithmetical sentences within some specific intervals of theories. To achieve rather delicate results technically we have to deal with more restricted kinds of progressions than in [9], the so-called *smooth progressions*. Nevertheless, many things have become easier than in Feferman's work. In particular, we do not have to apply a formalization of recursion theorem in arithmetic to construct recursive progressions but instead use simpler arithmetical fixed point arguments.

In [6] we mainly treat the most basic interval of fragments of arithmetic between the elementary arithmetic EA and Peano arithmetic PA. Various theories within this

[3]These formulas define initial segments of α.

interval have been extensively studied over the years using both model-theoretic and proof-theoretic methods. By now the zoo of considered fragments of PA within this interval has become rather large. The main fragments are defined by restricting the arithmetical complexity of the induction formula and by modifying in various ways the formulation of induction, for example, one often considers parameter-free forms of induction and the rule forms. Some schemata alternative to induction, such the collection principle, also give rise to new and important series of fragments of PA (see [14] for a survey). The use of progressions of iterated reflection principles allows for a systematic treatment of the most important fragments, and for a formulation of their basic relationships from a unified perspective. In this development, recursive progressions play the role of a standard measuring stick against which we compare various systems. This is what one would usually expect from a meaningful classification.

The approach taken in this paper is, therefore, rather far from philosophical concerns, but is aimed at finding optimal (the strongest and the most economical) formulations of various results in the proof theory of arithmetic. How it relates to the other existing notions of proof-theoretic ordinals is explained in the introduction to [6].

2 Introduction

Proof-Theoretic Ordinals: A Discussion Since the fundamental work of Gentzen in the late 30s it was understood that formal theories of sufficient expressive power can, in several natural ways, be associated ordinals. Informally, these ordinals can be thought of as a kind of quantitative measure of the "proof-theoretic strength" of a theory. On a more formal level, proof-theoretic ordinal of a theory is one of its main metamathematical characteristics, and its knowledge usually reveals much other information about the theory under consideration, in particular, the information of computational character. Thus, the calculation of proof-theoretic ordinals, or *ordinal analysis*, has become one of the central aims in the study of formal systems (cf. [28]).

Perhaps, the most traditional approach to ordinal analysis is the definition of the proof-theoretic ordinal of a theory T as the supremum of order types of primitive recursive well-ordering relations, whose well-foundedness is provable in T. Since the well-foundedness is a Π^1_1-concept, this ordinal is sometimes called the Π^1_1-*ordinal* of T (denoted $|T|_{\Pi^1_1}$). This definition provides a stable and convenient measure of proof-theoretic strength of theories. Moreover, it is not dependent on any special concepts of natural well-orderings, which is typical for the other proof-theoretic ordinals discussed below. However, it has the following drawbacks: (1) it is only applicable to theories in which the well-foundedness is expressible, e.g., it does not (directly) apply to first order Peano arithmetic; (2) it is a fairly rough measure. It is well-known that the Π^1_1-ordinal does not allow to distinguish between a theory and any of its extensions by true Σ^1_1-axioms: all such extensions share one and the same Π^1_1-ordinal.

An alternative approach to ordinal analysis makes use of the notion of *provably total computable function* of a theory T, that is, a Σ^0_1-definable function, whose totality (formulated as an arithmetical Π^0_2-sentence) is provable in T. The class of p.t.c.f. is an important computational characteristic of a theory. In typical cases, for sufficiently strong theories, such classes can be characterized recursion-theoretically using transfinite subrecursive hierarchies of fast growing functions. This yields what is usually called the proof-theoretic Π^0_2-*ordinal* of T. There are several choices for such hierarchies and the resulting ordinals depend on what hierarchy you actually take as a scale. Most popular are the Fast Growing (or the Schwichtenberg–Wainer) hierarchy, the Hardy hierarchy, and the Slow Growing hierarchy. However, under certain conditions, there are fixed relationships between these hierarchies, which allows to translate the results obtained for one of them into the others.

Π^0_2-analysis, although sharper, is more problematic than the Π^1_1 one. A well-known difficulty here is that the common hierarchies, most notably the Slow Growing one, depend on a particular choice of the ordinal notation system and a particular fundamental sequences assignment. Such a choice always presents a certain degree of arbitrariness and technical complication.[4] However, there is a payoff: first of all, Π^0_2-analysis allows to extract computational information from proofs. Another important aspect of Π^0_2-analysis is that it is closely related to constructing independent finite combinatorial principles for formal theories.

In this paper the Fast Growing hierarchy is taken as basic. This hierarchy corresponds to a natural jump hierarchy of subrecursive function classes (the Kleene hierarchy) and therefore is less ad hoc. Besides, it is more robust than the others, in particular, it is possible to give a formulation of the Fast Growing hierarchy which is independent of the fundamental sequences assignments (see Sect. 4 for the details). Whenever we speak about Π^0_2-ordinals of theories, we always mean the ordinals measured by this hierarchy.

Proof-theoretic Π^0_2-analysis deals with independent principles of complexity Π^0_2 (sentences, expressing the totality of fast-growing functions), but it fails to distinguish between the theories only different in true Π^0_1-axioms. However, the most prominent independent principle—Gödel's consistency assertion $\mathrm{Con}(T)$ for an axiom system T—has logical complexity Π^0_1. So, for example, $|\mathrm{PA}+\mathrm{Con}(\mathrm{PA})|_{\Pi^0_2} = |\mathrm{PA}|_{\Pi^0_2} = \epsilon_0$. In general, theories having the same Π^0_2-ordinal can be of quite different consistency strength.

Historically, there have been proposals to define proof-theoretic ordinals relevant to logical complexity Π^0_1. This level of logical complexity is characteristic for the "consistency strength" of theories and thus plays a role, e.g., in connection with

[4]It is a long standing open question, whether a natural ordinal notation system can be canonically chosen for sufficiently large constructive ordinals. It has to be noted, however, that the standard proof-theoretic methods, in practical cases, usually allow to define natural ordinal notation systems for suitable initial segments of the constructive ordinals, that is, they simultaneously allow for Π^1_1- and Π^0_2-analyses of a theory, whenever they work. Pohlers [27] calls this property *profoundness* of the ordinal analysis.

Hilbert's program. On the other hand, an independent interest in Π^0_1-analysis is its relationship with the concept of *relative interpretability* of formal theories. By the results of Orey, Feferman and Hájek, this notion (for large classes of theories) is equivalent to Π^0_1-conservativity. The proposals to define general notions of proof-theoretic Π^0_1-ordinals, however, generally fell victim to just criticism, see [18]. To refresh the reader's memory, we discuss one such proposal below.

Indoctrinated by Hilbert's program, Gentzen formulated his ordinal analysis of Peano arithmetic as a proof of consistency of PA by transfinite induction up to ϵ_0. Accordingly, a naive attempt at generalization was to define the Π^0_1-ordinal of a system T as the order type of the shortest primitive recursive well-ordering $\prec$ such that the corresponding scheme of transfinite induction $TI(\prec)$ proves $\mathrm{Con}(T)$.

This definition is inadequate for several reasons. The first objection is that the formula $\mathrm{Con}(T)$ may not be canonical, that is, it really depends on the chosen *provability predicate* for T rather than T itself. Feferman [8] gave examples of Σ_1-provability predicates externally numerating PA and satisfying Löb's derivability conditions such that the corresponding consistency assertions are not PA-provably equivalent. In Appendix 2 we consider another example of this sort, for which the two provability predicates correspond to sufficiently *natural* proof systems axiomatizing PA. This indicates that the intended Π^0_1-ordinal of a theory T can, in fact, be a function of its *provability predicate* (and possibly some additional data), rather than just of the set of axioms of T taken externally.[5] Two possible ways to avoid this problem are: (1) to restrict the attention to specific natural theories, for which the canonical provability predicates are known; (2) to stipulate that theories always come together with their own fixed provability predicates. In other words, if two deductively equivalent axiom systems are formalized with different provability predicates, they should be considered as different. As remarked above, the second option appears to be better than the first, and we stick to it in this paper.

The second objection is that the primitive recursive well-ordering may be sufficiently pathological, and then $TI(\prec)$ can already prove $\mathrm{Con}(T)$ for some well-ordering $\prec$ of type ω (as shown by Kreisel). This problem can be avoided, if we only consider *natural* primitive recursive well-orderings, which are known for certain initial segments of the constructive ordinals. This would make the definition work at least for certain classes of theories, whose ordinals are not too large. Notice that essentially the same problem appears in the definition of the proof-theoretic Π^0_2-ordinal described above.

The third objection is that, although $\mathrm{Con}(T)$ is a Π^0_1-formula, the logical complexity of the schema $TI(\prec)$ is certainly higher. Kreisel noticed that the formulation of Gentzen's result would be more informative, if one restricts the complexity of transfinite induction formulas to primitive recursive, or open, formulas (we denote this schema $TI_{p.r.}(\prec)$). That is, Gentzen's result can be recast as a reduction of ω-induction of arbitrary arithmetical complexity to open transfinite induction up to ϵ_0.

---

[5]This is also typical for the other attempts to define proof-theoretic ordinals "from above" (cf. Appendix 2 for a discussion).

This formulation allows to rigorously attribute to, say, PA the natural ordinal (notation system up to) ϵ_0. However, for other theories T this approach is not yet fully satisfactory, for it is easy to observe that $TI_{p.r.}(\prec)$ has logical complexity Π^0_2, which is higher than Π^0_1. So, the definition of $|T|_{\Pi^0_1}$ as the infimum of order types of natural primitive recursive well-orderings $\prec$ such that

$$\mathrm{PRA} + TI_{p.r.}(\prec) \vdash \mathrm{Con}(T),$$

in fact, reduces a Π^0_1-principle to a Π^0_2-principle. The opposite reduction, however, is not possible. Thus, the ordinals obtained are not necessarily 'the right ones'. For example, in this sense the ordinal of PA + Con(PA) happens to be the same number ϵ_0, whereas any decent Π^0_1-analysis should separate the system from PA. One can attempt to push down the complexity of $TI_{p.r.}(\prec)$ by formulating it as a transfinite induction *rule* and disallowing nested applications of the rule, but in the end this would look less natural than the approach proposed in this paper.

Proof-Theoretic Analysis by Iterated Reflection The aim of this paper is to present another approach to proof-theoretic Π^0_1- and, in general, Π^0_n-analysis for any $n \geq 1$. The treatment of arbitrary n is not substantially different from the treatment of $n = 1$. For $n = 2$ our definition is shown to agree with the usual Π^0_2-analysis w.r.t. the Fast Growing hierarchy.[6] The apparent advantage of the method is that for the 'problematic' cases, such as PA + Con(PA), one obtains meaningful ordinal assignments. For example, we will see that $|\mathrm{PA} + \mathrm{Con(PA)}|_{\Pi^0_1} = \epsilon_0 \cdot 2$, which is well above the Π^0_1- and Π^0_2-ordinal ϵ_0 of PA, as expected.

A basic idea of the proof-theoretic Π^0_n-analysis is that of *conservative approximation* of a given theory T by formulas of complexity Π^0_n whose behavior is well-understood. Many properties of T, e.g., its class of p.t.c.f., can be learned from the known properties of the conservative approximations. As suitable approximations we take progressions of transfinitely iterated reflection principles (of relevant logical complexity). In particular, progressions of iterated consistency assertions, which are equivalent to iterated Π^0_1-reflection principles, provide suitable approximations of complexity Π^0_1.

The choice of the reflection formulas as the approximating ones has the following two advantages. First of all, the hierarchies of reflection principles are natural analogs of the jump hierarchies in recursion theory (this analogy is made more precise in Sect. 4 of this paper). So, in a sense, they are more elementary than the other candidate schemata, such as transfinite induction. Second, and more important, they allow for a convenient *calculus*. That is, the proof-theoretic ordinals for many theories can be determined by rather direct calculations, once some basic rules of handling iterated reflection principles are established. The key tool for this kind of calculations is Schmerl's formula [32], which is generalized and provided a new proof in this paper.

[6] This can be considered as an evidence supporting our definition for the other n.

The idea of using iterated reflection principles for the classification of axiomatic systems goes back to the old works of Turing [37] and Feferman [9]. Given a base theory T, one constructs a transfinite sequence of extensions of T by iteratedly adding formalized consistency statements, roughly, according to the following clauses:

(T1) $T_0 = T$;
(T2) $T_{\alpha+1} = T_\alpha + \mathrm{Con}(T_\alpha)$;
(T3) $T_\alpha = \bigcup_{\beta<\alpha} T_\beta$, for α a limit ordinal.

By Gödel's Incompleteness Theorem, whenever the initial theory T is sound,[7] the theories T_α form a strictly increasing transfinite sequence of sound Π^0_1-axiomatized extensions of T. Choosing for T some reasonable minimal fragment of arithmetic (in this paper we work over the elementary arithmetic EA) this sequence can be used to associate an ordinal $|U|_{\Pi^0_1}$ to any theory U extending EA as follows:

$$|U|_{\Pi^0_1} := \sup\{\alpha : \mathrm{EA}_\alpha \subseteq U\}.$$

This definition provides interesting information only for those theories U which can be well approximated by the sequence EA_α. For such U one should be able to show that for $\alpha = |U|_{\Pi^0_1}$ the theory EA_α axiomatizes *all* arithmetical Π^0_1-consequences of U, that is,

$$U \equiv_{\Pi_1} \mathrm{EA}_\alpha. \tag{1}$$

(Here and below $T \equiv_{\Pi_n} U$ means that the theories T and U prove the same Π_n-sentences.) Thus, (1) can be viewed as an exact reduction of U to a purely Π^0_1-axiomatized theory EA_α, and in this sense $|U|_{\Pi^0_1}$ is called the proof-theoretic Π^0_1-ordinal of U. Theories U satisfying equivalence (1) are called Π^0_1-*regular*. Verifiability of (1) within, say, EA implies

$$\mathrm{EA} \vdash \mathrm{Con}(U) \leftrightarrow \mathrm{Con}(\mathrm{EA}_\alpha),$$

and thus, $|U|_{\Pi^0_1}$ can also be thought of as the ordinal measuring the consistency strength of the theory U.

The program as described above, however, encounters several technical difficulties. One familiar difficulty is the fact that the clauses (T1)–(T3) do not uniquely define the sequence of theories T_α, that is, the theory T_α depends on the formal representation of the ordinal α within arithmetic rather than on the ordinal itself.

For the analysis of this problem Feferman [9] considered families of theories of the form $(T_c)_{c\in\mathcal{O}}$ satisfying (T1)–(T3) along every path within $\mathcal{O}$, where $\mathcal{O}$ is Kleene's universal system of ordinal notation. Using an idea of Turing, he showed

[7] That is, if all theorems of T hold in the standard model of arithmetic.

that every true Π^0_1-sentence is provable in T_c for a suitable ordinal notation $c \in \mathcal{O}$ with $|c| = \omega + 1$. It follows that there are two ordinal notations $a, b \in \mathcal{O}$ with $|a| = |b| = \omega + 1$ such that T_a proves $\text{Con}(T_b)$, and this observation seems to break down the program of associating ordinals to theories as described above, at least in the general case.

However, a possibility remains that for *natural* (mathematically meaningful) theories U one can exhaust all Π^0_1-consequences of U using only specific *natural* ordinal notations, and a careful choice of such notations should yield proper ordinal bounds. This idea has been developed in the work of Schmerl [32], who showed among other things that for natural ordinal notations

$$\text{PA} \equiv_{\Pi_1} \text{PRA}_{\epsilon_0}.$$

This essentially means that $|\text{PA}|_{\Pi^0_1} = \epsilon_0$, which coincides with the ordinal associated to PA through other proof-theoretic methods.

The significant work of Schmerl, however, attracted less attention than it, in our opinion, deserved. Partially this could be explained by a rather special character of the results, as they were stated in his paper. At present, 20 years later, thanks to the development of provability logic and formal arithmetic, we know much more about the structure of the fragments of PA, as well as about the properties of provability predicates. One of the goals of this paper is to revise and put in the right context this work of Schmerl. We provide a simpler approach to defining and treating iterated reflection principles, which helps to overcome some technical problems and allows for further development of these methods.

Plan of the Paper In Sect. 3 we define progressions of iterated reflection principles and note some basic facts about them. This allows to rigorously define Π^0_n-ordinals of theories following the ideas presented in the introduction.

In Sect. 4 we relate, in a very general setup, the hierarchy of iterated Π^0_2-reflection principles and the Fast Growing hierarchy. This shows that our approach, for the particular case of logical complexity Π^0_2, agrees with the usual proof-theoretic Π^0_2-analysis and provides the expected kind of information about the classes of provably total computable functions. Proofs of some technical lemmata are postponed till Appendix 1.

Section 5 can be read essentially independently from the previous parts of the paper. It presents a new conservation result relating the uniform and local reflection schemata. In particular, it is shown that uniform Π_2-reflection principle is Σ_2-conservative over the local Σ_1-reflection principle. This yields as an immediate corollary the result in [15] on the relation between parametric and parameter-free induction schemata: $I\Sigma_n$ is Σ_{n+2}-conservative over $I\Sigma^-_n$. The results of that section also provide a clear proof of a particular case of Schmerl's formula, which already has some meaningful corollaries for fragments of PA. At the same time it serves as a basis for a generalization given in the further sections.

Section 6, aiming at a proof of Schmerl's formula, presents a few lemmata to the effect that some conservation results for noniterated reflection principles can be directly extended to iterated ones. In Sect. 7 we formulate and prove (a generalization of) Schmerl's formula. In particular, we formulate a general relationship between Π_1^0- and Π_2^0-ordinals of Π_2^0-regular theories: Π_1^0-ordinal of a theory is one ω-power higher than its Π_2^0-ordinal.

In Sect. 8 we apply the general methodology to calculate proof-theoretic ordinals of main fragments of PA, including forms of parameter-free induction and their combinations with the parametric ones. This section is mostly meant as an illustration of the possibilities of our techniques. In our opinion, the most interesting examples treated are Π_2^0-irregular theories such as parameter-free Π_1-induction schema $I\Pi_1^-$, PA $+$ Con(PA), and the like.

Appendix 1 supplies technical lemmata for the results of Sect. 4. Appendix 2 discusses an attempt to define proof-theoretic ordinals "from above", and the role of nonstandard provability predicates. In Appendix 3 ordinal Π_1^0-ordinals of the weak systems (not proving the totality of the superexponentiation function) are treated.

3 Constructing Iterated Reflection Principles

In defining iterated reflection principles we closely follow [3]. Our present approach is slightly more general, but the proofs of basic lemmas remain essentially the same, so we just fix the terminology and indicate some basic ideas.

Iterated Consistency Assertions We deal with first order theories formulated in a language containing that of arithmetic. Our basic system is Kalmar elementary arithmetic EA (or $I\Delta_0(\exp)$, cf. [14]). For convenience we assume that a symbol for the exponentiation function 2^x is explicitly present in the language of EA. EA^+ denotes the extension of EA by an axiom stating the totality of the superexponential function 2_x^x (or $I\Delta_0 +$ supexp). EA^+ is the minimal extension of EA where the cut-elimination theorem for first order logic is provable. Hence, it will often play the role of a natural metatheory for various arguments in this paper.

Elementary formulas are bounded formulas in the language of EA. A theory T is *elementary presented* if it is equipped with a *numeration*, that is, an elementary formula $\mathrm{Ax}_T(x)$ defining the set of axioms of T in the standard model of arithmetic.

By an *elementary linear ordering* $(D, \prec)$ we mean a pair of elementary formulas $x \in D$ and $x \prec y$ such that EA proves that the relation $\prec$ linearly orders the domain D. An elementary *well-ordering* is an elementary linear ordering, which is well-founded in the standard model.

Given an elementary linear ordering $(D, \prec)$, we use Greek variables α, β, γ, etc. to denote the elements of D (and the corresponding ordinals). Since D is elementary definable, these variables can also be used within EA.

An elementary formula $\mathrm{Ax}_T(\alpha, x)$ *numerates* a family of theories $(T_\alpha)_{\alpha \in D}$, if for each α the formula $\mathrm{Ax}_T(\bar{\alpha}, x)$ defines the set of axioms of T_α in the standard model. If such a formula Ax_T exists, the family $(T_\alpha)_{\alpha \in D}$ is called *uniformly elementary presented*.

From $\mathrm{Ax}_T(\alpha, x)$, as well as from a numeration of an individual theory, the (parametric) provability predicate $\Box_T(\alpha, x)$ and the consistency assertion $\mathrm{Con}(T_\alpha)$ are constructed in a standard way. Specifically, there is a canonical Σ^0_1-formula $P(X, x)$, with a set parameter X and a number parameter x, expressing the fact that x codes a formula logically provable from the set of (non-logical) axioms coded by X. Then $\Box_T(\alpha, x) := P(\{u : \mathrm{Ax}_T(\alpha, u)\}, x)$ and $\mathrm{Con}(T_\alpha) := \neg\Box_T(\alpha, \bot)$. Notice that $\Box_T(\alpha, x)$ is first order Σ_1, and $\mathrm{Ax}_T(\alpha, u)$ occurs in $\Box_T(\alpha, x)$ as a subformula [replacing the occurrences of the form $u \in X$ in $P(X, x)$].

As usual, we write $\Box_T(\alpha, \varphi)$ instead of $\Box_T(\alpha, \ulcorner\varphi\urcorner)$, and $\Box_T(\alpha, \varphi(\dot{x}))$ instead of $\Box_T(\alpha, \ulcorner\varphi(\dot{x})\urcorner)$. Here $\ulcorner\varphi(\dot{x})\urcorner$ denotes the standard elementary function (and the corresponding EA-definable term) that maps a number n to the code of the formula $\varphi(\bar{n})$.

Now we present progressions of iterated consistency assertions. Somewhat generalizing [3], we distinguish between explicit and implicit progressions. Both are defined by formalizing (in two different ways) the following variant of conditions (T1)–(T3): for all $\alpha \in D$,

$$T_\alpha \equiv T + \{\mathrm{Con}(T_\beta) : \beta \prec \alpha\}.$$

Suppose we are given an "initial" elementary presented theory T and an elementary linear ordering $(D, \prec)$. An elementary formula $\mathrm{Ax}_T(\alpha, x)$ *explicitly numerates* a progression based on iteration of consistency along $(D, \prec)$ if

$$\mathrm{EA} \vdash \mathrm{Ax}_T(\alpha, x) \leftrightarrow (\mathrm{Ax}_T(x) \vee \exists \beta \prec \alpha\; x = \ulcorner\mathrm{Con}(T_{\dot{\beta}})\urcorner). \tag{2}$$

A formula $\mathrm{Ax}_T(\alpha, x)$ *implicitly numerates* such a progression if

$$\mathrm{EA} \vdash \Box_T(\alpha, x) \leftrightarrow P(\{u : \mathrm{Ax}_T(u) \vee \exists \beta \prec \alpha\; u = \ulcorner\mathrm{Con}(T_{\dot{\beta}})\urcorner\}, x). \tag{3}$$

Obviously, every explicit numeration is implicit, but the converse is generally false. It is often technically more convenient to deal with implicit numerations, for it allows one to disregard the superfluous information about the exact axiomatization of theories T_α (see, e.g., our proof of Theorem 1).

An *explicit/implicit progression* based on iteration of consistency is any family of theories $(T_\alpha)_{\alpha \in D}$ presented by an explicit/implicit numeration. If $(D, \prec)$ is an elementary *well*-ordering and the initial theory T is Σ_1-sound, then any implicit progression based on iteration of consistency is a strictly increasing sequence of Σ_1-sound theories satisfying (T1)–(T3).[8]

Notice that the definition of explicit and implicit numerations is self-referential. This raises the questions about the existence and uniqueness of such progressions.

[8]This is essentially the only place in all the development below, where well-foundedness matters. Actually, for the progressions based on iteration of consistency, well-foundedness w.r.t. the Σ_2-definable subsets would be sufficient.

Lemma 1 (Existence) *For any elementary linear ordering $(D, \prec)$ and any initial theory T, there is an explicit progression based on iteration of consistency along $(D, \prec)$.*

Proof The definition (2) has the form of a fixed point equation. Indeed, the formula $\mathrm{Con}(T_\beta)$ is constructed effectively from $\mathrm{Ax}_T(\beta, x)$, essentially by replacing x by u and substituting the result into $\neg P(X, \bot)$ for $u \in X$. Hence, there is an elementary definable term con that outputs the Gödel number of $\mathrm{Con}(T_{\overline{\beta}})$ given the Gödel number of $\mathrm{Ax}_T(\overline{\beta}, x)$. Then Eq. (2) can be rewritten as follows:

$$\mathrm{EA} \vdash \mathrm{Ax}_T(\alpha, x) \leftrightarrow \big(\mathrm{Ax}_T(x) \vee \exists \beta \leq x\, (\beta \prec \alpha \wedge x = \mathrm{con}(\ulcorner \mathrm{Ax}_T(\dot{\beta}, x)\urcorner))\big). \quad (4)$$

Fixed point lemma guarantees that an elementary solution $\mathrm{Ax}_T(\alpha, x)$ exists. To see that the solution satisfies (2) it only has to be noted that, assuming the Gödel numbering we use is standard, provably in EA for any β,

$$\beta \leq \ulcorner \overline{\beta} \urcorner \leq \ulcorner \mathrm{Ax}_T(\overline{\beta}, x)\urcorner \leq \ulcorner \mathrm{Con}(T_{\overline{\beta}})\urcorner,$$

⊠

In the following, the words *progression based on iteration of consistency* will always refer to implicit progressions. We note the obvious *monotonicity* property of such progressions:

Lemma 2 $\mathrm{EA} \vdash \alpha \prec \beta \rightarrow (\Box_T(\alpha, x) \rightarrow \Box_T(\beta, x))$.

The next lemma shows that any progression based on iteration of consistency is uniquely defined by the initial theory and the elementary linear ordering.

Lemma 3 (Uniqueness) *Let U and V be elementary presented extensions of* EA, *$(D, \prec)$ an elementary linear ordering, $(U_\alpha)_{\alpha \in D}$ and $(V_\alpha)_{\alpha \in D}$ progressions based on iteration of consistency with the initial theories U and V, respectively. Then*

$$\mathrm{EA} \vdash \forall x\, (\Box_U(x) \leftrightarrow \Box_V(x)) \implies \mathrm{EA} \vdash \forall \alpha \forall x\, (\Box_U(\alpha, x) \leftrightarrow \Box_V(\alpha, x)).$$

The uniqueness property is a robust background for further treatment of recursive progressions. In particular, it allows one to consistently use combined expressions like $(T_\alpha)_\beta$ for the composition of progressions (along the same ordering). The proof of Lemma 3 employs a trick coming from the work of Schmerl [32], which will also be used for the other results below.

Lemma 4 (Reflexive Induction) *For any elementary linear ordering $(D, \prec)$, any theory T is closed under the following* reflexive induction *rule:*

$$\forall \alpha\, (\Box_T(\forall \beta \prec \dot{\alpha}\, A(\beta)) \rightarrow A(\alpha)) \vdash \forall \alpha\, A(\alpha).$$

Proof Assuming $T \vdash \forall\alpha\, (\Box_T(\forall\beta \prec \dot{\alpha}\, A(\beta)) \to A(\alpha))$ we derive:

$$\begin{aligned} T \vdash \Box_T \forall\alpha\, A(\alpha) &\to \forall\alpha\, \Box_T \forall\beta \prec \dot{\alpha}\, A(\beta) \\ &\to \forall\alpha\, A(\alpha). \end{aligned}$$

Löb's theorem for T then yields $T \vdash \forall\alpha\, A(\alpha)$, ⊠

Proof of Lemma 3 then easily follows by reflexive induction in EA. We refer the reader to [3] for the details. It is important to realize that, although the reflexive induction has formally nothing to do with the well-foundedness, it allows one to prove certain properties of progressions *as if* they were proved by transfinite induction, in agreement with the underlying intuition. This makes the uses of reflexive induction in this context quite simple and natural.

Remark 5 All the above results continue to hold, if one replaces EA by EA^+ or any other sound elementary presented extension of EA.

Iterated Reflection Principles Let T be an elementary presented theory. The *local* reflection principle for T is the schema

$$\mathrm{Rfn}(T): \qquad \mathrm{Prov}_T(\ulcorner\varphi\urcorner) \to \varphi, \quad \varphi \text{ a sentence.}$$

The *uniform* reflection principle is the schema

$$\mathrm{RFN}(T): \qquad \forall x\, (\mathrm{Prov}_T(\ulcorner\varphi(\dot{x})\urcorner) \to \varphi(x)), \quad \varphi(x) \text{ a formula.}$$

Partial reflection principles are obtained from the above schemata by imposing a restriction that φ belongs to one of the classes Γ of the arithmetical hierarchy (denoted $\mathrm{Rfn}_\Gamma(T)$ and $\mathrm{RFN}_\Gamma(T)$, respectively). See [3, 19, 35] for some basic information about reflection principles.

We shall also consider the following *metareflection rule*:

$$\mathrm{RR}_{\Pi_n}(T): \qquad \frac{\varphi}{\mathrm{RFN}_{\Pi_n}(T+\varphi)}.$$

We let Π_m-$\mathrm{RR}_{\Pi_n}(T)$ denote the above rule with the restriction that φ is a Π_m-sentence.

For $n \geq 1$, $\Pi_n(\mathbb{N})$ denotes the set of all true Π_n-sentences. $\mathrm{True}_{\Pi_n}(x)$ denotes a canonical truth definition for Π_n-sentences, that is, a Π_n-formula naturally defining the set of Gödel numbers of $\Pi_n(\mathbb{N})$-sentences in EA.

Let T be an elementary presented theory containing EA. The set of axioms of the theory $T + \Pi_n(\mathbb{N})$ can be defined, e.g., by the Π_n-formula $\mathrm{Ax}_T(x) \vee \mathrm{True}_{\Pi_n}(x)$. Then the formula

$$\Box_T^{\Pi_n}(x) := P(\{u : \mathrm{Ax}_T(u) \vee \mathrm{True}_{\Pi_n}(u)\}, x)$$

naturally represents the Σ_{n+1}-complete provability predicate for $T + \Pi_n(\mathbb{N})$, and $\mathrm{Con}^{\Pi_n}(T) := \neg\Box_T^{\Pi_n}\bot$ is the corresponding consistency assertion. (For $n = 0$ we stipulate that these formulas coincide with $\Box_T$ and $\mathrm{Con}(T)$, respectively.)

Relativized provability predicates $\Box_T^{\Pi_n}$, as well as the usual provability predicate $\Box_T$, satisfy Löb's derivability conditions. $\Box_T^{\Pi_n}$ is EA-provably Σ_{n+1}-complete, that is,

$$\mathrm{EA} \vdash \forall x\,(\sigma(x) \to \Box_T^{\Pi_n}\sigma(\dot{x})),$$

for any Σ_{n+1}-formula $\sigma(x)$. Besides, for $n \geq 0$ the following relationships are known (see [4]).

Lemma 6 *For any elementary presented theory T containing* EA, *the following schemata are equivalent over* EA*:*

(i) $\mathrm{Con}^{\Pi_n}(T)$;
(ii) $\mathrm{RFN}_{\Pi_{n+1}}(T)$;
(iii) $\mathrm{RFN}_{\Sigma_n}(T)$.

This shows that the uniform reflection principles are generalizations of the consistency assertions to higher levels of the arithmetical hierarchy. (Notice that the schema $\mathrm{RFN}_{\Pi_1}(T)$ is equivalent to the standard consistency assertion $\mathrm{Con}(T)$.)

Relativized local reflection principles are generally not equivalent to any of the previously considered schemata. They are defined as follows:

$$\mathrm{Rfn}_{\Sigma_m}^{\Pi_n}(T): \qquad \Box_T^{\Pi_n}\varphi \to \varphi, \quad \text{for } \varphi \in \Sigma_m,$$

and similarly for the local Π_m-reflection principle. Notice that the relativized analog of, say, $\mathrm{Rfn}_{\Sigma_m}(T)$ is actually $\mathrm{Rfn}_{\Sigma_{m+n}}^{\Pi_n}(T)$.

Progressions based on iteration of reflection principles are defined in an analogy with (2). If $\Phi(T)$ is any of the reflection schemata for T introduced above, then the progressions based on iteration of Φ along $(D, \prec)$ will be denoted $(\Phi(T)_\alpha)_{\alpha\in D}$. Thus, $T_\alpha \equiv \mathrm{Con}(T)_\alpha$; we also write $(T)_\alpha^n$ short for $\mathrm{RFN}_{\Pi_n}(T)_\alpha$.

Theories $\Phi(T)_\alpha$ are defined by formalizing the condition

$$\Phi(T)_\alpha \equiv T + \{\Phi(\Phi(T)_\beta) : \beta \prec \alpha\}.$$

This can be done as follows. Since the instances of the reflection principles are elementarily recognizable, with each of the above schemata Φ one can naturally associate an elementary formula Φ-code(e, x) expressing that *e is the code of a Σ_1-formula $\Box_U(v)$, and x is the code of an instance of $\Phi(U)$ formulated for $\Box_U$*. Then $\mathrm{Ax}_T(\alpha, x)$ is called an explicit numeration of a progression based on iteration of Φ, if

$$\mathrm{EA} \vdash \mathrm{Ax}_T(\alpha, x) \leftrightarrow (\mathrm{Ax}_T(x) \vee \exists \beta \prec \alpha\ \Phi\text{-code}(\ulcorner\Box_T(\dot{\beta}, v)\urcorner, x)),$$

and an implicit numeration, if

$$\mathrm{EA} \vdash \Box_T(\alpha, x) \leftrightarrow P(\{u : \mathrm{Ax}_T(u) \lor \exists \beta \prec \alpha\ \Phi\text{-code}(\ulcorner \Box_T(\dot{\beta}, v)\urcorner, u)\}, x).$$

Then the analogs of existence, monotonicity, and uniqueness lemmas hold for such progressions too, with similar proofs. We omit them.

Π^0_n-Ordinals Let an elementary well-ordering $(D, \prec)$ be fixed. All the definitions below are to be understood relative to this ordering. We define:

$$|T|_{\Pi^0_n} := \sup\{\alpha : (\mathrm{EA})^n_\alpha \subseteq T\}.$$

If the ordering $(D, \prec)$ is too short, that is, if for all $\alpha \in D$, $(\mathrm{EA})^n_\alpha \subseteq T$, we can set $|T|_{\Pi^0_n} := \infty$.

A theory T is *Π^0_n-regular*, if there is an $\alpha \in D$ such that

$$T \equiv_{\Pi_n} (\mathrm{EA})^n_\alpha. \tag{5}$$

Notice that Π^0_n-regular theories are Π^0_n-sound, because $(\mathrm{EA})^n_\alpha$ is. If the equivalence (5) is provable in a (meta)theory U, then T is called *U-provably Π^0_n-regular.* For $U \subseteq T$ in this case we have $\alpha = |T|_{\Pi^0_n}$, because the formalization of (5) implies

$$U \vdash \mathrm{RFN}_{\Pi_n}(T) \leftrightarrow \mathrm{RFN}_{\Pi_n}((\mathrm{EA})^n_\alpha),$$

and $T \nvdash \mathrm{RFN}_{\Pi_n}(T)$ then yields $(\mathrm{EA})^n_{\alpha+1} \nsubseteq T$.

Comparing this definition with the above discussion of proof-theoretic Π^0_1-ordinals we notice that it lacks the mentioned drawbacks (1) and (3).

Ad 1). Indeed, EA is a natural finitely axiomatizable theory, so it has a canonical provability predicate. Uniqueness theorem then guarantees that the progression $(\mathrm{EA})^n_\alpha$ for $\alpha \in D$ is uniquely defined. (There is no mentioning of the provability predicates *for* T in the definition of $|T|_{\Pi^0_n}$.)

Ad 2). As defined above, the Π^0_n-ordinal of a theory T is a function of a prefixed elementary well-ordering $(D, \prec)$. We shall see later that the analysis of natural theories requires imposing some additional natural structure on the well-ordering. (This situation is only slightly better than a restriction to concrete natural well-orderings.) As expected in view of the general problem of natural ordinal notations, at present we do not have an answer to the question what kind of structure is needed for the analysis of arbitrary theories. None of the existing definitions of proof-theoretic ordinals of logical complexity below Π^1_1 is free from this drawback, and this may even be unavoidable.

Ad 3). Equation (5) provides an exact reduction of a given Π^0_n-regular theory T to a purely Π^0_n-axiomatized theory $(\mathrm{EA})^n_\alpha$. This is the main advantage of the considered definition.

4 Iterated Π_2-Reflection and the Fast Growing Hierarchy

In this section we relate the hierarchies of iterated uniform Π_2-reflection principles and the hierarchies of fast growing functions. This shows that, under very general assumptions, the proof-theoretic analysis by iterated Π_2-reflection principles over EA provides essentially the same information as the usual Π_2^0-ordinal analysis. For the natural ordinal notation system up to ϵ_0 similar results can be deduced from the work of Sommer [36]. Our present approach is somewhat more general and also seems to be technically simpler, so we opted for an independent presentation.

Let $\mathcal{E}$ denote the class of elementary functions. For any set of functions $\mathcal{K}$, $\mathbf{C}(\mathcal{K})$ denotes the closure of $\mathcal{K} \cup \mathcal{E}$ under composition; $\mathbf{E}(\mathcal{K})$ denotes is the elementary closure of $\mathcal{K}$, that is, the closure of $\mathcal{K} \cup \mathcal{E}$ under composition and bounded recursion. If all the functions from $\mathcal{K}$ are monotone and have elementary graphs, then $\mathbf{C}(\mathcal{K}) = \mathbf{E}(\mathcal{K})$, see [2].

Let an elementary well-ordering $(D, \prec)$ be fixed. Throughout this section we assume that there is an element $0 \in D$ satisfying EA $\vdash \forall\alpha\,(0 = \alpha \vee 0 \prec \alpha)$.

A hierarchy of functions F_α for $\alpha \in D$ is defined recursively as follows:

$$F_\alpha(x) := \max\{2_x^x + 1\} \cup \{F_\beta^{(v)}(u) + 1 : \beta \prec \alpha,\ \beta, u, v \leq x\}. \qquad (*)$$

Since $(D, \prec)$ is well-founded, all F_α are well-defined. The functions F_α generate the hierarchy of function classes

$$\mathcal{F}_\alpha := \mathbf{E}(\{F_\beta : \beta \prec \alpha\}).$$

One easily verifies that for the initial elements $\alpha \in D$ the classes $\mathcal{F}_\alpha$ coincide with the classes of the familiar Grzegorczyk hierarchy: $\mathcal{F}_0 = \mathcal{E}$, $\mathcal{F}_1 = \mathcal{E}'$, ..., $\mathcal{F}_\omega$ = primitive recursive functions, ... The further classes are a natural extension of the Grzegorczyk hierarchy into the transfinite. Notice that this hierarchy is defined for an arbitrary (not necessarily natural) well-ordering and does not depend on the assignments of fundamental sequences.

A slight modification of this hierarchy has recently been proposed by Weiermann and studied in detail by Möllerfeld [22]. Building on some previous results, see [31] for an overview, he relates this hierarchy to some other natural hierarchies of function classes. Since our hierarchy has to be reasonably representable in EA, in some respects we need a sharper treatment than in [22].

Proofs of the following two lemmas will be given in the Appendix.

Lemma 7 *$F_\alpha(x) = y$ is an elementary relation of α, x, and y.*

Notice that a priori we only know that this relation is recursive. Let $F_\alpha(x) \simeq y$ be a natural elementary formula representing it.

Lemma 8 *The following properties are verifiable in* EA*:*

(i) $(x_1 \leq x_2 \wedge F_\alpha(x_1) \simeq y_1 \wedge F_\alpha(x_2) \simeq y_2) \to y_1 \leq y_2$;

(ii) $(\beta \prec \alpha \wedge F_\beta(x) \simeq y_1 \wedge F_\alpha(x) \simeq y_2) \to y_1 \leq y_2$*;*
(iii) A natural formalization of $(*)$ *(axioms (F1)–(F3) from Appendix 1).*

Let $F_\alpha\downarrow$ denote the formula $\forall x \exists y F_\alpha(x) \simeq y$, and let S_α denote the theory $\mathrm{EA} + \{F_{\bar\beta}\downarrow : \beta \prec \alpha\}$. Obviously, the theories $(S_\alpha)_{\alpha \in D}$ are uniformly elementary presented, e.g., one can define

$$\mathrm{Ax}_S(\alpha, x) :\leftrightarrow \mathrm{Ax}_{\mathrm{EA}}(x) \vee \exists \beta < x\, (\beta \prec \alpha \wedge x = \ulcorner \forall u \exists v\, F_{\dot\beta}(u) \simeq v \urcorner).$$

Our aim is to prove that $(S_\alpha)_{\alpha \in D}$ is deductively equivalent to the progression of iterated uniform Π_2-reflection principles over EA.

Notice that $S_0 \equiv \mathrm{EA}$, and S_α contains EA^+ for $\alpha \succ 0$.

Theorem 1 *Provably in* EA^+, $\forall \alpha\, S_\alpha \equiv (\mathrm{EA})^2_\alpha$.

As a corollary we obtain the following statement.

Corollary 9 *For all* $\alpha \in D$, $\mathcal{F}((\mathrm{EA})^2_\alpha) = \mathcal{F}_\alpha$.

Proof Obviously, symbols for all functions F_β for $\beta \prec \alpha$ can be introduced into the language of S_α. The corresponding definitional extension of S_α admits a purely universal axiomatization, because the graphs of all F_β are elementary. By Herbrand's theorem

$$\mathcal{F}(S_\alpha) = \mathbf{C}(\{F_\beta : \beta \prec \alpha\}),$$

which coincides with the class $\mathcal{F}_\alpha$, since all functions F_β are monotone and have elementary graphs. ⊠

Proof of Theorem 1 By the uniqueness lemma it is sufficient to establish within EA^+ that $(S_\alpha)_{\alpha \in D \setminus \{0\}}$ is an implicit progression based on iteration of uniform Π_2-reflection principles over EA^+. So, we show the following main

Lemma 10 *Provably in* EA^+,

$$\forall \alpha\, S_\alpha \equiv \mathrm{EA} + \{\mathrm{RFN}_{\Pi_2}(S_\beta) : \beta \prec \alpha\}.$$

Proof We formalize the proofs of the following two lemmas in EA^+. (Notice that the arguments are local, that is, they do not use any form of transfinite induction on α.) ⊠

Lemma 11 *Provably in* EA,

$$\forall \beta\, \mathrm{EA} + \mathrm{RFN}_{\Pi_2}(S_\beta) \vdash F_{\overline{\beta}}\downarrow.$$

Proof Let $F^{(u)}_\beta(x) \simeq y$ abbreviate

$$\exists s \in Seq\, [(s)_0 = x \wedge \forall i < u\, F_\beta((s)_i) \simeq (s)_{i+1} \wedge (s)_u = y].$$

From the assumption $\gamma \prec \overline{\beta}$ within $\mathrm{EA} + \mathrm{RFN}_{\Pi_2}(S_\beta)$ one can derive:

1. $\Box_{S_\beta} F_{\dot\gamma}\!\downarrow$ (by the definition of S_β)
2. $\forall u\, \Box_{S_\beta} \forall x \exists y\, F_{\dot\gamma}^{(\dot u)}(x) \simeq y$ (by elementary induction on u from 1.)
3. $\forall x, u \exists y F_\gamma^{(u)}(x) \simeq y$ (by $\mathrm{RFN}_{\Pi_2}(S_\beta)$ from 2.)

This shows that

$$\mathrm{EA} + \mathrm{RFN}_{\Pi_2}(S_\beta) \vdash \forall \gamma \prec \overline{\beta}\, \forall x, u \exists y F_\gamma^{(u)}(x) \simeq y. \tag{6}$$

On the other hand, by elementary induction on x one obtains

$$\mathrm{EA} \vdash \forall x\, \exists \gamma_0 \leq x\, \forall \gamma \leq x\, (\gamma \prec \beta \rightarrow \gamma \preceq \gamma_0).$$

Reasoning inside EA, from (6) for this particular γ_0 one obtains a y such that $F_{\gamma_0}^{(x)}(x) \simeq y$. We claim that y is as required. By (i) and (ii) of Lemma 8 and by (6), for all $u, \gamma \leq x$ such that $\gamma \prec \beta$ one has $\exists z \leq y\, F_\gamma^{(u)}(x) \simeq z$. By property (F3) from Appendix 1 we then obtain $F_\beta(x) \simeq y$. ⊠

Lemma 12 *Provably in* EA^+,

$$\forall \beta \prec \alpha\; S_\alpha \vdash \mathrm{RFN}_{\Pi_2}(S_\beta).$$

Proof Let S_β^* denote the definitional extension of S_β by function symbols for all the functions $\{F_\gamma : \gamma \prec \beta\}$. Clearly, S_β^* is a conservative extension of S_β, moreover, this can be shown in EA^+ uniformly in β. Thus, it is sufficient to prove the lemma for the theories S_β^*.

First of all, by a standard result (cf. [12] and [2], Proposition 5.11) based on the monotonicity of the functions F_β we obtain that S_α^* proves induction for bounded formulas in the extended language (and this is, obviously, formalizable).

Second, since Herbrand's theorem is formalizable in EA^+, we have that, for any elementary formula $\sigma(y, x)$,

$$S_\beta^* \vdash \exists y \sigma(y, \bar{n}) \Rightarrow S_\beta^* \vdash \sigma(t, \bar{n}),$$

for some closed term t in S_β^*. So, it is sufficient to establish in S_α^* the reflection principle for S_β^* for open formulas (in the language of S_β^*). This proof is very similar to the proof of Theorem 2 in [2], so we only sketch it.

The proof involves two main ingredients. First, we need a natural evaluation function for terms in the language of S_β^*, that is, a function $\mathrm{eval}_\beta(e, x)$ satisfying

$$\mathrm{eval}_\beta(\ulcorner t \urcorner, \langle \vec{x} \rangle) = t(\vec{x}),$$

for all such terms $t(\vec{x})$.

It is not difficult to see that $\mathrm{eval}_\beta \in \mathcal{F}_\alpha$ and is naturally definable in S^*_α. Essentially, it is sufficient to check that eval_β is bounded by an elementary function in F_β. Indeed, by monotonicity of all functions F_γ, every term $t(x)$ is bounded by some iterate of a function F_γ, for a suitable $\gamma \prec \beta$, which means that

$$\mathrm{eval}_\beta(e,x) \leq F^{(n(e))}_{\gamma(e)}(x),$$

where $\gamma(e)$ and $n(e)$ are elementary functions. For the natural coding of terms we can additionally assume that $n(e) < e$, for all e. Then we can estimate the evaluation function as follows:

$$F^{(n(e))}_{\gamma(e)}(x) \leq F^{(\max(e,x))}_{\gamma(e)}(\max(e,x)) \leq F_\beta(\max(e,x,\gamma(e))).$$

Therefore, eval_β is elementary in F_β and belongs to $\mathcal{F}_\alpha$. As a corollary we obtain that S^*_α proves $\Delta_0(\mathrm{eval}_\beta)$-induction.

The second ingredient is a proof in S^*_α of the reflection principle for S^*_β for open formulas. This is done straightforwardly by the induction on the length of a cut-free S^*_β-derivation using eval_β. This induction has $\Delta_0(\mathrm{eval}_\beta)$-form, hence it is formalizable in S_α. The whole argument (involving cut-elimination) is then formalizable in EA^+. Details can be found in [2], Sect. 7. This completes the proof of Lemma 12 and Theorem 1. ⊠

Remark 13 One can show that the hierarchy $(T)^2_\alpha$, for a given Π_2-axiomatized sound extension T of EA, corresponds to the so-called *Kleene hierarchy* over the class $\mathcal{F}(T)$. The Kleene hierarchy is essentially obtained by adding a canonical universal function for the previous class at successor stages and taking the unions at limit stages. For $T = \mathrm{EA}$ a standard result (elaborated for the kind of hierarchies considered here in [22]) shows that the Kleene hierarchy coincides with the classes $\mathcal{F}_\alpha$ introduced above.

5 Uniform Reflection Is Not Much Stronger Than Local Reflection

In this section we establish a relationship between the uniform reflection schema and a suitable version of reflection rule. This allows us to prove that for every elementary presented theory T containing EA the theory $\mathrm{EA} + \mathrm{RFN}_{\Sigma_1}(T)$ is Σ_2-conservative over $\mathrm{EA} + \mathrm{Rfn}_{\Sigma_1}(T)$. Since the arithmetical complexity of the schema $\mathrm{Rfn}_{\Sigma_1}(T)$ is $\mathcal{B}(\Sigma_1)$, a somewhat unexpected aspect of this result is that the Σ_2-consequences of $\mathrm{RFN}_{\Sigma_1}(T)$ can be axiomatized by a set of formulas of lower arithmetical complexity.

A relativization of this theorem allows us to obtain an alternative proof of the results of Kaye et al. [15] on the partial conservativity of the parametric induction schemata over the parameter-free ones. At the same time, we also obtain for free the well-known result of Parsons on the partial conservativity of induction schemata over the induction rules over EA. Further, this result leads to a more general version of Schmerl's theorem [32], which plays an important role in the present approach to proof-theoretic analysis.

Theorem 2 *Let T be an elementary presented theory containing* EA, *and let U be a Π_{n+1}-axiomatized extension of* EA *($n \geq 1$). Then $U + \mathrm{RFN}_{\Sigma_n}(T)$ is Π_n-conservative over $U + \Pi_n\text{-}RR_{\Pi_n}(T)$.*

Proof For the proof of this theorem it is convenient to give a sequential formulation of Π_n-$\mathrm{RR}_{\Pi_n}(T)$. Let Π_n-$\mathrm{RR}^G_{\Pi_n}(T)$ denote the following inference rule in the formalism of Tait calculus:

$$\frac{\Gamma, \varphi(s)}{\Gamma, \neg \mathrm{Prf}_T(t, \ulcorner \neg\varphi(\dot{s})\urcorner)},$$

for all terms t, s and formulas $\varphi(a) \in \Pi_n$, where $\ulcorner \psi(\dot{s})\urcorner$ denotes the result of substitution of a term s in the term $\ulcorner \psi(\dot{a})\urcorner$. Π_n-$\mathrm{RR}^g_{\Pi_n}(T)$ will denote the same rule with the restriction that Γ consists of Π_n-formulas.

The following lemma states that the terms $\ulcorner \psi(\dot{s})\urcorner$ have a natural commutation property.

Lemma 14 *For any term $s(\vec{x})$, where the list $\vec{x}$ exhausts all the variables of s, and any formula $\phi(a)$ (where s is substitutable in ϕ for a) there holds:*

$$\mathrm{EA} \vdash \forall \vec{x}\, \big(\Box_T \phi(s(\dot{\vec{x}})) \leftrightarrow \Box_T \phi(\dot{s})\big).$$

Proof Obviously,

$$\mathrm{EA} \vdash s(\vec{x}) = y \to (\varphi(s(\vec{x})) \leftrightarrow \varphi(y)).$$

Hence, by the provable Σ_1-completeness and Löb's conditions

$$\begin{aligned} \mathrm{EA} \vdash s(\vec{x}) = y &\to \Box_T(s(\dot{\vec{x}}) = \dot{y}) \\ &\to \Box_T\big(\varphi(s(\dot{\vec{x}})) \leftrightarrow \varphi(\dot{y})\big) \\ &\to \big(\Box_T\varphi(s(\dot{\vec{x}})) \leftrightarrow \Box_T\varphi(\dot{y})\big) \end{aligned} \tag{7}$$

On the other hand, by the definition of $\ulcorner \varphi(\dot{s})\urcorner$,

$$\mathrm{EA} \vdash s(\vec{x}) = y \to \big(\Box_T\varphi(\dot{s}) \leftrightarrow \Box_T\varphi(\dot{y})\big),$$

which together with (7) yields the claim. ⊠

Under the standard interpretation of a sequent as the disjunction of the formulas occurring in it the following lemma holds.

Lemma 15 *The rule Π_n-$RR^G_{\Pi_n}(T)$ is equivalent to the schema* $\mathrm{RFN}_{\Sigma_n}(T)$.

Proof For a reduction of Π_n-$\mathrm{RR}^G_{\Pi_n}(T)$ to $\mathrm{RFN}_{\Sigma_n}(T)$ consider an arbitrary Σ_n-formula $\sigma(a)$. In the formalism of Tait calculus $\neg\neg\sigma$ happens to be graphically the same as σ, so we can derive:

$$\dfrac{\dfrac{\dfrac{\sigma(a), \neg\sigma(a)}{\sigma(a), \neg\mathrm{Prf}_T(b, \ulcorner\sigma(\dot{a})\urcorner)}}{\sigma(a), \forall y \neg\mathrm{Prf}_T(y, \ulcorner\sigma(\dot{a})\urcorner)}}{\forall x\, (\Box_T\sigma(\dot{x}) \to \sigma(x)).}$$

For a reduction of Π_n-$\mathrm{RR}^G_{\Pi_n}(T)$ to $\mathrm{RFN}_{\Sigma_n}(T)$ notice that for any terms s, t and any Π_n-formula φ we have:

$$\begin{aligned} \mathrm{EA} + \mathrm{RFN}_{\Sigma_n}(T) \vdash \varphi(s) &\to \neg\Box_T\neg\varphi(\dot{s}) \\ &\to \forall y \neg\mathrm{Prf}_T(y, \ulcorner\neg\varphi(\dot{s})\urcorner) \\ &\to \neg\mathrm{Prf}_T(t, \ulcorner\neg\varphi(\dot{s})\urcorner), \end{aligned}$$

⊠

Let $\Diamond_T^{\Pi_n}\varphi$ denote $\neg\Box_T^{\Pi_n}\neg\varphi$. Notice that for any φ, $\Diamond_T^{\Pi_n}\varphi$ is EA-equivalent to $\mathrm{RFN}_{\Sigma_n}(T + \varphi)$.

Lemma 16 *For $n \geq 1$, the following rules are equivalent (and even congruent in the terminology of Beklemishev [2]):*

(i) Π_n-$RR_{\Pi_n}(T)$,
(ii) Π_n-$RR^g_{\Pi_n}(T)$,
(iii) $\dfrac{\Gamma, \varphi(s)}{\Gamma, \Diamond_T^{\Pi_{n-1}}\varphi(\dot{s})}$, *for* $\Gamma \cup \{\varphi\} \subseteq \Pi_n$.

Proof Reduction of (ii) to (iii) is obvious, because

$$\begin{aligned} \mathrm{EA} \vdash \Diamond_T^{\Pi_{n-1}}\varphi(\dot{s}) &\to \Diamond_T\varphi(\dot{s}) \\ &\to \neg\mathrm{Prf}_T(t, \ulcorner\neg\varphi(\dot{s})\urcorner). \end{aligned}$$

For a reduction of (iii) to (i) we reason as follows. Let $\vec{x}$ denote the list of all the free variables in Γ and s. Notice that, if $\Gamma \subseteq \Pi_n$, then the universal closure of $\bigvee\Gamma \vee \varphi(s) \in \Pi_n$ and we can construct the following derivation:

1. $\bigvee\Gamma(\vec{x}) \vee \varphi(s(\vec{x}))$
2. $\forall\vec{x}\left(\bigvee\Gamma(\vec{x}) \vee \varphi(s(\vec{x}))\right)$
3. $\Diamond_T^{\Pi_{n-1}}\forall\vec{x}\left(\bigvee\Gamma(\vec{x}) \vee \varphi(s(\vec{x}))\right)$ (by Π_n-$\mathrm{RR}_{\Pi_n}(T)$)
4. $\forall\vec{x}\ \Diamond_T^{\Pi_{n-1}}\left(\bigvee\Gamma(\dot{\vec{x}}) \vee \varphi(s(\dot{\vec{x}}))\right)$ (by Löb's conditions from 3)
5. $\Diamond_T^{\Pi_{n-1}}(\bigvee\Gamma(\dot{\vec{x}})) \to \bigvee\Gamma(\vec{x})$ (by provable Σ_n-completeness of $\Box_T^{\Pi_{n-1}}$)

6. $\bigvee \Gamma(\vec{x}) \vee \Diamond_T^{\Pi_{n-1}} \varphi(\dot{s(\vec{x})})$ (by 4, 5 and Löb's conditions)
7. $\bigvee \Gamma(\vec{x}) \vee \Diamond_T^{\Pi_{n-1}} \varphi(\dot{s})$ (by Lemma 14)

In order to reduce (i) to (ii), for any Π_n-formula φ and any Σ_{n-1}-formula $\sigma(x)$ we reason as follows:

$$\dfrac{\dfrac{\dfrac{\dfrac{\neg\sigma(x), \sigma(x) \quad \varphi}{\varphi \wedge \neg\sigma(x), \sigma(x)}}{\Diamond_T(\varphi \wedge \neg\sigma(\dot{x})), \sigma(x)}}{\neg\Box_{T+\varphi}\sigma(\dot{x}), \sigma(x)}}{\forall x\,(\Box_{T+\varphi}\sigma(\dot{x}) \to \sigma(x))} \quad \text{(by } \Pi_n\text{-RR}^g_{\Pi_n}(T) \text{ and logic)}$$

This gives the required proof of an arbitrary instance of $\mathrm{RFN}_{\Sigma_{n-1}}(T + \varphi)$ from a derivation of φ. ⊠

Resuming the proof of Theorem 2 we show that the standard cut-elimination procedure can be considered as a reduction of $\mathrm{RFN}_{\Sigma_n}(T)$ to Π_n-$\mathrm{RR}^g_{\Pi_n}(T)$. Consider a cut-free derivation of a sequent of the form

$$\neg U, \neg\mathrm{RFN}_{\Sigma_n}(T), \Pi, \tag{8}$$

where Π is a set of Π_n-formulas, $\neg U$ is a finite set of negated axioms of U, and $\neg\mathrm{RFN}_{\Sigma_n}(T)$ is a finite set of negated instances of $\mathrm{RFN}_{\Sigma_n}(T)$ of the form

$$\exists y \exists x\, [\mathrm{Prf}_T(y, \ulcorner\neg\varphi(\dot{x})\urcorner) \wedge \varphi(x)]$$

for some $\varphi(x) \in \Pi_n$. Let $R_\varphi(x, y)$ denote the formula in square brackets. We can also assume that the axioms of U have the form $\forall x_1 \ldots \forall x_m \neg A(x_1, \ldots, x_m)$ for some Π_n-formulas $A(\vec{x})$.

By the subformula property, any formula occurring in the derivation of a sequent Γ of the form (8) either (a) is a Π_n-formula, or (b) has the form $\neg\mathrm{RFN}_{\Sigma_n}(T)$, $\exists x R_\varphi(t, x)$ or $R_\varphi(t, s)$, for appropriate terms s, t, or (c) has the form

$$\exists x_{i+1} \ldots \exists x_m A(t_1, \ldots, t_i, x_{i+1}, \ldots, x_m)$$

for some $i < n$ and terms $t_1, \ldots, t_i$. Let Γ^- denotes the result of deleting all formulas of types (b) and (c) from Γ.

Lemma 17 *If a sequent Γ of the form (8) is cut-free provable, then Γ^- is provable from the axioms of U (considered as initial sequents) using the logical rules, including Cut, and the rule Π_n-$RR^g_{\Pi_n}(T)$.*

Proof goes by induction on the height of the derivation d of Γ. It is sufficient to consider the cases that a formula of the form (b) or (c) is introduced by the last inference in d. Besides, it is sufficient to only consider the formulas of the form

$R_\varphi(t,s)$ and $\exists x_m A(t_1,\dots,t_{m-1},x_m)$, because in all other cases after the application of $(\cdot)^-$ the premise and the conclusion of the rule coincide.

So, assume that the derivation d has the form

$$\frac{\mathrm{Prf}_T(t,\ulcorner\neg\varphi(\dot{s})\urcorner),\Delta \quad \varphi(s),\Delta}{R_\varphi(t,s),\Delta}\ (\wedge)$$

where $\varphi \in \Pi_n$. Then by the induction hypothesis we obtain $\Pi_n\text{-RR}^g_{\Pi_n}(T)$-derivations of the sequents

$$\mathrm{Prf}_T(t,\ulcorner\neg\varphi(\dot{s})\urcorner),\Delta^- \tag{9}$$

and

$$\varphi(s),\Delta^-. \tag{10}$$

Since Δ^- consists of Π_n-formulas, the rule $\Pi_n\text{-RR}^g_{\Pi_n}(T)$ is applicable to (10), and we obtain a derivation of

$$\neg\mathrm{Prf}_T(t,\ulcorner\neg\varphi(\dot{s})\urcorner),\Delta^-.$$

Applying the Cut-rule with the sequent (9) we obtain the required derivation of Δ^-. If the last inference in d has the form

$$\frac{A(t_1,\dots,t_{m-1},t_m),\Delta}{\exists x_m A(t_1,\dots,t_{m-1},x_m),\Delta}\ (\exists)$$

then by the induction hypothesis we obtain a $\Pi_n\text{-RR}^g_{\Pi_n}(T)$-derivation of

$$A(t_1,\dots,t_{m-1},t_m),\Delta^-.$$

Then a derivation of

$$\exists x_1\dots\exists x_m A(x_1,\dots,x_m),\Delta^-$$

is obtained by several applications of the rule $(\exists)$. The sequent Δ^- is now derived applying the Cut-rule with the axiom sequent $\forall x_1\dots\forall x_m\neg A(x_1,\dots,x_m)$. ⊠

Theorem 2 now follows immediately from Lemmas 16 and 17. ⊠

Proposition 18 *If T is a Π_{n+1}-axiomatized extension of* EA, *then $T+\mathrm{RFN}_{\Sigma_n}(T)$ is Π_n-conservative over $(T)^n_\omega$.*

This statement has been obtained (by other methods) for $T=\mathrm{PRA}$ in [32], and for $n=1$ and $T=\mathrm{EA}$ in [1].

Proof It is sufficient to notice that $(T)^n_\omega$ is closed under the rule $\Pi_n\text{-RR}_{\Pi_n}(T)$. ⊠

Proposition 19 *If U is a Π_{n+1}-axiomatized extension of* EA, *then $U + \mathrm{RFN}_{\Sigma_n}(T)$ is a Σ_{n+1}-conservative extension of $U + \mathrm{Rfn}_{\Sigma_n}^{\Pi_{n-1}}(T)$. In particular, if U is Π_2-axiomatized, then $U + \mathrm{RFN}_{\Sigma_1}(T)$ is Σ_2-conservative over $U + \mathrm{Rfn}_{\Sigma_1}(T)$.*

Proof Assume $U + \mathrm{RFN}_{\Sigma_n}(T) \vdash \sigma$ for a sentence $\sigma \in \Sigma_{n+1}$, then

$$U + \neg\sigma + \mathrm{RFN}_{\Sigma_n}(T) \vdash \bot,$$

and by Theorem 2

$$U + \neg\sigma + \Pi_n\text{-}\mathrm{RR}_{\Pi_n}(T) \vdash \bot.$$

Notice that the rule Π_n-$\mathrm{RR}_{\Pi_n}(T)$ is obviously reducible to the schema $\mathrm{Rfn}_{\Sigma_n}^{\Pi_{n-1}}(T)$. Hence, we obtain

$$U + \neg\sigma + \mathrm{Rfn}_{\Sigma_n}^{\Pi_{n-1}}(T) \vdash \bot,$$

and by deduction theorem

$$U + \mathrm{Rfn}_{\Sigma_n}^{\Pi_{n-1}}(T) \vdash \sigma,$$

⊠

Thus, from our characterization of parameter-free induction schemata (cf. [2] or Sect. 8 of this paper) we directly obtain an interesting conservation result due to Kaye et al. [15] (by a model-theoretic proof).

Corollary 20 *For $n \geq 1$, $I\Sigma_n$ is a Σ_{n+2}-conservative extension of $I\Sigma_n^-$.*

From Proposition 18 and the characterization of induction rules in terms of reflection principles (cf. [2]) one can also obtain the following theorem of Parsons [26].

Corollary 21 *For $n \geq 1$, $I\Sigma_n$ is Π_{n+1}-conservative over $I\Sigma_n^R$.*

Proposition 18 also implies that $I\Sigma_1$ is Π_2-conservative over $(\mathrm{EA})^2_\omega$. Together with Corollary 9 this implies that $\mathcal{F}(I\Sigma_1)$ coincides with $\mathcal{F}_\omega$, that is, with the class of primitive recursive functions. This well-known result was originally established by Parsons [25], G. Takeuti and G. Mints by other methods.

6 Extending Conservation Results to Iterated Reflection Principles

The definition of progressions based on iteration of reflection principles allows one to directly "extend by continuity" some basic conservation results for reflection principles to their transfinite iterations. In particular, this leads to a useful generalization of Schmerl's fine structure theorem, which will be discussed in the next section. Here we just state a number of such easy extension results. Throughout this section we fix an elementary well-ordering $(D, \prec)$ and an initial elementary presented theory T.

The following proposition generalizes Statement 3 of Theorem 1 in [3].

Proposition 22 *The following statements are provable in* EA*:*

(i) $\forall\alpha\ \mathrm{Rfn}_{\Sigma_1}(T)_\alpha \subseteq \mathrm{Rfn}(T)_\alpha$;
(ii) $\forall\alpha\ \mathrm{Rfn}(T)_\alpha \subseteq_{\mathcal{B}(\Sigma_1)} \mathrm{Rfn}_{\Sigma_1}(T)_\alpha$.

Proof We give an informal argument by reflexive induction on α. Since both (i) and (ii) are formalized as Π_2-formulas, we may actually argue in EA $+ B\Sigma_1$ (and then use Π_2-conservativity of the latter over EA). Denote $V^\alpha := \mathrm{Rfn}_{\Sigma_1}(T)_\alpha$ and $U^\alpha := \mathrm{Rfn}(T)_\alpha$.

(i) By the definition of (implicit) progressions, modulo provable equivalence every axiom of V^α is either an axiom of T, and in this case there is nothing to prove, or it is an instance of the schema $\mathrm{Rfn}_{\Sigma_1}(V^\beta)$ for some $\beta \prec \alpha$, that is, it has the form $\Box_{V^\beta}\sigma \to \sigma$ for a sentence $\sigma \in \Sigma_1$. By the reflexive induction hypothesis we have

$$\mathrm{EA} \vdash \Box_{V^\beta}\sigma \to \Box_{U^\beta}\sigma,$$

whence

$$\mathrm{EA} \vdash (\Box_{U^\beta}\sigma \to \sigma) \to (\Box_{V^\beta}\sigma \to \sigma).$$

Thus, $U^\alpha \vdash \Box_{V^\beta}\sigma \to \sigma$, by the definition of U^α, that is, every axiom of V^α is provable in U^α, as required. ($B\Sigma_1$ then implies that every *theorem* of V^α is provable in U^α. In the following we shall not mention such uses of $B\Sigma_1$.)

(ii) Assume $U^\alpha \vdash \delta$, where δ is a $\mathcal{B}(\Sigma_1)$-sentence. By the definition of U^α and the formalized deduction theorem there exist $\beta_1, \ldots, \beta_m \prec \alpha$ and sentences $\varphi_1, \ldots, \varphi_m$ such that

$$T \vdash \bigwedge_{i=1}^{m}(\Box_{U^{\beta_i}}\varphi_i \to \varphi_i) \to \delta.$$

By provable monotonicity, stipulating $\beta := \max_\prec\{\beta_1, \ldots, \beta_m\}$, we obtain

$$T \vdash \bigwedge_{i=1}^{m}(\Box_{U^\beta}\varphi_i \to \varphi_i) \to \delta. \tag{11}$$

Lemmas 4 and 5 in [3] then yield $\mathcal{B}(\Sigma_1)$-sentences $\psi_1, \dots, \psi_m$ such that

$$T \vdash \bigwedge_{i=1}^{m} (\Box_{U^\beta} \psi_i \to \psi_i) \to \delta. \tag{12}$$

(Note that these sentences together with a proof of (12) are constructed elementarily from the proof (11).) The reflexive induction hypothesis implies

$$\mathrm{EA} \vdash \Box_{U^\beta} \psi_i \to \Box_{V^\beta} \psi_i,$$

whence

$$T \vdash \bigwedge_{i=1}^{m} (\Box_{V^\beta} \psi_i \to \psi_i) \to \delta.$$

Thus, $V^\alpha \vdash \delta$, by the definition of V^α. ⊠

The following generalization of Statements 1 and 2 of Theorem 1 in [3] is similarly proved.

Proposition 23 *For all $n > 1$, the following statements are provable in* EA*:*

(i) $\forall \alpha\, \mathrm{Rfn}(T)_\alpha \equiv_{\Sigma_n} \mathrm{Rfn}_{\Sigma_n}(T)_\alpha$*;*
(ii) $\forall \alpha\, \mathrm{Rfn}(T)_\alpha \equiv_{\Pi_n} \mathrm{Rfn}_{\Pi_n}(T)_\alpha$.

We also state without proof an obvious relativization of Proposition 22.

Proposition 24 *For all $n \geq 1$, the following statements are provable in* EA*:*

$$\forall \alpha\, \mathrm{Rfn}^{\Pi_n}(T)_\alpha \equiv_{\mathcal{B}(\Sigma_{n+1})} \mathrm{Rfn}^{\Pi_n}_{\Sigma_{n+1}}(T)_\alpha.$$

The next proposition generalizes Proposition 19. Its proof is based on a formalization of Theorem 2, which is possible in EA^+, but not in EA itself. (The nonelementarity is only due to the application of cut-elimination in that proof.)

Proposition 25 *If T is a Π_{n+1}-axiomatized extension of* EA*, then the following is provable in* EA^+ :

$$\forall \alpha\, \mathrm{RFN}_{\Sigma_n}(T)_\alpha \equiv_{\Sigma_{n+1}} \mathrm{Rfn}^{\Pi_{n-1}}_{\Sigma_n}(T)_\alpha.$$

Proof Inclusion ($\supseteq$) is obvious. We give a proof of ($\subseteq_{\Sigma_{n+1}}$) for $n = 1$ (for $n > 1$ the proof is no different, but the notation would be heavier to read). Denote $U^\alpha := \mathrm{RFN}_{\Sigma_1}(T)_\alpha$; $V^\alpha := \mathrm{Rfn}_{\Sigma_1}(T)_\alpha$. We give an informal argument by reflexive induction on α in EA^+.

Assume $U^{\alpha} \vdash \sigma$, where $\sigma \in \Sigma_2$. By the definition of U^{α}, for some $\beta \prec \alpha$ we have

$$T + \mathrm{RFN}_{\Sigma_1}(U^{\beta}) \vdash \sigma.$$

Notice that by the monotonicity of V^{β}

$$T + \mathrm{RFN}_{\Sigma_1}(V^{\beta}) \vdash \mathrm{RFN}_{\Sigma_1}(\mathrm{EA}),$$

that is, $T + \mathrm{RFN}_{\Sigma_1}(V^{\beta})$ contains EA^{+}. On the other hand, by the reflexive induction hypothesis

$$\mathrm{EA}^{+} \vdash \forall x \in \Sigma_1\, (\Box_{U^{\beta}}(x) \leftrightarrow \Box_{V^{\beta}}(x)),$$

whence

$$\mathrm{EA}^{+} \vdash \mathrm{RFN}_{\Sigma_1}(U^{\beta}) \leftrightarrow \mathrm{RFN}_{\Sigma_1}(V^{\beta}).$$

Therefore, $\mathrm{RFN}_{\Sigma_1}(U^{\beta})$ is contained in $T + \mathrm{RFN}_{\Sigma_1}(V^{\beta})$, and thus,

$$T + \mathrm{RFN}_{\Sigma_1}(V^{\beta}) \vdash \sigma.$$

It follows that

$$T + \neg\sigma + \mathrm{RFN}_{\Sigma_1}(V^{\beta}) \vdash \bot,$$

and by (formalized) Theorem 2

$$T + \neg\sigma + \Pi_1\text{-}\mathrm{RR}_{\Pi_1}(V^{\beta}) \vdash \bot,$$

and

$$T + \neg\sigma + \mathrm{Rfn}_{\Sigma_1}(V^{\beta}) \vdash \bot.$$

Thus,

$$T + \mathrm{Rfn}_{\Sigma_1}(V^{\beta}) \vdash \sigma,$$

that is, $V_{\alpha} \vdash \sigma$. ⊠

7 Schmerl's Formula

Our approach to Schmerl's formula borrows a general result from [3] relating the hierarchies of iterated local reflection principles and of iterated consistency assertions over an arbitrary initial theory T. This result and the results below hold under the assumption that $(D, \prec)$ is a *nice* elementary well-ordering. A nice well-ordering is an elementary well-ordering equipped with elementary terms representing the ordinal constants and functions $0, 1, +, \cdot, \omega^x$. These functions should provably in EA satisfy some minimal obvious axioms NWO listed in [3]. Besides, there should be an elementary EA-provable isomorphism between natural numbers (with the usual order) and the ordinals $\prec \omega$. Under these assumptions on the class of well-orderings we have the following theorem [3].

Proposition 26 EA *proves that, for all* α, β *such that* $\alpha \succeq 1$, *there holds*

$$(\text{Rfn}(T)_\alpha)_\beta \equiv_{\Pi_1} T_{\omega^\alpha \cdot (1+\beta)}.$$

In particular, for all $\alpha \succeq 1$, $\text{Rfn}(T)_\alpha \equiv_{\Pi_1} T_{\omega^\alpha}$.

A proof is obtained by elementary reflexive induction and for the nontrivial inclusion ($\subseteq_{\Pi_1}$) it only uses provability logic. Since the relativized provability predicates satisfy the same provability logic, essentially the same argument also yields the following theorem.

Proposition 27 *For any* $n \geq 1$, *provably in* EA,

$$\forall \alpha \succeq 1\ \forall \beta\ \ (\text{Rfn}^{\Pi_{n-1}}(T)_\alpha)_\beta \equiv_{\Pi_n} (T)^n_{\omega^\alpha \cdot (1+\beta)}.$$

Proof One only has to notice that provably in EA, for all α,

$$\text{Con}^{\Pi_{n-1}}(T)_\alpha \equiv \text{RFN}_{\Pi_n}(T)_\alpha \equiv (T)^n_\alpha,$$

which can be verified by a straightforward reflexive induction on α. ⊠

Now we can combine this result with Proposition 25 and obtain a generalization of Schmerl's theorem. (Schmerl [32] proves this statement for specially defined progressions corresponding to $(\text{PRA})^n_\alpha$).

Theorem 3 (Schmerl's Formula) *For all* $n \geq 1$, *if* T *is an elementary presented* Π_{n+1}*-axiomatized extension of* EA, *the following holds provably in* EA^+:

$$\forall \alpha \succeq 1\ (T)^{n+1}_\alpha \equiv_{\Pi_n} (T)^n_{\omega^\alpha}.$$

Proof It is sufficient to notice that for all $\alpha \succeq 1$,

$$(T)^{n+1}_\alpha \equiv_{\Sigma_{n+1}} \text{Rfn}^{\Pi_{n-1}}_{\Sigma_n}(T)_\alpha \equiv_{\Pi_n} (T)^n_{\omega^\alpha},$$

by Propositions 25 and 27, respectively. ⊠

Next we observe an easy but useful extension lemma.

Lemma 28 *Let U, V be elementary presented extensions of* EA, *and Γ be one of the classes Σ_{k+1}, Π_{k+1} or $\mathcal{B}(\Sigma_k)$, for $k \geq n \geq 1$. Then*

$$\mathrm{EA} \vdash U \subseteq_\Gamma V \quad \Longrightarrow \quad \mathrm{EA} \vdash \forall\alpha\ (U)^n_\alpha \subseteq_\Gamma (V)^n_\alpha.$$

Proof Reasoning by reflexive induction on α assume $U_\alpha \vdash \varphi$ for a sentence $\varphi \in \Gamma$. Then for some $\beta \prec \alpha$,

$$U + \mathrm{RFN}_{\Pi_n}((U)^n_\beta) \vdash \varphi.$$

By reflexive induction hypothesis

$$U + \mathrm{RFN}_{\Pi_n}((V)^n_\beta) \vdash \varphi,$$

whence

$$U \vdash \neg\mathrm{RFN}_{\Pi_n}((V)^n_\beta) \vee \varphi.$$

The latter formula is in Γ (modulo logical equivalence), therefore

$$V \vdash \neg\mathrm{RFN}_{\Pi_n}((V)^n_\beta) \vee \varphi,$$

and thus $V_\alpha \vdash \varphi$. ⊠

Remark 29 This lemma also holds for EA^+ in place of EA, with the same proof.

The ordinal functions $\omega_n(\alpha)$ are introduced as usual:

$$\begin{cases} \omega_0(\alpha) & := \alpha \\ \omega_{k+1}(\alpha) & := \omega^{\omega_k(\alpha)} \end{cases}$$

Denote $\omega_n := \omega_n(1)$, $\quad \epsilon_0 := \sup\{\omega_n : n < \omega\}$.

Our next theorem generalizes Theorem 3 to mixed hierarchies (for T = PRA established by Schmerl).

Theorem 4 *For all $n, m \geq 1$, if T is an elementary presented Π_{n+1}-axiomatized extension of* EA, *the following statements hold provably in* EA^+ :

(i) $\forall\alpha \succeq 1\ (T)^{n+m}_\alpha \equiv_{\Pi_n} (T)^n_{\omega_m(\alpha)}$;
(ii) $\forall\alpha \succeq 1\ ((T)^{n+m}_\alpha)^n_\beta \equiv_{\Pi_n} (T)^n_{\omega_m(\alpha)\cdot(1+\beta)}$.

Proof Part (i) follows by m-fold application of Theorem 3. For a proof of (ii) notice that by (i) and Proposition 25

$$(T)^{n+m}_\alpha \equiv_{\Pi_{n+1}} (T)^{n+1}_{\omega_{m-1}(\alpha)} \equiv_{\Sigma_{n+1}} \mathrm{Rfn}^{\Pi_{n-1}}_{\Sigma_n}(T)_{\omega_{m-1}(\alpha)}.$$

Lemma 28 and Proposition 27 imply that

$$((T)^{n+m}_{\alpha})^n_{\beta} \equiv_{\Pi_{n+1}} ((T)^{n+1}_{\omega_{m-1}(\alpha)})^n_{\beta} \equiv_{\Sigma_{n+1}} (\mathrm{Rfn}^{\Pi_{n-1}}_{\Sigma_n}(T)_{\omega_{m-1}(\alpha)})^n_{\beta} \equiv_{\Pi_n} (T)^n_{\omega^{\omega_{m-1}(\alpha)}\cdot(1+\beta)},$$

which yields the required formula

$$((T)^{n+m}_{\alpha})^n_{\beta} \equiv_{\Pi_n} (T)^n_{\omega_m(\alpha)\cdot(1+\beta)},$$

⊠

This theorem directly applies to theories T like EA or PRA. Following Schmerl's idea it is also possible to derive from it similar formulas for iterated reflection principles over PA, which we obtain in the next section.

Corollary 30 *If T is a Π^0_2-regular theory, then it is Π^0_1-regular and $|T|_{\Pi^0_1}$ is one ω-power higher than $|T|_{\Pi^0_2}$.*

Proof If T is a Π^0_2-regular theory and $|T|_{\Pi^0_2} = \alpha$, then $T \equiv_{\Pi_2} (\mathrm{EA})^2_{\alpha}$. By Theorem 3 $(\mathrm{EA})^2_{\alpha} \equiv_{\Pi_1} \mathrm{EA}_{\omega^{\alpha}}$, which means that $T \equiv_{\Pi_1} \mathrm{EA}_{\omega^{\alpha}}$, that is, T is Π^0_1-regular and $|T|_{\Pi^0_1} = \omega^{\alpha}$. ⊠

8 Ordinal Analysis of Fragments

Now we have at our disposal all necessary tools to give an ordinal analysis of arithmetic and its fragments. The general methodology bears similarities with the traditional Π^1_1-ordinal analysis, see [28]. To determine the ordinal of a given formal system T one first finds a suitable *embedding* of T into the hierarchy of reflection principles over EA. Then one applies Schmerl's formula for a *reduction* of the reflection principles axiomatizing T to iterated reflection principles of lower complexity. From this point of view the use of Schmerl's formula substitutes the use of cut-elimination for ω-logic. Notice that, although the meaning of the ordinals is different, ordinal bounds are essentially the same in both approaches. Also notice that in the present approach the embedding part is more informative. In particular, this allows to obtain upper and lower bounds for proof-theoretic ordinals simultaneously.

The following embedding results are known ($n \geq 1$).

(E1) Leivant [20] and Ono [23] show that $I\Sigma_n$ is equivalent to $\mathrm{RFN}_{\Pi_{n+2}}(\mathrm{EA})$ over EA, that is,

$$I\Sigma_n \equiv (\mathrm{EA})^{n+2}_1.$$

Notice that this is sharper than the related results in [32] and the original result by Kreisel and Lévy [19] stating that

$$\mathrm{PA} \equiv \mathrm{EA} + \mathrm{RFN}(\mathrm{EA}).$$

(E2) For the closures of EA under Σ_n- and Π_{n+1}-induction rules we have the following characterization (cf. [2]):

$$I\Sigma_n^R \equiv I\Pi_{n+1}^R \equiv (\mathrm{EA})_\omega^{n+1}.$$

(E3) Parameter-free induction schemata have been characterized in [5]:

(a) $I\Sigma_n^- \equiv \mathrm{EA} + \mathrm{Rfn}_{\Sigma_{n+1}}^{\Pi_n}(\mathrm{EA})$.
(b) $I\Pi_{n+1}^- \equiv \mathrm{EA} + \mathrm{Rfn}_{\Sigma_{n+2}}^{\Pi_n}(\mathrm{EA})$.
(c) $\mathrm{EA}^+ + I\Pi_1^- \equiv \mathrm{EA}^+ + \mathrm{Rfn}_{\Sigma_2}(\mathrm{EA}) \equiv \mathrm{EA}^+ + \mathrm{Rfn}_{\Sigma_2}(\mathrm{EA}^+)$.

Over EA the schema $I\Pi_1^-$ is equivalent to the local Σ_2-reflection principle for EA formulated for the predicate of *cut-free provability*, see [5] and Appendix 3 for more details.

Remark 31 The upper bound results only require the embeddings (E1)–(E3) from left to right, that is, the provability of the induction principles from suitable forms of the reflection principles. All these embeddings are very easy to prove using a trick by Kreisel [19]. For example, in order to prove $I\Sigma_n \subseteq (\mathrm{EA})_1^{n+2}$ let $\sigma(x, a)$ be any Σ_n-formula and consider the formula $\psi(x, a) := \sigma(0, a) \wedge \forall u\, (\sigma(u, a) \to \sigma(u + 1, a)) \to \sigma(x, a)$. ψ is logically equivalent to a Π_{n+2}-formula. By an elementary induction on m it is easy to see that, for all m, k, $\mathrm{EA} \vdash \psi(\overline{m}, \overline{k})$, and this fact is formalizable in EA. Hence $\mathrm{EA} \vdash \forall x, a\, \Box_{\mathrm{EA}} \psi(\dot{x}, \dot{a})$, and applying uniform Π_{n+2}-reflection yields

$$\mathrm{EA} + \mathrm{RFN}_{\Pi_{n+2}}(\mathrm{EA}) \vdash \forall x, a\, \psi(x, a),$$

which is equivalent to an instance of $I\Sigma_1$. The proofs of the corresponding embeddings in (E2) and (E3) are rather similar.

The embeddings (E1)–(E3) from right to left are increasingly more difficult to prove (but, if one is only interested in the upper bound results, this is not strictly necessary). The simplest argument for PA, due to Kreisel and Lévy [19], goes from an EA-proof of a formula φ to a cut-free derivation in Tait calculus of a sequent consisting of φ and some negated instances of the axioms of EA. All formulas occurring in this derivation are Π_n, where n is the maximum of the logical complexity of φ and 2 (since all the axioms of EA are Π_1). Then one can use a truth predicate for Π_n-formulas and prove by induction on the depth of the cut-free derivation that all sequents occurring in it are true. This induction argument is clearly formalizable in PA.

Remark 32 All the embedding results mentioned in (E1)–(E3) can be naturally formalized in EA. This is obvious for the finitely axiomatized systems in (E1) and can be checked for (E2) and (E3) as well.

Now we obtain some corollaries, starting from the simplest analysis of $I\Sigma_n$ and PA.

Proposition 33 *For all* $n \geq 1$,

(i) $I\Sigma_n \equiv_{\Pi_2} (\mathrm{EA})^2_{\omega_n} \equiv_{\Pi_1} \mathrm{EA}_{\omega_{n+1}}$, *hence* $|I\Sigma_n|_{\Pi^0_1} = \omega_{n+1}$;
(ii) $\mathrm{PA} \equiv_{\Pi_n} (\mathrm{EA})^n_{\epsilon_0}$, *hence* $|\mathrm{PA}|_{\Pi^0_n} = \epsilon_0$.

Proof Statement (i) follows from (E1) and Theorem 4. Statement (ii) similarly follows from the equivalence

$$\mathrm{PA} \equiv \bigcup_{m\geq 0} (\mathrm{EA})^{n+m}_1 \equiv_{\Pi_n} \bigcup_{m\geq 0} (\mathrm{EA})^n_{\omega_m(1)},$$

which holds for any $n \geq 1$. ⊠

Remark 34 Part (i) of this proposition is formalizable in EA^+. Part (ii) will be formalizable in EA^+ under some additional assumptions on the choice of a nice well-ordering. For example, we can extend NWO by a binary function symbol for $\omega_n(\alpha)$ and a constant symbol ϵ_0 with the obvious defining axioms. The extended theory will be denoted $\mathrm{NWO}(\epsilon_0)$, and we require that the nice well-ordering interprets these axioms by elementary functions.

For the induction rules from (E2) we obtain the same bounds ($n \geq 1$).

Proposition 35 $I\Sigma^R_n \equiv_{\Pi_2} (\mathrm{EA})^2_{\omega_n} \equiv_{\Pi_1} \mathrm{EA}_{\omega_{n+1}}$.

For parameter-free induction schemata we have, by the conservation results for local reflection principles in [2, 5] and Sect. 5 of this paper

Proposition 36 $(\mathrm{EA})^{n+2}_1 \equiv_{\Sigma_{n+2}} I\Sigma^-_n \equiv_{\mathcal{B}(\Sigma_{n+1})} I\Pi^-_{n+1}$.

It follows that, for $n \geq 1$, the theories $I\Sigma^-_n$, $I\Pi^-_{n+1}$ and $I\Sigma_n$ have the same Π^0_2- and Π^0_1-ordinals.

In [5] the theory $I\Sigma_n + I\Pi^-_{n+1}$ is analyzed ($n \geq 1$). On the basis of (E3)(b) it is shown that

$$I\Sigma_n + I\Pi^-_{n+1} \equiv_{\Pi_{n+1}} (I\Sigma_n)^{n+1}_\omega \equiv ((\mathrm{EA})^{n+2}_1)^{n+1}_\omega.$$

Applying Theorem 4 yields the following corollary, which determines its Π^0_2- and Π^0_1-ordinals.

Proposition 37 $I\Sigma_n + I\Pi^-_{n+1} \equiv_{\Pi_2} (\mathrm{EA})^2_{\omega_n(2)} \equiv_{\Pi_1} \mathrm{EA}_{\omega_{n+1}(2)}$.

Now we consider the exceptional case of the parameter-free Π_1-induction schema. We first analyze the system $\mathrm{EA}^+ + I\Pi^-_1$, the weaker theory $\mathrm{EA} + I\Pi^-_1$ will be treated in Appendix 3.

Notice that by (E3)(c) the theory $\mathrm{EA}^+ + I\Pi_1^-$ is certainly not Π_2-conservative (not even Π_1-conservative) over EA^+. Yet, the next proposition shows that its class of provably total computable functions coincides with that of EA^+. This means that $\mathrm{EA}^+ + I\Pi_1^-$ is not Π_2-regular (its Π_2^0-ordinal equals 1).

Proposition 38

(i) $\mathcal{F}(\mathrm{EA}^+ + I\Pi_1^-) = \mathcal{F}(\mathrm{EA}^+) = \mathcal{F}_1$;
(ii) $\mathcal{F}(\mathrm{EA} + I\Pi_1^-) = \mathcal{F}_0 = \mathcal{E}$.

Proof By (E3)(c), $\mathrm{EA}^+ + I\Pi_1^-$ is contained in $\mathrm{EA}^+ + \mathrm{Rfn}(\mathrm{EA}^+)$ and similarly

$$\mathrm{EA} + I\Pi_1^- \subseteq \mathrm{EA} + \mathrm{Rfn}(\mathrm{EA}).$$

Feferman [8] noticed that $\mathrm{Rfn}(T)$ is provable in T together with all true Π_1-sentences. Yet, it is equally well-known that adding any amount of true Π_1-axioms to a sound theory does not increase its class of provably total computable functions. This proves both parts (i) and (ii). ⊠

Proposition 39 $\mathrm{EA}^+ + I\Pi_1^-$ *is* Π_1^0*-regular and* $|\mathrm{EA}^+ + I\Pi_1^-|_{\Pi_1^0} = \omega^2$.

Proof Recall that by Proposition 26 (for this particular case established by Goryachev [13]) $T + \mathrm{Rfn}(T)$ is Π_1-conservative over T_ω. Hence, by Theorem 4,

$$\mathrm{EA}^+ + I\Pi_1^- \equiv_{\Pi_1} (\mathrm{EA}^+)_\omega \equiv ((\mathrm{EA})_1^2)_\omega \equiv_{\Pi_1} \mathrm{EA}_{\omega^2}.$$

Thus, the theory is Π_1^0-regular with the ordinal ω^2. ⊠

Now we consider the extensions of PA by reflection principles. The following proposition holds for nice well-orderings satisfying $\mathrm{NWO}(\epsilon_0)$. Notice that the function ϵ_0^α can be expressed in $\mathrm{NWO}(\epsilon_0)$ by the term $\omega^{\epsilon_0 \cdot \alpha}$.

Proposition 40 *For each* $n \geq 1$, *provably in* EA^+,

(i) $(\mathrm{PA})_\alpha^n \equiv_{\Pi_n} (\mathrm{EA})_{\epsilon_0\cdot(1+\alpha)}^n$;
(ii) $(\mathrm{PA})_\alpha^{n+1} \equiv_{\Pi_n} (\mathrm{PA})_{\epsilon_0^\alpha}^n$, *if* $\alpha \succeq 1$.

Proof By formalized Proposition 33(ii), provably in EA^+,

$$\mathrm{PA} \equiv_{\Pi_{n+1}} (\mathrm{EA})_{\epsilon_0}^{n+1}.$$

By Lemma 28 and Theorem 4(ii) we then obtain:

$$(\mathrm{PA})_\alpha^n \equiv_{\Pi_{n+1}} ((\mathrm{EA})_{\epsilon_0}^{n+1})_\alpha^n \equiv_{\Pi_n} (\mathrm{EA})_{\epsilon_0\cdot(1+\alpha)}^n.$$

Formula (ii) follows from (i) and Theorem 3, because for $\alpha \succeq 1$ one has

$$(\mathrm{EA})_{\epsilon_0\cdot(1+\alpha)}^{n+1} \equiv_{\Pi_n} (\mathrm{EA})_{\omega^{\epsilon_0\cdot(1+\alpha)}}^n \equiv (\mathrm{EA})_{\epsilon_0\cdot(1+\epsilon_0^\alpha)}^n \equiv_{\Pi_n} (\mathrm{PA})_{\epsilon_0^\alpha}^n,$$

⊠

As a particular case we obtain that PA + Con(PA) is a Π_1^0-regular theory with the ordinal $\omega \cdot 2$.

Corollary 41 PA + Con(PA) $\equiv_{\Pi_1}$ $\mathrm{EA}_{\epsilon_0 \cdot 2}$.

Since Con(PA) is a true Π_1-sentence, $|\mathrm{PA} + \mathrm{Con}(\mathrm{PA})|_{\Pi_2^0} = \epsilon_0$, therefore PA + Con(PA) is not Π_2^0-regular.

As another example of this sort let us compute the ordinals of the following Π_1^0-regular, but Π_2^0-irregular, theories.

Proposition 42

(i) $|I\Sigma_1 + \mathrm{Con}(\mathrm{PA})|_{\Pi_1^0} = \epsilon_0 + \omega^\omega$;
(ii) $|\mathrm{EA}^+ + \mathrm{Con}(I\Sigma_1)|_{\Pi_1^0} = \omega^\omega + \omega$.

Proof For (i) notice that by Theorem 4 provably in EA^+,

$$(I\Sigma_1)^1_{\epsilon_0} \equiv ((\mathrm{EA})^3_1)^1_\omega \equiv_{\Pi_1} \mathrm{EA}_{\omega_2(1)\cdot(1+\epsilon_0)} \equiv \mathrm{EA}_{\epsilon_0} \equiv_{\Pi_1} \mathrm{PA}.$$

Therefore,

$$\mathrm{EA}^+ \vdash \mathrm{Con}(\mathrm{PA}) \leftrightarrow \mathrm{Con}((I\Sigma_1)^1_{\epsilon_0}).$$

It follows that

$$I\Sigma_1 + \mathrm{Con}(\mathrm{PA}) \equiv I\Sigma_1 + \mathrm{Con}((I\Sigma_1)^1_{\epsilon_0}) \equiv (I\Sigma_1)^1_{\epsilon_0+1}.$$

For the latter theory we have

$$(I\Sigma_1)^1_{\epsilon_0+1} \equiv ((\mathrm{EA})^3_1)^1_{\epsilon_0+1} \equiv_{\Pi_1} \mathrm{EA}_{\omega_2\cdot(\epsilon_0+1)} \equiv \mathrm{EA}_{\epsilon_0+\omega^\omega}.$$

For (ii) reasoning in a similar way we obtain

$$\mathrm{EA}^+ + \mathrm{Con}(I\Sigma_1) \equiv ((\mathrm{EA})^2_1)_{\omega^\omega+1} \equiv_{\Pi_1} \mathrm{EA}_{\omega\cdot(1+\omega^\omega+1)} \equiv \mathrm{EA}_{\omega^\omega+\omega},$$

⊠

9 Conclusion and Further Work

This paper demonstrates the use of reflection principles for the ordinal analysis of fragments of Peano arithmetic. More importantly, reflection principles provide a uniform definition of proof-theoretic ordinals for any arithmetical complexity Π_n^0, in particular, for the complexity Π_1^0.

The results of this paper are further developed in our later paper [7], where the notion of *graded provability algebra* is introduced. It provides an abstract algebraic framework for proof-theoretic analysis and links canonical ordinal notation systems with certain algebraic models of provability logic. We hope that this further development will shed additional light on the problem of canonicity of ordinal notations.

Appendix 1

Let $(D, \prec)$ be an elementary well-ordering. Define

$$\alpha[x] := \max_{\prec}\{\beta \leq x : \beta \prec \alpha\}$$
$$\beta \prec_x \alpha :\Leftrightarrow (\beta \leq x \wedge \beta \prec a).$$

Recall that the functions F_α are defined as follows:

$$F_\alpha(x) := \max\{2^x_x + 1\} \cup \{F^{(v)}_\beta(u) + 1 : \beta \prec \alpha,\ u, v, \beta \leq x\}.$$

For technical convenience we also define $F_{-1}(x) = 2^x$ and $\alpha[x] = -1$, if there is no $\beta \prec_x \alpha$.

Lemma 43 *For all* α, β, x, y,

(i) $x \leq y \to F_\alpha(x) \leq F_\alpha(y)$;
(ii) $\beta \prec \alpha \to F_\beta(x) \leq F_\alpha(x)$.

Proof Part (i) is obvious. Part (ii) follows from the fact that

$$\gamma \prec_x \beta \prec \alpha \Rightarrow \gamma \prec_x \alpha,$$

⊠

Lemma 44 *For all* α, x, $F_\alpha(x) = F^{(x)}_{\alpha[x]}(x) + 1$.

Proof This is obvious for $\alpha[x] = -1$. Otherwise, from Part (i) of the previous lemma we obtain

$$u, v \leq x \to F^{(v)}_\beta(u) \leq F^{(x)}_\beta(x).$$

Part (ii) a nd Part (i) by an elementary induction on y then yield

$$\beta \prec \alpha \to F^{(y)}_\beta(x) \leq F^{(y)}_\alpha(x).$$

Hence, if $u, v \leq x$ and $\beta \prec_x \alpha$, then $\beta \prec_x \alpha[x]$ or $\beta = \alpha[x]$, and in both cases

$$F_{\beta}^{(v)}(u) \leq F_{\alpha[x]}^{(x)}(x),$$

⊠

We now define evaluation trees. An *evaluation tree* is a finite tree labeled by tuples of the form $\langle \alpha, x, y \rangle$ satisfying the following conditions:

1. If there is no $\beta \prec_x \alpha$, then $\langle \alpha, x, y \rangle$ is a terminal node and $y = 2_x^x + 1$.
2. Otherwise, there are x immediate successors of $\langle \alpha, x, y \rangle$, and their labels have the form

$$\langle \alpha[x], x, y_1 \rangle, \ \langle \alpha[x], y_1, y_2 \rangle, \ldots, \langle \alpha[x], y_{x-1}, y_x \rangle$$

and $y = y_x + 1$.

Obviously, the relation x *is a code of an evaluation tree* is elementary.

Lemma 45

(i) If a node of an evaluation tree is labeled by $\langle \alpha, x, y \rangle$, then $F_\alpha(x) = y$.
(ii) If $F_\alpha(x) = y$, then there is an evaluation tree with the root labeled by $\langle \alpha, x, y \rangle$.

Proof Part (i) is proved by transfinite induction on α. If $\alpha[x] = -1$, the statement is obvious. Otherwise, $\alpha[x] \prec \alpha$, hence by the induction hypothesis at the immediate successor nodes one has $F_{\alpha[x]}(y_i) = y_{i+1}$ for all $i < x$ (where we also put $y_0 := x$). It follows that $y_x = F_{\alpha[x]}^{(x)}(x)$, whence $y = y_x + 1 = F_\alpha(x)$.

Part (ii) obviously follows from the definition of F_α, ⊠

Now we observe that, for any evaluation tree T, whose root is labeled by $\langle \alpha, x, y \rangle$, the value $\max(\alpha, y)$ is a common bound to the following parameters:

(a) each γ, u, v such that $\langle \gamma, u, v \rangle$ occurs in T;
(b) the number of branches at every node of T;
(c) the depth of T.

Ad (a): If $\langle \gamma, u, v \rangle$ is an immediate successor of $\langle \alpha, x, y \rangle$, then $\gamma = \alpha[x] \leq x$, and $u, v \leq y$ by the monotonicity of F.

Statement (b) follows from (a) and the fact that the number of branches at a node $\langle \gamma, u, v \rangle$ equals u.

Statement (c) follows from the observation that if $\langle \gamma, u, v \rangle$ is an immediate successor of $\langle \alpha, x, y \rangle$, then $y > v$.

An immediate corollary of (a)–(c) is that the code of the evaluation tree T is bounded by the value $g(\max(\alpha, y))$, for an elementary function g. Hence we obtain

Proof of Lemma 7 Using Lemma 45 the relation $F_\alpha(x) = y$ can be expressed by formalizing the statement that *there is an evaluation tree with the code* $\leq$

$g(\max(\alpha, y))$, *whose root is labeled by* $\langle \alpha, x, y \rangle$. All quantifiers here are bounded, hence the relation $F_\alpha(x) = y$ is elementary. ⊠

Inspecting the definition of the relation $F_\alpha(x) = y$ notice that the proofs of the monotonicity properties and bounds on the size of the tree only required elementary induction (transfinite induction is not used). Hence, these properties together with the natural defining axioms for F_α can be verified in EA. This yields a proof of Lemma 8. Here we just formally state the required properties of F_α formalizable in EA.

F1. $(\forall \beta \leq x \neg \beta \prec \alpha) \rightarrow [F_\alpha(x) \simeq y \leftrightarrow y = 2^x_x + 1]$
F2. $F_\alpha(x) \simeq y \wedge \beta \leq x \wedge \beta \prec \alpha \rightarrow \exists z \leq y \, \exists u, v \leq x \, F^{(v)}_\beta(u) \simeq z$
F3. $\forall \beta, u, v \leq x(\beta \prec \alpha \rightarrow \exists y \leq z \, F^{(v)}_\beta(u) \simeq y) \rightarrow \exists y \leq z \, F_\alpha(x) \simeq y.$

Here, as usual, $F^{(x)}_\beta(x) \simeq y$ abbreviates

$$\exists s \in Seq \, [(s)_0 = x \wedge \forall i < x \, F_\beta((s)_i) \simeq (s)_{i+1} \wedge (s)_x = y].$$

Appendix 2

One can roughly classify the existing definitions of proof-theoretic ordinals in two groups, which I call the definitions “from below” and “from above”. Informally speaking, proof-theoretic ordinals defined from below measure the strength of the principles of certain complexity Γ that are *provable* in a given theory T. In contrast, the ordinals defined from above measure the strength of certain characteristic for T *unprovable* principles of complexity Γ. For example, $\mathrm{Con}(T)$ is such a characteristic principle of complexity Π^0_1.

The standard Π^1_1- and Π^0_2-ordinals are defined from below, and so are the Π^0_n-ordinals introduced in this paper. The notorious ordinal of the shortest natural primitive recursive well-ordering $\prec$ such that $TI_{p.r.}(\prec)$ proves $\mathrm{Con}(T)$ (apart from the already discussed feature of logical complexity mismatch) is a typical definition from above.

All the usual definitions of proof-theoretic ordinals can also be reformulated in the form “from above”. Let a natural elementary well-ordering be fixed. For the case of Π^0_n-ordinals the corresponding approach would be to let

$$|T|^\vee_{\Pi^0_n} := \min\{\alpha : \mathrm{EA}^+ + \mathrm{RFN}_{\Pi_n}((\mathrm{EA})^n_\alpha) \vdash \mathrm{RFN}_{\Pi_n}(T)\}.$$

(Notice that for $n > 1$ the theory on the left hand side of $\vdash$ can be replaced by $(\mathrm{EA})^n_{\alpha+1}$.)

In a similar manner one can transform the definition of the Π^0_2-ordinal via the Fast Growing hierarchy into a definition “from above”. The class of p.t.c.f. of T has a natural indexing, e.g., we can take as indices of a function f the pairs $\langle e, p \rangle$ such that e is the usual Kleene index (= the code of a Turing machine) of f, and p is the code of a T-proof of the Π^0_2-sentence expressing the totality of the function $\{e\}$. With this natural indexing in mind one can write out a formula defining the universal function $\varphi_T(e,x)$ for the class of unary functions in $\mathcal{F}(T)$. Then the Π^0_2-sentence expressing the totality of φ_T would be the desired characteristic principle. (It is not difficult to show that the totality of φ_T formalized in this way is EA^+-equivalent to $\mathrm{RFN}_{\Pi_2}(T)$.) The Π^0_2-ordinal of T can then be defined as follows:

$$|T|^{\vee}_{\Pi^0_2} := \min\{\alpha : \varphi \in \mathcal{F}_{\alpha+1}\}.$$

Notice that the proof-theoretic ordinals of T defined "from above" not only depend on the externally taken set of theorems of T, but also on the way T is formalized, that is, essentially on the provability predicate or the *proof system* for T. For example, in the above definition the universal function $\varphi_T(e,x)$ depends on the Gödel numbering of proofs in T. In practice, for most of the natural(ly formalized) theories the ordinals defined "from below" and those "from above" coincide:

Proposition 46 *If T is EA^+-provably Π^0_n-regular and contains EA^+, then $|T|^{\vee}_{\Pi^0_n} = |T|_{\Pi^0_n}$.*

Proof Let $\alpha = |T|_{\Pi^0_n}$. By provable regularity,

$$\mathrm{EA}^+ \vdash \mathrm{RFN}_{\Pi_n}(T) \leftrightarrow \mathrm{RFN}_{\Pi_n}((\mathrm{EA})^n_\alpha),$$

hence $\mathrm{EA}^+ + (\mathrm{EA})^n_{\alpha+1} \vdash \mathrm{RFN}_{\Pi_n}(T)$. On the other hand, by Gödel's theorem

$$\mathrm{EA}^+ + (\mathrm{EA})^n_\alpha \subseteq T \nvdash \mathrm{RFN}_{\Pi_n}(T),$$

⊠

The following example demonstrates that, nonetheless, there are reasonable (and naturally formalized) proof systems for which these ordinals are different, so sometimes the ordinal defined from above bears essential additional information.

Consider some standard formulation of PA, it has a *natural* provability predicate $\Box_{\mathrm{PA}}$. The system PA^* is obtained from PA by adding Parikh's inference rule:

$$\frac{\Box_{\mathrm{PA}}\varphi}{\varphi},$$

where φ is any sentence. For the reasons of semantical correctness, Parikh's rule is admissible in PA, so PA^* proves the same theorems as PA. However, as is well known, the equivalence of the two systems cannot be established within PA (otherwise, PA^* would have a speed-up over PA bounded by a p.t.c.f. in PA, which was disproved by Parikh [24]).[9] Below we analyze the situation from the point of view of the proof-theoretic ordinals.

Notice that PA^* is a reasonable proof system, and it has a natural Σ_1 provability predicate $\Box_{\mathrm{PA}^*}$. Lindström [21] proves the following relationship between the provability predicates in PA and PA^*:

Lemma 47 $\mathrm{EA} \vdash \forall x\, (\Box_{\mathrm{PA}^*}(x) \leftrightarrow \exists n\, \Box_{\mathrm{PA}}\Box^n_{\mathrm{PA}}(\dot{x}))$, *where* $\Box^n_{\mathrm{PA}}$ *means* n *times iterated* $\Box_{\mathrm{PA}}$.

Notation: The right hand side of the equivalence should be understood as the result of substituting in the external $\Box_{\mathrm{PA}}$ the elementary term for the function $\lambda n, x. \ulcorner \Box^n_{\mathrm{PA}}(\bar{x}) \urcorner$.

Proof (Sketch) The implication ($\leftarrow$) holds, because PA^* is provably closed under Parikh's rule, that is,

$$\mathrm{PA} \vdash \Box^n_{\mathrm{PA}}\varphi \quad \Rightarrow \quad \mathrm{PA}^* \vdash \varphi,$$

by n applications of the rule, and this is obviously formalizable.

The implication ($\rightarrow$) holds, because the predicate $\exists n\, \Box_{\mathrm{PA}}\Box^n_{\mathrm{PA}}(\dot{x})$ is provably closed under PA, *modus ponens* and Parikh's rule:

$$\mathrm{PA} \vdash \varphi \Rightarrow \mathrm{PA} \vdash \Box^1_{\mathrm{PA}}\varphi,$$

$$\mathrm{PA} \vdash \Box^n_{\mathrm{PA}}\varphi, \quad \mathrm{PA} \vdash \Box^m_{\mathrm{PA}}(\varphi \to \psi) \Rightarrow \mathrm{PA} \vdash \Box^{\max(n,m)}_{\mathrm{PA}}\psi,$$

$$\mathrm{PA} \vdash \Box^n_{\mathrm{PA}}(\Box_{\mathrm{PA}}\varphi) \Rightarrow \mathrm{PA} \vdash \Box^{n+1}_{\mathrm{PA}}\varphi,$$

and this is formalizable. ⊠

Corollary 48 $\mathrm{EA} \vdash \mathrm{Con}(\mathrm{PA}^*) \leftrightarrow \mathrm{Con}(\mathrm{PA}_\omega)$.

Proof By induction, $\mathrm{Con}(\mathrm{PA}_n)$ is equivalent to $\neg\Box_{\mathrm{PA}}\Box^n_{\mathrm{PA}}\bot$, moreover this equivalence is formalizable in EA (with n a free variable). Hence, $\mathrm{Con}(\mathrm{PA}_\omega)$ is equivalent to $\forall n\, \neg\Box_{\mathrm{PA}}\Box^n_{\mathrm{PA}}\bot$, which yields the claim by the previous lemma. ⊠

Applying this to Π^0_1-ordinals defined from above, we observe

Proposition 49

(i) $|\mathrm{PA}|^{\vee}_{\Pi^0_1} = \epsilon_0$;
(ii) $|\mathrm{PA}^*|^{\vee}_{\Pi^0_1} = \epsilon_0 \cdot \omega$.

Proof Statement (i) follows from the EA^+-provable Π^0_1-regularity of PA.

[9] An even simpler argument: otherwise one can derive from $\Box_{\mathrm{PA}}\Box_{\mathrm{PA}}\bot$ the formulas $\Box^*_{\mathrm{PA}}\bot$ and $\Box_{\mathrm{PA}}\bot$, which yields $\mathrm{PA} \vdash \Box_{\mathrm{PA}}\Box_{\mathrm{PA}}\bot \to \Box_{\mathrm{PA}}\bot$ contradicting Löb's theorem.

To prove (ii) we first obtain

$$\mathrm{PA}_\omega \equiv_{\Pi_1} \mathrm{EA}_{\epsilon_0\cdot\omega}$$

by Proposition 40(i). Formalization of this in EA^+ yields

$$\mathrm{EA}^+ + \mathrm{EA}_{\epsilon_0\cdot\omega+1} \vdash \mathrm{Con}(\mathrm{PA}_\omega) \vdash \mathrm{Con}(\mathrm{PA}^*),$$

by Corollary 48. By Schmerl's formula it is also clear that

$$\mathrm{EA}^+ + \mathrm{EA}_{\epsilon_0\cdot\omega} \subseteq ((\mathrm{EA})^2_1)_{\epsilon_0\cdot\omega} \nvdash \mathrm{Con}(\mathrm{PA}^*),$$

which proves Statement (ii). ⊠

Observe that the Π^0_1-ordinals of PA and PA^* defined from below both equal ϵ_0, because PA^* is deductively equivalent to PA. Hence, for PA^* the Π^0_1-ordinals defined from below and from above are different—this reflects the gap between the power of axioms of this theory and the effectiveness of its proofs.

Despite the fact that, as we have seen, the proof-theoretic ordinals defined from above may have some independent meaning, it seems that those from below are more fundamental and better behaved.

Appendix 3

Here we discuss the role of the metatheory EA^+ that was taken as basic in this paper. On the one hand, it is the simplest choice, and if one is interested in the analysis of strong systems, there is no reason to worry about it. Yet, if one wants to get meaningful ordinal assignments for theories not containing EA^+, such as $\mathrm{EA} + I\Pi^-_1$ or $\mathrm{EA} + \mathrm{Con}(I\Sigma_1)$, the problem of weakening the metatheory has to be addressed. For example, somewhat contrary to the intuition, it can be shown (see below) that $\mathrm{EA} + \mathrm{Con}(I\Sigma_1)$ is not a Π^0_1-regular theory in the usual sense.

These problems can be handled, if one reformulates the hierarchies of iterated consistency assertions using the notion of *cut-free provability* and formalizes Schmerl's formulas in EA using *cut-free conservativity*. Over EA^+ these notions provably coincide with the usual ones, so they can be considered as reasonable generalizations of the usual notions in the context of the weak arithmetic EA. The idea of using cut-free provability predicates for this kind of problems comes from Wilkie and Paris [39]. Below we briefly sketch this approach and consider some typical examples.

A formula φ is *cut-free provable* in a theory T (denoted $T \vdash^{\mathrm{cf}} \varphi$), if there is a finite set T_0 of (closed) axioms of T such that the sequent $\neg T_0, \varphi$ has a cut-free proof in predicate logic. Similarly, φ is *rank k provable*, if for some finite $T_0 \subseteq T$, the sequent $\neg T_0, \varphi$ has a proof with the ranks of all cut-formulas bounded by k.

If T is elementary presented, its cut-free $\Box_T^{\mathrm{cf}}$ and rank k provability predicates can be naturally formulated in EA. It is known that EA^+ proves the equivalence of the ordinary and the cut-free provability predicates. On the other hand, EA can only prove the equivalence of the cut-free and the rank k provability predicates for any fixed k, but not the equivalence of the cut-free and the ordinary provability predicates.

The behavior of $\Box_T^{\mathrm{cf}}$ in EA is very much similar to that of $\Box_T$, e.g., $\Box_T^{\mathrm{cf}}$ satisfies the EA-provable Σ_1-completeness and has the usual provability logic—this essentially follows from the equivalence of the bounded cut-rank and the cut-free provability predicates in EA.

Visser [38], building on the work H. Friedman and P. Pudlák, established a remarkable relationship between the predicates of ordinary and cut-free provability: if T is a finite theory, then[10]

$$\mathrm{EA} \vdash \forall x\, (\Box_T \varphi(\dot{x}) \leftrightarrow \Box_{\mathrm{EA}}^{\mathrm{cf}} \Box_T^{\mathrm{cf}} \varphi(\dot{x})). \tag{13}$$

In particular, for EA itself Visser's formula (13) attains the form

$$\mathrm{EA} \vdash \forall x\, (\Box_{\mathrm{EA}} \varphi(\dot{x}) \leftrightarrow \Box_{\mathrm{EA}}^{\mathrm{cf}} \Box_{\mathrm{EA}}^{\mathrm{cf}} \varphi(\dot{x})).$$

This can be immediately generalized (by reflexive induction) to progressions of iterated cut-free consistency assertions. Let $\mathrm{Con}^{\mathrm{cf}}(T)$ denote $\neg\Box_T^{\mathrm{cf}}\bot$ and let a nice well-ordering be fixed.

Proposition 50 *The following holds provably in* EA*:*

(i) If $\alpha \prec \omega$*, then* $\mathrm{EA}_\alpha \equiv \mathrm{Con}^{\mathrm{cf}}(\mathrm{EA})_{2\cdot\alpha}$;
(ii) If α *is a limit ordinal, then* $\mathrm{EA}_\alpha \equiv \mathrm{Con}^{\mathrm{cf}}(\mathrm{EA})_\alpha$;
(iii) If $\alpha = \omega \cdot \beta + n + 1$*, where* $\beta \succ 0$ *and* $n \prec \omega$*, then*

$$\mathrm{EA}_\alpha \equiv \mathrm{Con}^{\mathrm{cf}}(\mathrm{EA})_{\omega\cdot\beta+2n+1}.$$

We omit a straightforward proof by Visser's formula. We call a theory T Π_1^0*-cf-regular*, if for some α,

$$T \equiv_{\Pi_1} \mathrm{Con}^{\mathrm{cf}}(\mathrm{EA})_\alpha. \tag{14}$$

The situation with the cut-free reflection principles of higher arithmetical complexity is even easier.

[10] A. Visser works in a relational language and uses efficient numerals, but this does not seem to be essential for the general result over EA.

Proposition 51 *Let T be a finite extension of* EA. *Then for any $n > 1$,*

$$\mathrm{EA} \vdash \mathrm{RFN}^{\mathrm{cf}}_{\Pi_n}(T) \leftrightarrow \mathrm{RFN}_{\Pi_n}(T).$$

Proof We only show the implication ($\rightarrow$), the opposite one is obvious. For any $\varphi(x) \in \Pi_n$ the formula $\Box_T\varphi(\dot{x})$ implies $\Box^{\mathrm{cf}}_T\Box^{\mathrm{cf}}_T\varphi(\dot{x})$. Applying $\mathrm{RFN}^{\mathrm{cf}}_{\Pi_n}(T)$ twice yields $\varphi(x)$. ⊠

This equivalence carries over to the iterated principles. Let $(T)^{n,\mathrm{cf}}_\alpha$ denote $\mathrm{RFN}^{\mathrm{cf}}_{\Pi_n}(T)_\alpha$. Using the fact that for successor ordinals α the theories $(T)^{n,\mathrm{cf}}_\alpha$ are finitely axiomatizable we obtain by reflexive induction in EA using Proposition 51 for the induction step:

Proposition 52 *For any $n > 1$, provably in* EA,

$$\forall\alpha\ \ (T)^{n,\mathrm{cf}}_\alpha \equiv (T)^n_\alpha.$$

We say that T is *cut-free Π_n-conservative* over U, if for every $\varphi \in \Pi_n$, $T \vdash^{\mathrm{cf}} \varphi$ implies $U \vdash^{\mathrm{cf}} \varphi$. Let $T \equiv^{\mathrm{cf}}_{\Pi_n} U$ denote a natural formalization in EA of the mutual cut-free Π_n-conservativity of T and U. Externally $\equiv^{\mathrm{cf}}_{\Pi_n}$ is the same as $\equiv_{\Pi_n}$, so the difference between the two notions only makes sense in formalized contexts.

Analysis of the proof of Schmerl's formula reveals that we deal with an elementary transformation of a cut-free derivation into a derivation of a bounded cut-rank. To see this, the reader is invited to check the ingredient proofs of Theorem 2 and Propositions 19 and 25. All these elementary proof transformations are verifiable in EA, which yields the following formalized variant of Schmerl's formula (we leave out all the details).

Proposition 53 *For all $n \geq 1$, if T is an elementary presented Π_{n+1}-axiomatized extension of* EA, *the following holds provably in* EA*:*

$$\forall\alpha \succeq 1\ (T)^{n+m}_\alpha \equiv^{\mathrm{cf}}_{\Pi_n} (T)^n_{\omega_m(\alpha)}$$

We notice that this relationship holds for the ordinary as well as for the cut-free reflection principles, because, even for $n = 0$, the ordinal on the right hand side of the equivalence is a limit (if $m > 0$).

Now we consider a few examples. The following proposition shows that the theory $\mathrm{EA} + I\Pi_1^-$ is both Π_1^0-cf-regular and Π_1^0-regular with the ordinal ω.

Proposition 54 *Provably in* EA,

$$\mathrm{EA} + I\Pi_1^- \equiv_{\Pi_1} \mathrm{Con}^{\mathrm{cf}}(\mathrm{EA})_\omega \equiv_{\Pi_1} \mathrm{EA}_\omega.$$

Proof The logics of the ordinary and the cut-free provability for EA coincide, so by the usual proof the schema of local reflection w.r.t. the cut-free provability is Π_1-conservative over $\mathrm{Con}^{\mathrm{cf}}(\mathrm{EA})_\omega$. But the former contains $\mathrm{EA} + I\Pi_1^-$ by the comment following (E3)(c), moreover this inclusion is easily formalizable in EA (both in the usual and in the cut-free version). This proves the first equivalence. The second one follows from Proposition 50(ii). ⊠

Consider the theories $\mathrm{EA} + \mathrm{Con}^{\mathrm{cf}}(I\Sigma_1)$ and $\mathrm{EA} + \mathrm{Con}(I\Sigma_1)$. We show that the first one is Π^0_1-cf-regular with the ordinal $\omega^\omega + 1$, and the second is Π^0_1-cf-regular with the ordinal $\omega^\omega + 2$. Notice, however, that only the *first* theory is Π^0_1-regular in the usual sense: by Proposition 50(iii), $\mathrm{EA}_{\omega^\omega+1} \equiv \mathrm{Con}^{\mathrm{cf}}(\mathrm{EA})_{\omega^\omega+1}$, whereas

$$\mathrm{EA}_{\omega^\omega+2} \equiv \mathrm{Con}^{\mathrm{cf}}(\mathrm{EA})_{\omega^\omega+3} \not\equiv \mathrm{Con}^{\mathrm{cf}}(\mathrm{EA})_{\omega^\omega+2}.$$

Proposition 55 *Provably in* EA,

(i) $\mathrm{EA} + \mathrm{Con}^{\mathrm{cf}}(I\Sigma_1) \equiv \mathrm{Con}^{\mathrm{cf}}(\mathrm{EA})_{\omega^\omega+1}$;
(ii) $\mathrm{EA} + \mathrm{Con}(I\Sigma_1) \equiv \mathrm{Con}^{\mathrm{cf}}(\mathrm{EA})_{\omega^\omega+2}$.

Proof Part (i) follows from the (obvious) formalizability of the equivalence of $I\Sigma_1$ and $(\mathrm{EA})^3_1$ in EA and Proposition 53. One can show that the two systems are also cut-free equivalent, provably in EA.

To prove Part (ii) recall that by Visser's formula (13), provably in EA, $\mathrm{Con}(I\Sigma_1)$ is equivalent to $\mathrm{Con}^{\mathrm{cf}}(\mathrm{EA} + \mathrm{Con}^{\mathrm{cf}}(I\Sigma_1))$, and hence to $\mathrm{Con}^{\mathrm{cf}}(\mathrm{EA})_{\omega^\omega+2}$ by Part (i). ⊠

The following facts are also worth noticing. Statement (i) below implies that $I\Sigma_n$ is not EA-provably Π^0_1-regular (and thus the original Schmerl's formula is not formalizable in EA). In contrast, Statement (ii) implies that incidentally PA itself *is* EA-provably Π^0_1-regular.

Proposition 56

(i) $\mathrm{EA} \nvdash \mathrm{Con}(\mathrm{EA}_{\omega_{n+1}}) \to \mathrm{Con}(I\Sigma_n)$;
(ii) $\mathrm{PA} \equiv_{\Pi_1} \mathrm{EA}_{\epsilon_0}$, provably in EA, hence $\mathrm{EA} \vdash \mathrm{Con}(\mathrm{EA}_{\epsilon_0}) \to \mathrm{Con}(\mathrm{PA})$.

Proof Fact (i) has just been proved for $I\Sigma_1$: $\mathrm{Con}(I\Sigma_n)$ is Π^0_1-conservative over $\mathrm{Con}^{\mathrm{cf}}(\mathrm{EA})_{\omega_{n+1}+2}$, whereas $\mathrm{Con}(\mathrm{EA}_{\omega_{n+1}})$ is equivalent to $\mathrm{Con}^{\mathrm{cf}}(\mathrm{EA})_{\omega_{n+1}+1}$ by Proposition 50(iii). Hence, (i) follows by Löb's principle for the cut-free provability.

To prove (ii) we formalize the following reasoning in EA: Assume $\pi \in \Pi_1$ and $\mathrm{PA} \vdash \pi$. Then for some n, $I\Sigma_n \vdash \pi$. Since $I\Sigma_n$ is finitely axiomatized, by (13) we obtain that $\exists n\, \mathrm{EA} \vdash^{\mathrm{cf}} \Box^{\mathrm{cf}}_{I\Sigma_n} \pi$, therefore by Proposition 53

$$\exists n\, \mathrm{EA} \vdash^{\mathrm{cf}} \Box^{\mathrm{cf}}_{\mathrm{EA}_{\omega_{n+1}}} \pi,$$

which can be weakened to

$$\exists n\, \mathrm{EA}_{\epsilon_0} \vdash \Box^{\mathrm{cf}}_{\mathrm{EA}_{\omega_{n+1}}} \pi. \tag{15}$$

On the other hand, we notice that (provably in EA) for every fixed $\beta \prec \epsilon_0$ and $\pi \in \Pi_1$,

$$\mathrm{EA}_{\epsilon_0} \vdash \Box^{\mathrm{cf}}_{\mathrm{EA}_\beta} \pi \to \pi,$$

by the cut-free version of the Σ_1-completeness principle, and applying this to (15) yields $\mathrm{EA}_{\epsilon_0} \vdash \pi$. ⊠

Corollary 57 $|\mathrm{EA} + \mathrm{Con(PA)}|_{\Pi^0_1} = \epsilon_0 + 1$.

Acknowledgements The bulk of this paper was written during my stay in 1998–1999 as Alexander von Humboldt fellow at the Institute for Mathematical Logic of the University of Münster. Discussions with and encouragements from W. Pohlers, A. Weiermann, W. Burr, and M. Möllerfeld have very much influenced both the ideological and the technical sides of this work. I would also like to express my cordial thanks to W. and R. Pohlers, J. and D. Diller, H. Brunstering, M. Pfeifer, W. Burr, A. Beckmann, B. Blankertz, I. Lepper, and K. Wehmeier for their friendly support during my stay in Münster.

Supported by Alexander von Humboldt Foundation, INTAS grant 96-753, and RFBR grants 98-01-00282 and 15-01-09218.

References

1. L.D. Beklemishev, Remarks on Magari algebras of PA and $I\Delta_0 + \mathrm{EXP}$, in *Logic and Algebra*, ed. by P. Agliano, A. Ursini (Marcel Dekker, New York, 1996), pp. 317–325
2. L.D. Beklemishev, Induction rules, reflection principles, and provably recursive functions. Ann. Pure Appl. Log. **85**, 193–242 (1997)
3. L.D. Beklemishev, Notes on local reflection principles. Theoria **63**(3), 139–146 (1997)
4. L.D. Beklemishev, A proof-theoretic analysis of collection. Arch. Math. Log. **37**, 275–296 (1998)
5. L.D. Beklemishev, Parameter free induction and provably total computable functions. Theor. Comput. Sci. **224**(1–2), 13–33 (1999)
6. L.D. Beklemishev, Proof-theoretic analysis by iterated reflection. Arch. Math. Log. **42**(6), 515–552 (2003)
7. L.D. Beklemishev, Provability Algebras and Proof-Theoretic Ordinals, I. Ann. Pure Appl. Log. **128**, 103–124 (2004)
8. S. Feferman, Arithmetization of metamathematics in a general setting. Fundam. Math. **49**, 35–92 (1960)
9. S. Feferman, Transfinite recursive progressions of axiomatic theories. J. Symb. Log. **27**, 259–316 (1962)
10. S. Feferman, Systems of predicative analysis. J. Symbol. Log. **29**, 1–30 (1964)
11. S. Feferman, Turing in the land of $\mathcal{O}$(z), in *The Universal Turing Machine A Half-Century Survey*, ed. by R. Herken. Series Computerkultur, vol. 2 (Springer, Vienna, 1995), pp. 103–134
12. H. Gaifman, C. Dimitracopoulos, Fragments of Peano's arithmetic and the MDRP theorem, in *Logic and Algorithmic (Zurich, 1980)*. Monograph. Enseign. Math., vol. 30 (University of Genève, Genève, 1982), pp. 187–206
13. S. Goryachev, On interpretability of some extensions of arithmetic. Mat. Zametki **40**, 561–572 (1986) [In Russian. English translation in Math. Notes, 40]
14. P. Hájek, P. Pudlák, *Metamathematics of First Order Arithmetic* (Springer, Berlin, 1993)
15. R. Kaye, J. Paris, C. Dimitracopoulos, On parameter free induction schemas. J. Symb. Log. **53**(4), 1082–1097 (1988)
16. G. Kreisel, Ordinal logics and the characterization of informal concepts of proof, in *Proceedings of International Congress of Mathematicians* (Edinburgh, 1958), pp. 289–299

17. G. Kreisel, Principles of proof and ordinals implicit in given concepts, in *Intuitionism and Proof Theory*, ed. by R.E. Vesley A. Kino, J. Myhill (North-Holland, Amsterdam, 1970), pp. 489–516
18. G. Kreisel, Wie die Beweistheorie zu ihren Ordinalzahlen kam und kommt. Jahresbericht der Deutschen Mathematiker-Vereinigung **78**(4), 177–223 (1977)
19. G. Kreisel, A. Lévy, Reflection principles and their use for establishing the complexity of axiomatic systems. Zeitschrift f. math. Logik und Grundlagen d. Math. **14**, 97–142 (1968)
20. D. Leivant, The optimality of induction as an axiomatization of arithmetic. J. Symb. Log. **48**, 182–184 (1983)
21. P. Lindström, The modal logic of Parikh provability. Technical Report, Filosofiska Meddelanden, Gröna Serien 5, Univ. Göteborg (1994)
22. M. Möllerfeld, *Zur Rekursion längs fundierten Relationen und Hauptfolgen* (Diplomarbeit, Institut für Mathematische Logik, Westf. Wilhelms-Universität, Münster, 1996)
23. H. Ono, Reflection principles in fragments of Peano Arithmetic. Zeitschrift f. math. Logik und Grundlagen d. Math. **33**(4), 317–333 (1987)
24. R. Parikh, Existence and feasibility in arithmetic. J. Symb. Log. **36**, 494–508 (1971)
25. C. Parsons, On a number-theoretic choice schema and its relation to induction, in *Intuitionism and Proof Theory*, ed. by A. Kino, J. Myhill, R.E. Vessley (North Holland, Amsterdam, 1970), pp. 459–473
26. C. Parsons, On n-quantifier induction. J. Symb. Log. **37**(3), 466–482 (1972)
27. W. Pohlers, A short course in ordinal analysis, in *Proof Theory, Complexity, Logic*, ed. by A. Axcel, S. Wainer (Oxford University Press, Oxford, 1993), pp. 867–896
28. W. Pohlers, Subsystems of set theory and second order number theory, in *Handbook of Proof Theory*, ed. by S.R. Buss (Elsevier/North-Holland, Amsterdam, 1998), pp. 210–335
29. W. Pohlers, *Proof Theory. The First Step into Impredicativity* (Springer, Berlin, 2009)
30. M. Rathjen, Turing's "oracle" in proof theory, in *Alan Turing: His Work and Impact*, ed. by S.B. Cooper, J. van Leuven (Elsevier, Amsterdam, 2013), pp. 198–201
31. H.E. Rose, *Subrecursion: Functions and Hierarchies* (Clarendon Press, Oxford, 1984)
32. U.R. Schmerl, A fine structure generated by reflection formulas over primitive recursive arithmetic, in *Logic Colloquium'78*, ed. by M. Boffa, D. van Dalen, K. McAloon (North Holland, Amsterdam, 1979), pp. 335–350
33. K. Schütte, Eine Grenze für die Beweisbarkeit der transfiniten Induktion in der verzweigten Typenlogik. Arch. Math. Log. **7**, 45–60 (1965)
34. K. Schütte, Predicative well-orderings, in *Formal Systems and Recursive Functions*, ed. by J.N. Crossley, M.A.E. Dummet. Studies in Logic and the Foundations of Mathematics (North-Holland, Amsterdam), pp. 280–303 (1965)
35. C. Smoryński, The incompleteness theorems, in *Handbook of Mathematical Logic*, ed. by J. Barwise (North Holland, Amsterdam, 1977), pp. 821–865
36. R. Sommer, Transfinite induction within Peano arithmetic. Ann. Pure Appl. Log. **76**(3), 231–289 (1995)
37. A.M. Turing, System of logics based on ordinals. Proc. London Math. Soc. **45**(2), 161–228 (1939)
38. A. Visser, Interpretability logic, in *Mathematical Logic*, ed. by P.P. Petkov (Plenum Press, New York, 1990), pp. 175–208
39. A. Wilkie, J. Paris, On the scheme of induction for bounded arithmetic formulas. Ann. Pure Appl. Log. **35**, 261–302 (1987)

Part III
Philosophical Reflections

Alan Turing and the Foundation of Computer Science

Juraj Hromkovič

Abstract The goal of this article is to explain to non-specialists what computer science is about and what is the contribution of Alan Turing to general science. To do this well-understandable for everybody, we introduce mathematics as a language describing objects and relations between objects in an exact and unambiguously interpretable way, and as a language of correct, well-verifiable reasoning. Starting with the dream of Gottfried Wilhelm Leibniz, who wanted to automatize some part of the work of mathematicians including also reasoning, we finish with Alan Turing's exact axiomatic definition of the limits of automation of the intellectual work, i.e., with the birth date of computer science.

Keywords Models of computation • Philosophy of mathematics • Turing machines

MSC 2010: 68Q05, 68Q30, 03D10, 00A30

1 "What is Mathematics?" or The Dream of Leibniz

What is mathematics? What is the role of mathematics in science? These are the right questions if one wants to understand the contribution of Alan Turing to science. Already from the early beginning, numbers, calculations, equations, and geometry were used to describe the world surrounding us, to solve real tasks, and to do research in order to more and more understand the nature of our world. The development of mathematics inside philosophy (science) was very similar to the development of a language. One creates a vocabulary, word by word, and uses this vocabulary to study objects, relationships, simply everything that is accessible by the language in the current state of its development.

J. Hromkovič (✉)
Department of Computer Science, ETH Zurich, Zurich, Switzerland
e-mail: juraj.hromkovic@inf.ethz.ch

G. Sommaruga, T. Strahm (eds.), *Turing's Revolution*,
DOI 10.1007/978-3-319-22156-4_10

This seems to be same as when one observes the development of any natural language. What is the difference? The current language of mathematics is exact in the following sense. It is so accurate that it cannot be more accurate if one accepts its axioms that are the definitions of the basic notions. If one introduces a new word (notion) to the language of mathematics, then one must provide an exact definition of the meaning of this word. The definition has to be so precise that everybody understanding the language of mathematics interprets the meaning of the defined notion in exactly the same way and can unambiguously decide for every object whether it satisfies this definition or not. To get this high degree of precision was a long process which already started with the birth of mathematics. One often forgets that the search for the precise meaning of basic terms of fundamental concepts is at least as important as proving new theorems and that, in several cases, this took much more effort during periods of several 100 years. Mathematics became a language that avoids any kind of misinterpretation if one understands it properly. Because of that, one can measure a crucial characteristic of the development stage of a scientific discipline by the degree of the ability to use the language of mathematics in its research and in expressing its discoveries.

For instance, numbers as an abstraction were used to express and compare sizes and amounts for all kinds of measurements. The concept of number and calculating with numbers was considered so important that the Pythagoreans even believed that the whole world and all its principles could be explained by using natural numbers only. Their dream was definitely broken when one geometrically constructed the number $\sqrt{2}$ and showed that it cannot be expressed in any finite arithmetic calculation over the natural numbers (cannot be expressed as an arithmetic formula over the natural numbers). Equations were discovered in order to express relationships between different sizes in order to be able to estimate unknown sizes without measurements by calculating them from the known sizes. Euclidean geometry was developed in order to describe the three-dimensional space in which we live and move, and geometry became crucial for the measurement of areas, for architecture, for building houses, bridges, monuments etc. Each new concept in mathematics offered a new notion (word) for the language of mathematics. And any new word increased the expressive power of mathematics and made mathematics more powerful as a set of instruments for discovering our world.

The crucial point in the development of mathematics was and is its trustability which is absolute if one accepts the axioms of mathematics. If one is able to express the reality correctly in the language of mathematics or in a "mathematical model," then all products of calculations and of the research work on this model inside of mathematics offer true statements about the modeled reality. This unlimited trustability makes mathematics so valuable and, in fact, irreplaceable for science.

Leibniz [7, 8] was the first who formulated the role of mathematics in the following way. Mathematics offers an instrument to automatize the intellectual work of humans. One expresses a part of reality as a mathematical model, and then one calculates using arithmetics. One does not need to argue anymore that each step of the calculation correctly represents the objects or relationships under investigation. One simply calculates and takes the result of this calculation as a truth about the investigated part of reality. This reduction of complicated discussions about the matter of our interest to simple, but absolutely trustable calculations fascinated Leibniz and in his eyes it was an "automation" of human work. His dream was to achieve the same for reasoning. He wanted to have a formal system for reasoning as arithmetics is a formal system of calculations with numbers. With his famous sentence "Gentlemen, let us calculate," he hoped to solve all problems related to taking decisions in human society without complex, controversial discussions. The idea was simply to put all the known facts into a formal reasoning system and then to calculate. The result of the calculation would be absolutely trustable and there would be no political fighting in the process of decision making. A nice and a little bit naive dream. But we see through his dream what mathematics is for Leibniz and what his expectations were on the service of mathematics for science and for the whole human civilization. And to some extent he is right. Mathematics with its trustable way of reasoning can provide a lot of knowledge that may even essentially contribute to optimizing decisions in all areas of life if one is able to express the corresponding matters in the language of mathematics and if one has enough knowledge as an "input" for the mathematical calculation.

Leibniz tried, but was not able to build a formal system of reasoning, in spite of the fact that he had some good and interesting ideas. But the mathematicians in the next centuries made big progress and developed formal systems of reasoning that we call logic nowadays. And these systems satisfy the expectations of Leibniz to reduce formal reasoning to "logical calculation." However, one cannot put everything into this system and hope to get a solution to any question asked. This is the topic of the following sections.

2 "Is Mathematics Perfect?" or The Dream of Hilbert

After logic as a formal system of reasoning was developed, it became clear that each proof (formal argumentation) in mathematics is verifiable and this verification process can even be automatized. Let us explain this more carefully in what follows.

The end of the nineteenth century and the beginning of the twentieth century were the time of the industrial revolution. Engineers automatized a big portion of human physical work and made humans partially free from monotone physical activities, and so more free for intellectual growth. Life improved a lot during this time, and not only society, but also scientists became very optimistic about what can be reached and automatized. One of the leading experts in mathematics at that time was Hilbert [4]. He believed that searching for a solution can be automatized for any kind of problem. His dream was that the hardest and most intellectual part of the work of mathematicians, namely creating proofs of theorems, can be automatized. But the notion of "automation" developed in the meantime had a more precise meaning than in the dream of Leibniz. Let us describe this more carefully.

We start with the notion of a problem. A problem is a collection of (usually infinitely many) problem instances that have something in common. This something in common is the specification of a problem, for instance, solving quadratic equations is a problem. Each particular quadratic equation is an instance of this problem and, for sure, there are infinitely many different quadratic equations. A method for solving quadratic equations is a finite general description of the way to proceed (calculate) in order to get the solution for any of this infinitely many particular quadratic equations. The main point is the contrast between finiteness and infinity in this game. The infinite variety of particular cases of a problem can be reduced to a finite description of a method to manage each of these infinitely many problem instances. Clearly, one has to get a deep insight into the problem to be able to express its infinite dimension in finite terms.

The second crucial point is that, if someone discovers a method (called an algorithm by Hilbert) for solving a problem, then this problem becomes automatable. If one has a method for solving a problem, one can apply it and calculate the solution for any problem instance without understanding how this method was discovered and why it works. The description of a method (an algorithm) consists of simple calculation steps everybody can perform. Today, we would say that every computer as a device without intellect can perform it. Hilbert viewed the work of mathematicians as that of researchers who attempt to discover methods for solving problems and so to automatize intellectual work. He believed that, for each problem (task), there exists a method for solving it and so it is only a question of our intellectual effort to be able to discover methods and so to automatize everything. Hilbert believed in the perfection of mathematics in the following sense.

1. Every knowledge (fact, observation) expressible as a claim (statement) in the language of mathematics can be analyzed inside of mathematics and determined to be true or false.
2. For each problem expressible in the language of mathematics, there exists a method for solving it. Particularly, there must exist a method that, for a given true theorem, creates a proof that confirms its correctness.

Hence, in its kernel, the dream of Hilbert was the same as the dream of Leibniz: Mathematics has the capability of being a perfect tool for investigating this world and of solving all kinds of problems that occur in human society. The role of mathematicians is to explore the potential of mathematics and so to serve society.

3 The Incompleteness Theorem of Gödel as the End of the Dreams of Leibniz and Hilbert

We imagined that mathematics is a language and that the expressive power of this language is limited by its vocabulary. In other words, there are objects and relationships we are able to describe in mathematics and there are matters not expressible in mathematical terms. Mathematics is a perfect language with respect to unambiguous interpretation. But for this kind of perfection we pay by the descriptive power. Natural languages are much more powerful in their expressive (descriptive) power. However, they cannot be applied as instruments in the way mathematics is used. This is not surprising. Natural languages are built in order to communicate about everything we are able to experience (observe) and it does not matter whether we know what it is or not. For instance, a visual description may be enough to assign a new word to a new object. In mathematics, the situation is completely different. First, you name something only if you are able to completely express the nature (the essence) of this something. Secondly, you create notions for objects you never met and probably never will meet in reality, because they are only an auxiliary means to better understand the real world. We will discuss later how this is possible.

New concepts, and thus new words, are created in two different ways in mathematics. The fundamental way is to create new concepts based on our intuition and experience. We strongly believe that we understand the essence of an object and express it in terms of a mathematical definition. This kind of mathematical definition is called axioms. The axioms form the fundamental vocabulary of mathematics and unambiguously determine the expressive power of mathematics. All other words (concepts) are described by means of the fundamental axiomatic notions. This second kind of words simplifies the formulations in the language of mathematics. But these words do not increase the expressive power of mathematics, because one can always replace them by their description in terms of the fundamental vocabulary. Axioms of mathematics fix fundamental concepts and notions such as numbers and arithmetic calculations, correct reasoning, probability and correct calculation with probabilities, infinity, geometry, etc. One has to understand that the milestones in the development of mathematics were done by fixing new axioms and that the way to try to understand fundamental notions is usually much longer and much harder than the discovery of new theorems. To provide the formal definition of infinity took more than 2000 years, the development of the concept of probability took 300 years, etc.

Another crucial point is to recognize that inserting new axioms into mathematics increases also the argumentation power of mathematics. One can investigate and analyze things that mathematics without this axiom was unable to study. A nice example is the concept of infinity. To understand our finite world, one may say that it is not necessary to use the term of infinity. Everything we experience we consider to be finite, physics considers that the universe is finite, and we never touched infinity outside the artificial world of mathematics. Any size we ever need to calculate has a concrete, finite size. Why to enrich mathematics by the notion of "infinity" and the concept of comparing sizes of infinite objects? Because we would be unable to discover more than a fraction of the knowledge about our finite world without using infinity as an instrument. Even the classical physics of Newton would not exist without the concept of infinity. One cannot even express the notion of actual speed without the concept of infinity.

Why do we discuss this matter? Because we need to realize that one can measure the power of mathematics with respect to two parameters—the expressive power (about what one can speak in the language of mathematics) and the argumentation power (about what one can argue in mathematics). A part of the dream of Hilbert was that mathematics is perfect (he called it complete) in the sense that every statement expressible in the language of mathematics can be proved to be correct or false in mathematics (by logical calculus). With his Incompleteness Theorem, Gödel proved that the expressive power of mathematics is larger than its argumentation power. This means that there exist true statements in mathematics that cannot be proven to be true in the language of mathematics. Even worse. This situation will hold forever. If one adds a new axiom to mathematics in order to increase its argumentation power and to prove claims that were unprovable in the old mathematics, then the expressive power of mathematics will grow as well in such a way that new valid claims will be expressible that cannot be proved in the new mathematics. Hence, the gap between the expressive power of mathematics and the argumentation power of mathematics will persist forever.

The Incompleteness Theorem of Gödel is one of the main scientific discoveries of the twentieth century. This proven imperfection of mathematics must not be viewed as a negative result only or as a weakness of the formal language created by human beings. If everything including proving theorems would be solvable by mathematical methods, then everything would be automatable and one could ask what is the role of the human intellect.

Another important message we learnt from Gödel is that there is not one perfect mathematics and that we are asked to develop mathematics forever by adding new axioms and so increasing both its expressive power as well as its argumentation power.

4 Alan Turing and the Foundation of Informatics

Following the statement of the Incompleteness Theorem, it became clear that there does not exist any method for proving all valid mathematical claims, because, for some claims, no proofs exist. But the result of Gödel left a fundamental problem open. Does there exist a method for calculating proofs of provable claims? Do there exist methods for solving concrete problems for which up to now no method was discovered? At this point, we have to repeat that a "mathematical" problem is a collection of infinitely many instances of this problem. To provide a solution method means to find a finite description how to proceed in order to solve any of the infinitely many problem instances. In other words, discovering a method for solving a problem means to be able to reduce the infinite diversity of this problem to a finite size given by the description of the method.

Alan Turing wanted to show that there exist problems whose infinite diversity given by their infinitely many problem instances cannot be reduced to any finite size. But the question is: "How to prove a result like that? How to prove the non-existence of a solution method (an algorithm) for a concrete problem?" Up to that time, nobody needed a formal, mathematical definition of the notion of a "method." A good, intuitive understanding of the term "method" for a problem as a well-understandable description how to work in order to solve any instance of the given problem was sufficient. However, in order to try to prove the non-existence of a method for a problem, one has to know exactly what a method is. You cannot prove that something does not exist, if you do not know precisely what this something is. Therefore, there was the need to offer an exact, mathematical definition of the notion of a method. This means to transfer our intuition about the meaning of the word "method" into an exact description in the language of mathematics. And this is nothing else than to try to extend the current mathematics by a new axiom.

To formalize the meaning of the word "mathematical method" (called an algorithm by Hilbert) was the main contribution of Alan Turing. For this new definition, one prefers to use the new word "algorithm" instead of method in order to distinguish it from the broad use of the term "method" in our natural language. With the new concept of algorithm included in the language of mathematics, one is able to investigate the limits of automation. If a problem admits an algorithm, then solving the problem can be automatized. If there does not exist any algorithm for the given problem, then the problem is considered to be not automatically solvable. This new axiom of mathematics enabled to prove, for many problems, that they are not algorithmically solvable, and a very successful investigation of the border between algorithmic solvability and algorithmic unsolvability was initiated. An important consequence of the study of the limits of algorithmic solvability (automation) is that it is definitely not sufficient to be able to express a real-world problem in terms of mathematics. If this problem is not algorithmically solvable, then it may be an intellectual challenge to try to solve a particular instance of this problem.

To create an axiom is usually a long process and to get a new axiom accepted may take even more time. As already mentioned above, for the axiomatic definition of infinity, people needed 2000 years, for the concept of probability approximately 300 years, for correct reasoning in the form of a formal logic more than 2000 years, and we do not know how many years were needed for an axiomatic definition of geometry. The 6 years that started in 1930 with the Incompleteness Theorem of Gödel and finished in 1936 [9] with the presentation of the so-called "Turing Machine" is in fact a very short time. More time was needed to accept Turing's definition of an algorithm. This was done by trying to define the notion of an algorithm by many different formal approaches. The experience that all these approaches resulted in the same notion of algorithmic solvability (in the same limit on automation) led to the famous "Church-Turing Thesis" stating that all these formal definitions of an algorithm correspond to our intuition about the meaning of this notion.

What was the difficulty in formalizing the meaning of the word "method" ("algorithm")? One can always say that a description of a method must consist of simple actions for which nobody has doubt about that they can be simply executed. For instance, arithmetic operations or fundamental logical operations are definitely of this kind. But the question is whether we have already a set of instructions (activities) that is complete in the sense that each algorithm (method) can be expressed in terms of these instructions. If one misses a fundamental instruction that is not expressible by the instructions in our list, then all algorithms that need this instruction would not be considered as algorithms. In this way, some algorithmically solvable problems would be considered to be unsolvable and our concept of algorithm would be misleading. To express our intuition in terms of a complete set of allowed fundamental actions (instructions, operations, computational steps) was the main problem of Turing and all others who tried to formally fix the concept of algorithm.

Note that Turing did not focus on a technical device such as a computer when trying to define algorithms. He analyzed the intellectual work of a person (more precisely, a mathematician) that uses a pencil to write symbols on a paper or erases symbols from this paper. Everything written can be viewed as a sequence of symbols (text) over a finite alphabet that is chosen before starting the work. We omit to discuss details at this point, because it can be found well-explained in the classical literature.

We have to point out that the formal definition of the notion "algorithm" significantly increased the expressive power of the language of mathematics and enabled to investigate the limits of automation. The concept of algorithm also strengthened the argumentation power. The original proof of the Incompleteness Theorem of Gödel was long enough to be published as a book [3]. We can finish this article by showing that some true claims cannot be proved in the language of current mathematics. Due to using the concept of algorithm, our proof will only have the length of one page.

Chaitin and Kolmogorov [1, 2, 6] introduced the notion of Kolmogorov complexity of finite objects in order to measure the information content of finite objects. The Kolmogorov complexity of a finite binary string x, denoted by $K(x)$, is the binary code of the shortest program in a fixed programming language that generates x. This definition is ingenious, because it takes into account all infinitely many possible compression strategies and takes the best one for the compression of x. One can easily prove that there is no algorithm that, for a given x, computes $K(x)$. For a detailed explanation, see, for instance, [5].

Let us fix Σ_{math} as an alphabet over which all mathematical proofs can be written. Let $P_1, P_2, \ldots$ be the linear order of all mathematical proofs given by the canonical ordering, in which shorter proofs come before longer proofs and the proofs of the same length are lexicographically ordered. For any positive integer n, let w_n, with $K(w_n) \geq n$, be the first binary string for which there exists a j_n such that P_{j_n} is the proof of "$K(w_n) \geq n$;" and none of the proofs $P_1, P_2, \ldots, P_{j_n-1}$ proves for any binary string that its Kolmogorov complexity is at least n. Then, for each positive integer n, we give the following algorithm A_n that generates w_n.

Algorithm $\boldsymbol{A_n}$
Input: n

Generate all strings over Σ_{math} in the canonical ordering. Verify for each generated string whether it is a mathematical proof and whether this proof proves that a string x has $K(x) \geq n$. Halt, if the first proof of the fact $K(x) \geq n$ is found, for some x.

Output: x

Since w_n is the word with the first proof in the canonical order of the fact "$K(w_n) \geq n$," for each n, A_n generates w_n.

We observe that all algorithms A_n are almost the same for all n, they only differ in the input n that can be written using $\lceil \log_2(n+1) \rceil$ bits. The length of the rest can be bounded by c bits and c is independent of n. Hence, we proved

$$K(w_n) \leq c + \lceil \log_2(n+1) \rceil.$$

But following the definition of w_n, we have $K(w_n) \geq n$ and thus

$$c + \lceil \log_2(n+1) \rceil \geq K(w_n) \geq n$$

holds for all n. But this can be true only for finitely many n. The conclusion is that there exists an n_0 such that, for all $n \geq n_0$, there does not exist any proof that $K(x) \geq n$ for any word x.

5 Conclusion

Alan Turing enriched mathematics by a new axiom that increased the expressive power as well as the argumentation power of mathematics. Introducing the concept of algorithms enabled to study the limits of automation of intellectual work. In this way, a new scientific discipline, called computer science or informatics, was founded.

References

1. G.J. Chaitin. On the length of programs for computing finite binary sequences. J. Assoc. Comput. Mach. **13**(4), 547–569 (1966)
2. G.J. Chaitin, On the length of programs for computing finite binary sequences, statistical considerations. J. Assoc. Comput. Mach. **16**(1), 145–159 (1969)
3. K. Gödel, *Collected Works*, vol. 1–5 (Oxford University Press, Oxford, 1986/2003)
4. D. Hilbert, Die logischen Grundlagen der Mathematik. Math. Ann. **88**, 151–165 (1923)
5. J. Hromkovič, *Theoretical Computer Science* (Springer, New York, 2004)
6. A.N. Kolmogorov, Three approaches to the definition of the concept "quantity of information. Problemy Peredachi Informatsii **1**(1), 3–11 (1965). Russian Academy of Sciences
7. G.W. Leibniz, *Discourse on Metaphysics, Principles on Nature and Grace*. The Monadology (1686)
8. G.W. Leibniz, *Philosophical Essays*, edited and translated by R. Ariew and D. Garber (Hackett Publishing Company, Indianapolis, 1989)
9. A.M. Turing, On computable numbers, with an application to the Entscheidungsproblem. Proc. Lond. Math. Soc. **42**(2), 230–265 (Oxford University Press, Oxford, 1936)

Proving Things About the Informal

Stewart Shapiro

Abstract There has been a bit of discussion, of late, as to whether Church's thesis is subject to proof. The purpose of this article is to help clarify the issue by discussing just what it is to be a proof of something, and the relationship of the intuitive notion of proof to its more formal explications. A central item on the agenda is the epistemological role of proofs.

Keywords Church's thesis • Proof • Epistemology • Vagueness • Kleene • Church • Wang • Gandy • Godel • Sieg • Gandy • Lakatos • Foundationalism • Self-evidence

MSC Class: 03A05

1 The Received View: No Proof

For a long time, it was widely, if not universally, believed that Church's thesis is not subject to rigorous mathematical demonstration or refutation. This goes back to Alonzo Church's [1] original proposal:

> We now define the notion . . . of an *effectively calculable* function of positive integers by identifying it with the notion of a recursive function of positive integers . . . This definition is thought to be justified by the considerations which follow, so far as positive justification can ever be obtained for the selection of a formal definition to correspond to an intuitive one. (Church [1, §7])

The argument here is simple. Since Church's thesis is the identification of an intuitive notion (effective computability) with a precisely defined one (recursiveness) there is no sense to be made of establishing, mathematically, that the identification, or "definition" is correct.

S. Shapiro (✉)
The Ohio State University, Columbus, OH, USA
e-mail: shapiro.4@osu.edu

G. Sommaruga, T. Strahm (eds.), *Turing's Revolution*,
DOI 10.1007/978-3-319-22156-4_11

The founders, Church, Stephen Cole Kleene, Alan Turing, et al., were thinking of computability in terms of what can be done by a suitably idealized human following an algorithm (see, for example, [2]).[1] So the notion of computability seems to invoke psychological or other spatio-temporal notions. Emil Post [3] was explicit about this

> . . . for full generality a complete analysis would have to be made of all the possible ways in which the human mind could set up finite processes.
> . . . we have to do with a certain activity of the human mind as situated in the universe. As activity, this logico-mathematical process has certain temporal properties; as situated in the universe it has certain spatial properties.

So the theme underlying Church's claim is that there just is no hope of rigorously proving things when we are dealing with the messy world of space, time, and human psychology.

2 Vagueness: Does Church's Thesis Even Have a Truth Value?

In a retrospective article, Kleene [4, pp. 317–319] broached another, related matter:

> Since our original notion of effective calculability of a function (or of effective decidability of a predicate) is a somewhat vague intuitive one, [Church's thesis] cannot be proved . . .
> While we cannot prove Church's thesis , since its role is to delimit precisely an hitherto vaguely conceived totality, we require evidence that it cannot conflict with the intuitive notion which it is supposed to complete; i.e., we require evidence that every particular function which our intuitive notion would authenticate as effectively calculable is . . . recursive. The thesis may be considered a hypothesis about the intuitive notion of effective calculability, or a mathematical definition of effective calculability; in the latter case, the evidence is required to give the theory based on the definition the intended significance.

Kleene says that the intuitive notion of computability is *vague*. Recursiveness is presumably not vague. It may be too much of an anachronism to insist that Kleene was using the word "vague" in the sense it has taken in contemporary philosophy and linguistics, but the connection is, or may be illuminating. Proving Church's thesis would be like trying to prove that an American philosopher is rich if and only if she has a net worth of at least $783,299.45.

Kleene's central premise, it seems, is that we cannot prove things, mathematically, about "vague intuitive" notions. The analogy with ordinary vague terms, like "rich", "bald", and "thin" suggests that the identification with a precise notion (like having a net worth of at least $783,299.45) does not even have a truth value, or else

[1] Some later writers invoke a different notion of computability, which relates to mechanical computing devices. Prima facie, this latter notion is an idealization from a physical notion. So, for present purposes, the issues are similar.

it must be false, strictly speaking. Vague things cannot match up, precisely, with precise ones.

One can, of course, sharpen vague properties, for various purposes. Kleene speaks of "completing" a vague notion, presumably replacing it with a sharp one. As he notes, one can sometimes justify the chosen sharpening, or completion, at least partially, but there is no question of *proving*, with mathematical resources, that a given sharpening is the one and only correct one.

To make this case, however, we would need an argument that computability is vague. Are there any borderline computable functions? Can we construct a sorites series from a clearly computable function to a clearly non-computable function? I find it hard to imagine what such a series would look like.

In conversations with Hao Wang, Kurt Gödel challenged the conclusion based on the supposed vagueness of computability. Gödel invoked his famous (or infamous) analogy between mathematical intuition and sense perception:

> If we begin with a vague intuitive concept, how can we find a sharp concept to correspond to it faithfully? The answer is that the sharp concept is there all along, only we did not perceive it clearly at first. This is similar to our perception of an animal first far away and then nearby. We had not perceived the sharp concept of mechanical procedures before Turing, who brought us to the right perspective. And then we do perceive clearly the sharp concept.
>
> If there is nothing sharp to begin with, it is hard to understand how, in many cases, a vague concept can uniquely determine a sharp one without even the *slightest* freedom of choice.
>
> "Trying to see (i.e., understand) a concept more clearly" is the correct way of expressing the phenomenon vaguely described as "examining what we mean by a word" (quoted in Wang [5, pp. 232 and 233], see also Wang [6, pp. 84–85])

Gödel's claim seems to be that there is a precise property that somehow underlies the intuitive notion of computability. One way to put it is that computability only *appears* to be vague. The subsequent analysis of the intuitive notion convinced us, or at least convinced him, that there is a unique, sharp notion that underlies the intuitive one.

It is not clear how much "freedom of choice", to use Gödel's term, there was concerning the development of computability—in the sharpening of (what appears to be) a vague, intuitive notion. There are, after all, other notions of computability: finite state computability, push-down computability, computability in polynomial time, etc. For that matter, it is not clear what it is to have "freedom of choice" when examining certain concepts—and possibly sharpening them.

Our main question stands. When put in Gödelian terms, how does one *prove* that a proposed sharpening of an apparently vague notion is the uniquely correct one—that it captures the precise concept that was "there all along". Does mathematics extend that far? How?

3 Theses

Computability is not the first instance of an intuitive, or pre-theoretic notion that was eventually "defined" in more explicit and rigorous terms, to just about everyone's satisfaction. Think of "area", "measure", "continuous", and "smooth". Gödel mentions two other cases of initially vague, or at least intuitive notions for which, at least with hindsight, we can see that the formal development could not have gone in any other way:

> The precise concept meant by the intuitive idea of velocity is d*s*/d*t*, and the precise concept meant by "size" . . . is clearly equivalent with the Peano measure in the cases where either concept is applicable. In these cases the solutions again are *unquestionably* unique, which here is due to the fact that only they satisfy certain axioms which, on closer inspection, we find to be undeniably implied by the concept we had. For example, congruent figures have the same area, a part has no larger size than the whole, etc. (quoted in Wang [5, p. 233])

If "intuitive" notions are somehow disqualified from the pristine sphere of mathematical demonstration, there would not have been any mathematics, or at least no mathematical proofs, before the advent of formal deductive systems. If we are to make sense of mathematics before, say, 1900, and of much mathematics today, we had better allow intuitive, pre-theoretic notions into the fold. Our question concerns the epistemological process involved in accepting (or rejecting) a proposed definition.

In general, how do intuitive, mathematical notions relate to their rigorously defined successors? Is there a sense in which one can sometimes *prove* that the definitions are correct? Or is always more like a premise, or a presupposition of the significance of the treatment, as Kleene suggests?

4 The Opposition: It Is Possible to Prove These Things

Even from the beginning, it was only a *near* consensus that Church's thesis is not subject to mathematical demonstration. In a letter to Kleene, Church noted that Gödel opposed the prevailing views on this matter[2]:

> In regard to Gödel and the notions of recursiveness and effective calculability, the history is the following. In discussion with him . . . it developed that there is no good definition of effective calculability. My proposal that lambda-definability be taken as a definition of it he regarded as thoroughly unsatisfactory . . .
>
> His [Gödel's] only idea at the time was that it might be possible, in terms of effective calculability as an undefined term, to state a set of axioms which would embody the generally accepted properties of this notion, and to do something on that basis.

It is not easy to interpret these second-hand remarks, but Gödel seems to have demanded a *proof*, or something like a proof, of Church's thesis or of some related

[2]The letter, which was dated November 29, 1935, is quoted in Davis [12, p. 9].

thesis, contrary to the emerging consensus. His proposal seems to be that one begin with a conceptual analysis of computability. This might suggest axioms for the notion, axioms one might see to be self-evident. Then, perhaps, one could rigorously prove that every computable function is recursive, and vice versa, on the basis of those axioms. Something along these lines might count as a demonstration, or proof, of Church's thesis. As indicated in the passage from the Wang conversations, Gödel may have held that something similar underlies, or could underlie, the other "theses", concerning velocity and area.

Gödel himself was soon convinced that (something equivalent to) Church's thesis is true. Turing's landmark paper was a key event. In a 1946 address (published in Davis [7, pp. 84–88]), Gödel remarked that

> Tarski has stressed in his lecture (and I think justly) the great importance of the concept of general recursiveness (or Turing's computability). It seems to me that this importance is largely due to the fact that with this concept one has for the first time succeeded in giving an absolute definition of an interesting epistemological notion . . . (Gödel [8, p. 84])

Although it is not clear whether Gödel regarded Turing's original [9] treatment of computability as constituting, or perhaps suggesting, or leading to, a rigorous *proof* of Church's thesis—perhaps along the foregoing lines—it seems that Gödel was satisfied with Turing's treatment (see Kleene [10, 11], Davis [12], and my review of those, Shapiro [13]), and convinced that Turing computability is the uniquely correct explication of computability. In a 1964 postscript to Gödel [14], he wrote that "Turing's work gives an analysis of the concept of 'mechanical procedure' (alias 'algorithm' or 'computation procedure' or 'finite combinatorial procedure'). This concept is shown to be equivalent to that of a Turing machine". The word "shown" here might be read as "proved", but perhaps we should not be so meticulous in interpreting this remark.

In any case, the once near consensus that Church's thesis is not subject to mathematical proof or refutation was eventually challenged by prominent philosophers, logicians, and historians. Robin Gandy [15] and Elliott Mendelson [16] claimed that Church's thesis *is* susceptible of rigorous, mathematical proof; and Gandy went so far as to claim that Church's thesis has in fact been *proved*. He cites Turing's [9] study of a human following an algorithm as the germ of the proof. Gandy referred to (a version of) Church's thesis as "Turing's theorem". Wilfried Sieg [17] reached a more guarded, but similar conclusion: "Turing's theorem" is the proposition that if f is a number-theoretic function that can be computed by a being (or device) satisfying certain determinacy and finiteness conditions, then f can be computed by a Turing machine. Turing's original paper [9] contains arguments that humans satisfy some these conditions, but, apparently, Sieg [17] considered this part of Turing's text to be less than a rigorous proof. Sieg [18, 19] claims to have completed the project of rigorously demonstrating the equivalence of computability and recursiveness, via a painstaking analysis of the notions involved:

> My strategy . . . is to bypass theses altogether and avoid the fruitless discussion of their (un-)provability. This can be achieved by *conceptual analysis*, i.e., by sharpening the

> informal notion, formulating its general features axiomatically, and investigating the axiomatic framework . . .
>
> The detailed conceptual analysis of effective calculability yields rigorous characterizations that dispense with theses, reveal human and machine calculability as axiomatically given mathematical concepts, and allow their systematic reduction to Turing computability. (Sieg [18, pp. 390 and 391])

Sieg's use of the term "sharpening" raises the aforementioned matter of vagueness. It is indeed intriguing that a conceptual analysis, followed by a rigorous demonstration, can somehow clear away vagueness, showing—with full mathematical rigor—that there was a unique concept there all along, as Gödel contended.

The Gödel/Mendelson/Gandy/Sieg position(s) generated responses (e.g., Folina [20], Black [21]), and the debate continues. As the foregoing sketch indicates, the issues engage some of the most fundamental questions in the philosophy of mathematics. What is it to prove something? What counts as a proof? For that matter, what is mathematics about?

5 So What Is It to Prove Something?

Let us turn to some prominent formal explications of the notion of proof. Considering those will, I think, shed light on our question concerning Church's thesis.

From some perspectives, a proof is a sequence of statements that can be "translated", at least in principle, into a derivation in Zermelo-Fraenkel set theory (ZF, or ZFC). Call this a *ZFC-proof.* The staunch naturalist Penelope Maddy [22, p. 26] sums up this foundational role for set theory: "if you want to know if a given statement is provable or disprovable, you mean (ultimately), from the axioms of the theory of sets."

With this proposed explication in mind, we'd begin with a series of statements in ordinary natural language, supplemented with mathematical terminology—say the contents of the compelling Sieg analysis. We'd translate the statements into the language of ZFC—with membership being the only non-logical term—and show that the translation of our conclusion, Church's thesis, follows from the axioms of ZFC.

The issues would then focus on the translation from the informal language into that of set theory. What should be preserved by a decent translation from an informal mathematical text into the language of set theory? Despite the use of the word "translation", it is surely too much to demand that the ZFC-version has the same *meaning* as the original text. How can we capture the meaning of, say, a statement about bounded mechanical procedures, by using only the membership symbol? Presumably, the central mathematical and logical relations should be preserved in the translation. But what are those? Can we leave this matter at an intuitive level, claiming that we know a good "translation" when we see it? Surely, that would beg all of the questions.

The general matter of the envisioned "translation" is broached in Yiannis Moschovakis text [23, pp. 33–34], concerning another "thesis", taken from real analysis:

> A typical example of the method . . . is the "identification" of the . . . geometric line . . . with the set . . . of real numbers, via the correspondence which "identifies" each point . . . with its coordinate . . . What is the precise meaning of this "identification"? *Certainly not that points are real numbers*. Men have always had direct geometric intuitions about points which have nothing to do with their coordinates . . . What we mean by the "identification" . . . is that the correspondence . . . gives a **faithful representation** of [the line] in [the real numbers] which allows us to give arithmetic definitions for all the useful geometric notions and to study the mathematical properties of [the line] **as if points were real numbers**.

Set theory provides a target for embedding just about any mathematical domain. Moschovakis continues:

> . . . we . . . discover within the universe of sets *faithful representations* of all the mathematical objects we need, and we will study set theory . . . **as if all mathematical objects were sets**. The delicate problem in specific cases is to formulate precisely the correct definition of "faithful representation" and to prove that one such exists.

With computability and recursive function theory, we have an instance of what Moschovakis calls the "delicate problem". A ZFC-proof of Church's thesis would begin with set-theoretic "definitions" of "computability" and at least one of the notions of "recursiveness", "Turing computability", "λ-definability", etc. There are, or easily could be, good, or at least well-established formulations of the latter in ZFC. That much would be a routine exercise. However, formulating the notion of "computability" in the language of ZFC would be another story. How could we assure ourselves that the proposed set-theoretic predicate really is an accurate formulation of the intuitive, pre-theoretic notion of computability? Would we prove that? How? In ZFC?

In effect, a statement that a proposed predicate in the language of set theory (with its single primitive) does in fact coincide with computability would be the same sort of thing as Church's thesis, in that it would propose that a precisely defined—now set-theoretic—property is equivalent to an intuitive one. We would be back where we started, philosophically. Clearly, in claiming that Church's thesis is capable of proof, Mendelson and Gandy (not to mention Gödel) are not asserting the existence of a ZFC-proof for Church's thesis (see Mendelson [16, p. 223]).

Let us turn to another, perhaps earlier, explication of proof. We begin by separating questions concerning the underlying logic of a proof from questions concerning its premises. Define a *derivation* to be a sequence of well-formed-formulas in a formal language, constructed according to certain rules, specified in advance. There are philosophical issues concerning how the primitive inference-rules in the formal deductive system relate to the intuitive notion of a correct inference, an argument that is immediately or intuitively compelling. That is a matter much like Church's thesis, relating a precisely defined notion to an intuitive, in this case epistemic, notion: an argument is intuitively valid just in case a translation of it into the formal language is sanctioned as such in the formal deductive system.

I propose to just take it for granted, for the sake of argument, that we do have a good explication of suitably idealized deductive discourse. So let us just assume that we do know what it is for a premise-conclusion pair to be valid, and that translation into one of the standard formal deductive systems matches that. But, of course, not all valid deductive argumentations constitute *proofs*. Which discourses establish their conclusions with full mathematical rigor? With our proposed explication, our question becomes which formal derivations, in decent deductive systems, constitute or correspond to *proofs*?

A common view is that a given formal derivation constitutes a *proof* only if all of the (undischarged) premises are, or correspond to, or are translations of, either self-evident or previously established propositions.[3] Call such a derivation a *formal proof*.

Presumably, a formal proof of Church's thesis would begin with a standard formalization of number-theory. One would add a predicate for computability, which applies to number-theoretic functions (or sets of natural numbers), to the formal language. And one would supply some axioms for this new predicate.

Present action would then focus on these purported new axioms. They should either be unquestionably true—self-evident—or previously established. Then we would produce a sequence of formulas, each of which would either be an axiom (of number theory or of computability), or would follow from previous lines by the unquestionably valid, formally correct rules of the deductive system. The last line would be Church's thesis.

This model seems to fit Gödel's early suggestion to Church that logicians should try to "state a set of axioms which would embody the generally accepted properties of [effective calculability] and . . . do something on that basis". The axioms should constitute intuitively evident truths about computability; the "something" that we would do "on that basis" would be to produce a formalized, or at least formalizable derivation. The model also seems to fit the deep analysis in Sieg [18, 19]. Arguably, Sieg's conceptual analysis does give us evident truths about computability, perhaps even *self*-evident truths. It would be no harder to formalize Sieg's text than it would any sufficiently informal mathematical argument.

On the present explication, whether an unformalized text constitutes a proof presumably depends on how close the text is to a formalization, and how plausible the resulting formalized axioms and premises are, with respect to the intuitive notions invoked in it. This perspective might allow one to think of Turing's text as a formalizable proof—or the germ of one—despite the fact that some people were not convinced, and despite the fact that most mathematicians did not, and many still do not, recognize it as a proof or a proof-germ. The uncertainty is charged to the connection between an informal text in ordinary language and a formalization thereof.

[3]Clearly, this is not a sufficient condition. A one or two line argument whose sole premise and conclusion is a previously established proposition does not constitute a proof of that proposition.

This example, and thousands of others like it, raises a general question of how we tell whether a given piece of live mathematical reasoning corresponds to a given actual or envisioned formal proof. Clearly, there can be some slippage here: such is informal reasoning, and such is vagueness. How does one guarantee that the stated axioms or premises of the formal proof are in fact necessary for the intuitive, pre-theoretic notions invoked in the informal text? That is, how do we assure ourselves that the formalization is faithful? This question cannot be settled by a *formal* derivation. That would start a regress, at least potentially. We would push our problem to the axioms or premises of *that* derivation. Moreover, any formalization of a real-life argument reveals missing steps, or gaps, and plugging those would sometimes require theses much like Church's thesis.

In a collection of notes entitled "What does a mathematical proof prove?" (published posthumously in [24, pp. 61–69]), Imre Lakatos makes a distinction between the pre-formal development, the formal development, and the post-formal development of a branch of mathematics. Present concern is with the last of these. Lakatos observes that even after a branch of mathematics has been successfully formalized, there are residual questions concerning the relationship between the formal deductive system (or the definitions in ZFC) and the original, intuitive mathematical ideas. How can we be sure that the formal system accurately reflects the original mathematical ideas? These questions cannot be settled with a derivation in a further formal deductive system, not without begging the question or starting a regress—there would be new, but similar, questions concerning the new deductive system.

I take it as safe to claim that if there is to be a mathematical proof of Church's thesis, the proof cannot be *fully* captured with either a ZFC-proof or a formal proof or a ZFC-proof. There would be an intuitive or perhaps philosophical residue concerning the adequacy of the formal or ZFC surrogates to the original, pre-theoretic notion of computability. So if one *identifies* mathematical proof with formal proof or with ZFC-proof, then one can invoke *modus tolens* and accept the conclusion that Church's thesis is not subject to mathematical proof. But then there would not have been many proofs before the advent of formal deductive systems or the codification of ZFC. It is more natural, however, to hold that (1) Church's thesis is not *entirely* a formal or ZFC matter, but (2) this does not make Church's thesis any less mathematical, nor does it rule out a proof of Church's thesis.

6 Epistemology

The issue would turn on the status of either the set-theoretic definitions or the undischarged premises of the formal (or formalizable) derivation. What epistemic property must they have for the discourse to constitute a *proof* of its conclusion? Clearly, we cannot prove everything. What is required of the principles that, in any given derivation, are not proved?

There is a longstanding view that results from mathematics are absolutely certain. For some theorists, this is tacit; for others it is quite explicit. The idea is that for mathematics at least, humans are capable of proceeding, and should proceed, by applying infallible methods. We begin with self-evident truths, called "axioms", and proceed by self-evidently valid steps, to all of our mathematical knowledge. In practice (or performance) we invariably fall short of this, due to slips of the pen, faulty memory, and the practical need to take shortcuts, but in some sense we are indeed capable of error-free mathematics, and, moreover, all of the mathematics we have, or at least all of the correct mathematical we have, could be reproduced, in principle, via rigorous deductions from self-evident axioms, either in ZFC or in some other formalized theory.

Lakatos [24] calls this the "Euclidean" model for mathematical knowledge, but it might be better to steer away from any exegetical assumptions concerning the historical Euclid. Call it the *foundationalist* model, since it fits the mold of foundationalist epistemology. Indeed, mathematics, so conceived, is probably the model for foundationalism in epistemology, and for traditional rationalism.

In the Gibbs lecture, Gödel [25, p. 305] comes close to endorsing a foundationalist model for mathematics:

> [The incompletability of mathematics] is encountered in its simplest form when the axiomatic method is applied, not to some hypothetico-deductive system such as geometry (where the mathematician can assert only the conditional truth of the theorems), but to mathematics proper, that is, to the body of those mathematical propositions which hold in an absolute sense, without any further hypothesis . . . [T]he task of axiomatizing mathematics proper differs from the usual conception of axiomatics insofar as the axioms are not arbitrary, but must be correct mathematical propositions, and moreover, evident without proof.

Roger Penrose [26, Chap. 3] also seems endorse something in the neighborhood of the foundationalist model, with his central notion of "unassailable knowledge". He admits that even ideal subjects may be subject to error, but he insists that errors are "correctable", at least for anything we might call mathematics.

Although, as noted, the foundationalist model has advocates today, it is under serious challenge, partly due to the actual development of mathematics (see Shapiro [27]). Nowadays, the standard way to establish new theorems about, say, the natural numbers, is not to derive them from self-evident truths about natural numbers—such as the first-order Peano postulates—but rather to embed the natural numbers in a richer structure, and take advantage of the more extensive expressive and deductive resources. The spectacular resolution of Fermat's last theorem, via elliptical function theory, is not atypical in this respect.

I submit that as the embedding structures get more and more complex and powerful—as our ability to postulate gets more and more bold—we eventually lose our "unassailable" confidence that the background theories are themselves based on self-evident axioms (if we ever had such confidence). Even the consistency of the powerful theories can come into question. This seems to compromise the ideal of proceeding by self-evidently valid steps.

To be sure, the overwhelming bulk of mathematics today can be recast in axiomatic set theory, and sometimes is. That is why set theory can play the foundational role mentioned in the previous section. The iterative hierarchy is rich enough to embed any mathematical structure we may encounter, and the axioms of set theory seem to be rich enough to prove most (if not all) of what we would like to prove about various mathematical structures, once suitable definitions are provided.

Set theory has a single language, in fact only one non-logical symbol, and it has a fixed axiomatization, ZFC. But it is highly contentious to think of those axioms as themselves self-evident. Do we really have unassailable rational insight that every set has a unique powerset? Or of the axiom of choice? Or of replacement? Or, for that matter, of infinity?[4]

Moreover, Zermelo-Fraenkel set theory is clearly not sufficient for *all* mathematical knowledge, even all unassailable mathematical knowledge. Surely, we know that some Gödel sentence for ZFC is true, or that ZFC is consistent. There is a fruitful search for new axioms for set theory, in part to help decide propositions known to be independent of ZFC. Do we require that any new axioms added to ZFC must themselves be self-evident? The bold large cardinal candidates explored by contemporary set theorists hardly meet that norm.

All of this tells against the foundationalist model. I would think that a more holistic epistemology would make better sense for contemporary mathematics. The idea is that successful theories support each other, and there just is no foundational bedrock, at least not in any epistemological sense (see Shapiro [27]).

Of course, from the holistic perspective, as from any other, there is a difference between an intuitively evident proposition, an entrenched thesis, a well-confirmed working hypothesis, and a conjecture that serves as a premise to an argument. For the holist, however, the difference among these things is a matter of degree, and not a difference in kind. Being "evident", or being sufficiently evident to serve as an axiom, is not an epistemological natural kind.

Moreover, what is evident, or intuitively evident, or even what is regarded as *self*-evident, is itself highly context sensitive. What is evident to one generation of mathematicians is not always so evident to the previous generation. What is evident to a teacher is not always so evident to her students, at least at first. In general, what a given competent mathematician finds evident depends heavily on his background beliefs, on what other theories he has digested, and on what can reasonably be taken for granted in the situation at hand. In the foregoing sketch, I did not mean to raise (or endorse) doubts concerning the truth of, say, the axioms of ZFC (Boolos [28] notwithstanding). But I do doubt that they have the kind of *self*-evidence that traditional axioms are supposed to have. Some of the axioms, at least, are not the sort of thing that any rational thinker should accept, just upon grasping the very

[4] Boolos [28] is a delightful challenge to some of the more counter-intuitive consequences of ZFC, suggesting that we need not believe everything the theory tells us. If anything even remotely close to the conclusions of that paper is correct, then, surely, the axioms of ZFC fall short of the ideal of self-evident certainty.

concept of set. The history of our subject shows otherwise, assuming of course that at least some of the original skeptics were sufficiently rational. As Friedrich Waismann [29] once put it, "Mathematicians at first distrusting the new ideas . . . then got used to them". The axioms of set theory are accepted—with hindsight—in large part because of the fruit that the theory produced in organizing and extending mathematical knowledge. I'd go so far as to say that the axioms are *evident* only with hindsight.

Of course, this is not the place to settle broad questions concerning mathematical knowledge, but the general issues are centrally relevant to the matter at hand, whether it makes sense to speak of a proof of Church's thesis, and similar theses. I see no reason to withhold the honorific of "proof" or even "rigorous proof" to, say, the insightful development in Sieg [18, 19]. As noted, it is no more than an exercise to formalize the deductive aspects of the text. And the undischarged premises—say the principles of determinateness and finitude—are sufficiently evident, at least at this point in history. What makes them evident, however, is not that they are *self*-evident, the sort of thing that any rational thinker should accept upon properly grasping the concept of computability. As with set theory, the early history of Church's thesis makes this hard to maintain. What makes the premises compelling is the result of seventy-five years of sharpening, developing, and, most important, *using* recursive function theory, Turing computability, and the like in our theorizing.

This is not to say that the proof begs any questions, although some of the early critics of Church's thesis would certainly see it that way. So would any contemporary thinker who doubts the truth of Church's thesis. That is the way of deductive reasoning. If one wants to reject the conclusion of a valid argument, one will find a premise or premises to reject, possibly claiming that the question is begged. In general, and in some sense yet to be specified, when it comes to deduction, one does not get out any more than one puts in.

I presume that contemporary defenders of the original claim that Church's thesis is not susceptible of proof would also claim that the question is begged in the Sieg text. Otherwise, they would have to concede defeat (assuming that Sieg's argumentation is valid). The matter of whether a given argument begs the question is itself context-sensitive, and is often indeterminate. It depends on what various interlocutors are willing to take for granted.

I admit that, in the present intellectual climate, it is unlikely that there is anyone who was convinced of the truth of Church's thesis through, say, Sieg's analysis and argumentation—whatever its status. That is, I find it unlikely that there are theorists who were, at first, neutral or skeptical of Church's thesis, and were won over by the Sieg texts. But I take this to be more a statement about the present intellectual climate, and its history, than of the status of the argumentation. It is reasonably clear, nowadays, what an author can take her interlocutors to grant concerning the notions of computation, algorithm, and the like.

I submit, however, that Sieg's discourse and, for that matter, Turing's are no more question-begging than any other deductive argumentation. The present theme is more to highlight the holistic elements that go into the choice of premises, both in deductions generally and in discourses that are rightly called "proofs", at least in certain intellectual contexts.

Acknowledgments This paper is a sequel to some early sections in Shapiro [30] and Shapiro [31]. Thanks to the audience at the Conference on Church's thesis, Trends in Logic conference, in Krakow, Poland, in 2011, and the Conference on Turing, held in Zurich, in 2012.

References

1. A. Church, An unsolvable problem of elementary number theory. Am. J. Math. **58**, 345–363 (1936); reprinted in [7], pp. 89–107
2. J. Copeland, The Church-Turing thesis. Stanf. Encycl. Philos. (1997) http://plato.stanford.edu/entries/church-turing/
3. E. Post, Absolutely unsolvable problems and relatively undecidable propositions, in [7], pp. 338–433
4. S. Kleene, *Introduction to Metamathematics* (Amsterdam, North Holland, 1952)
5. H. Wang, *A Logical Journey: From Gödel to Philosophy* (The MIT Press, Cambridge, MA, 1996)
6. H. Wang, *From Mathematics to Philosophy* (Routledge and Kegan Paul, London, 1974)
7. M. Davis, *The Undecidable* (The Raven Press, Hewlett, NY, 1965)
8. K. Gödel, Remarks before the Princeton bicentennial conference on problems in mathematics (1946); [7], pp. 84–88
9. A. Turing, On computable numbers, with an application to the *Entscheidungsproblem Proc. Lond. Math. Soc.* **42**, 230–265 (1936) reprinted in [7], pp. 116–153
10. S. Kleene, Origins of recursive function theory. Ann. Hist. Comput. **3**(1), 52–67 (1981)
11. S. Kleene, Reflections on Church's thesis. Notre Dame J. Formal Log. **28**, 490–498 (1987)
12. M. Davis, Why Gödel didn't have Church's thesis. Inf. Control Academic Press **54**, 3–24 (1982)
13. S. Shapiro, Review of Kleene (1981), Davis (1982), and Kleene (1987). J. Symb. Log. **55**, 348–350 (1990)
14. K. Gödel, On undecidable propositions of formal mathematical systems (1934); [7], pp. 39–74
15. R. Gandy, The confluence of ideas in 1936, in *The Universal Turing Machine*, ed. by R. Herken (Oxford University Press, New York, 1988), pp. 55–111
16. E. Mendelson, Second thoughts about Church's thesis and mathematical proofs. J. Philos. **87**, 225–233 (1990)
17. W. Sieg, Mechanical procedures and mathematical experience, in *Mathematics and Mind*, ed. by A. George (Oxford University Press, Oxford, 1994), pp. 71–140
18. W. Sieg, Calculations by man and machine: conceptual analysis, in *Reflections on the Foundations of Mathematics: Essays in Honor of Solomon Feferman*, in ed. by W. Sieg, R. Sommer, C. Talcott, (Association for Symbolic Logic, A. K. Peters, Ltd., Natick, MA, 2002), pp. 390–409
19. W. Sieg, Calculations by man and machine: mathematical presentation, in *In the Scope of Logic, Methodology and Philosophy of Science 1*, ed. by P. Gärdenfors, J Woleński, K. Kijania-Placek (Kluwer Academic, Dordrecht, 2002), pp. 247–262
20. J. Folina, Church's thesis: prelude to a proof. Philos. Math. **6** (3), 302–323 (1998)
21. R. Black, Proving Church's thesis. Philo. Math. (III) **8**, 244–258 (2000)
22. P. Maddy, *Naturalism in Mathematics* (Oxford University Press, Oxford, 1997)
23. Y. Moschovakis, *Notes on Set Theory* (Springer, New York, 1994)
24. I. Lakatos, *Mathematics, Science and Epistemology*, ed. by J. Worrall, G. Currie (Cambridge University Press, Cambridge, 1978)

25. K. Gödel, Some basic theorems on the foundations of mathematics and their implications, in *Collected Works 3* (Oxford University Press, Oxford, 1951, 1995), pp. 304–323
26. R. Penrose, *Shadows of the Mind: A Search for the Missing Science of Consciousness* (Oxford University Press, Oxford, 1994)
27. S. Shapiro, We hold these truths to be self evident: but what do we mean by that? Rev. Symb. Log. **2**, 175–207 (2009)
28. G. Boolos, Must we believe in set theory? in *Logic, Logic, and Logic*, ed, by G. Boolos (Harvard University Press, Cambridge, MA, 1998), pp. 120–132
29. F. Waismann, *Lectures on the Philosophy of Mathematics*, edited and with an Introduction by W. Grassl (Rodopi, Amsterdam, 1982)
30. S. Shapiro, Computability, proof, and open-texture, in *Church's Thesis After 70 Years*, ed. by A. Olszewski, J. Woleński, R. Janusz (Ontos, Frankfurt, 2006), pp. 420–455
31. S. Shapiro, The open-texture of computability, in *Computability: Gödel, Turing, Church, and Beyond*, ed. by J. Copeland, C. Posy, O. Shagrir (The MIT Press, Cambridge, MA, 2013), pp. 153–181

Why Turing's Thesis Is Not a Thesis

Robert Irving Soare

Abstract In 1936 Alan Turing showed that any effectively calculable function is computable by a Turing machine. Scholars at the time, such as Kurt Gödel and Alonzo Church, regarded this as a convincing *demonstration* of this claim, not as a mere hypothesis in need of continual reexamination and justification. In 1988 Robin Gandy said that Turing's analysis "proves a theorem." However, Stephen C. Kleene introduced the term "thesis" in 1943 and in his book in 1952. Since then it has been known as "Turing's Thesis." Here we discuss whether it is a thesis, a definition, or a theorem. This is important to determine what Turing actually accomplished.

Keywords Logic in the philosophy of science • Turing machines and related notions

AMS Classifications: 03A10, 03D10

1 Introduction

1.1 Precision of Thought and Terminology

Charles Sanders Peirce in *Ethics of Terminology* [39, p. 129] wrote that science advances upon precision of thought, and precision of thought depends upon precision of terminology. He described the importance of language for science this way.

> the woof and warp of all thought and all research is symbols, and the life of thought and science is the life inherent in symbols; so that it is wrong to say that a good language is *important* to good thought, merely; for it is of the essence of it.

R.I. Soare (✉)
Department of Mathematics, The University of Chicago, 5734 University Avenue, Chicago, IL 60637-1546, USA
e-mail: soare@uchicago.edu
http://www.people.cs.uchicago.edu/~soare/

G. Sommaruga, T. Strahm (eds.), *Turing's Revolution*,
DOI 10.1007/978-3-319-22156-4_12

> Next would come the consideration of the increasing value of precision of thought as it advances.
> thirdly, the progress of science cannot go far except by collaboration; or, to speak more accurately, no mind can take one step without the aid of other minds.

1.2 The Main Points of this Paper

This paper is about precision of thought and terminology in computability theory and in the work of Alan Turing. The main points of the paper are the following.

1. We use the term *effectively calculable* as it was used in the 1930s by Turing to denote a function on the integers which could be calculated by an idealized *human being* according to a definite rule. Sieg [47] and [53], building on Gandy [11], coined the term *computor* for this idealized human being and he called such a function *computorable*. We shall restrict the term "effectively calculable" to computorable functions, and we shall use the two terms interchangeably. Gödel and Church used the term for a function which is mechanically computable.
2. Turing [65] made several monumental achievements. (a) He gave a detailed description (definition) of the informal notion of an effectively calculable function in [65, Sect. 9]. (b) He defined a mathematical model which he called an automatic machine (a-machine), now called a *Turing machine*. (c) He demonstrated the following version called *Turing's Thesis (TT)*.

   Thesis 1.1 (Turing Thesis (TT)) *A function on the integers is effectively calculable iff it is Turing computable.*

   This is "Thesis T" in [12]. This thesis can be extended to a function on finite symbols by coding symbols using integers.
3. Thesis 1.1 is the restricted (computorable) form which was demonstrated in [65] and the form in which Church and Gödel understood it. It has been later stated in the following more general form.

   Thesis 1.2 *A function on the integers is computable by a machine iff it is Turing computable.*

   This is "Thesis M" in [12]. We do *not* consider this form Thesis 1.2 because the notion of a "machine" is not defined as precisely as that of a computorable function. However, several of our references consider the more general notion of an algorithm or machine which gives a computation.
4. The term *thesis* from the Greek and Latin refers to a proposition laid down to be debated or defended. It is not a fact. A theorem, such as Gödel's Completeness Theorem, cannot be a thesis.
5. In the 1930s neither Gödel, Church, Turing, nor even Kleene, referred to TT as a "thesis" but rather as a *definition* or *characterization*. The term "thesis" entered the lexicon of computability only in [29] and especially in Kleene's very influential book in [30]. Kleene's use of the term "thesis" in [30] did not accord

with the dictionary and scientific use of the term at that time, but it became firmly established anyway.

6. Church's Thesis (CT), so named by Kleene [30], is Church's assertion [2] that a function is effectively calculable iff it is Herbrand-Gödel recursive. This is *extensionally* equivalent to Turing's Thesis 1.1 because the recursive functions coincide formally with the Turing computable ones. However, because of our precise use of terms and concepts, we wish to make an *intensional* distinction between CT and TT because Church's demonstration of CT was less convincing than Turing's demonstration of TT. Therefore, in this paper we do *not* identify the two, and we do *not* use the term Church–Turing Thesis (CTT) introduced by Kleene [30].

2 What is a Thesis?

The English term[1] "thesis" comes from the Greek word θέασις, meaning "something put forth." In logic and rhetoric it refers to a "proposition laid down or stated, especially as a theme to be discussed and proved, or to be maintained against attack."[2] It can be a hypothesis presented *without* proof, or it can be an assertion put forward with the intention of defending and debating it. For example, a PhD thesis is a dissertation prepared by the candidate and defended by him before the faculty in order to earn a PhD degree.

In music, a thesis is a downbeat anticipating an upbeat. In Greek philosophy dialectic is a form of reasoning in which propositions (theses) are put forward, followed by counter-propositions (antitheses), and a debate to resolve them. In general, a thesis is something laid down which expects a response, often a counter response.

The Harvard online dictionary says that a thesis is "not a topic; nor is it a fact; nor is it an opinion." A theorem such as the Gödel Completeness Theorem is not a thesis. It is a fact with a proof in a formal axiomatic system, ZFC set theory.

However, we use the term "proof" in a more general sense in our everyday world to evaluate what is true and what is not. For example, in a trial, a jury must decide whether the prosecution has *proved* the defendant guilty "beyond a reasonable doubt." This requires the evaluation of evidence which cannot be strictly formulated with well-formed formulas, but whose truth is often just as important in our daily lives as that of formal mathematical statements. We develop a sense of judgment for a demonstration and we depend upon evidence and logic to verify its validity. We often refer informally to that verification as a "proof" of the assertion.

The conclusion is that a *theorem or fact* is a proposition which is laid down and which may possibly need to be demonstrated, but about which there will be no

[1]See the *Oxford English Dictionary*, the *Websters International Dictionary*, and Wikipedia.

[2]*Oxford English Dictionary*.

debate. It will be accepted in the future without further question. A *thesis* is a weaker proposition which invites debate, discussion, and possibly repeated verification. Attaching the term "thesis" to such a proposition invites continual reexamination. It signals that to the reader that the proposition may not be completely valid, but rather it should continually be examined more critically.

This is not a question of mere semantics, but about what Turing actually *achieved*. If we use the term "thesis" in connection with Turing's work, then we are continually suggesting some doubt about whether he really gave an authentic characterization of the intuitively calculable functions. The central question of this paper is to consider whether Turing proved his assertion beyond any reasonable doubt or whether it is merely a thesis, in need of continual verification.

3 The Concept of Effectively Calculable

The concept of an effectively calculable function emerged during the 1920s and 1930s until the work of Turing [65] gave it a definitive meaning. Kleene wrote in [33, p. 46] that the objective was to find a decision procedure for validity "of a given logical expression by a finite number of operations" as stated in [25, pp. 72–73]. Hilbert characterized this as the fundamental problem of mathematical logic. Davis in [8, p. 108] wrote, "This was because it seemed clear to Hilbert that with the solution of this problem, the *Entscheidungsproblem*, it should be possible, at least in principle, to settle all mathematical questions in a purely mechanical manner." Von Neumann doubted that such a procedure existed but had no idea how to prove it when he wrote the following in [70, pp. 11–12].

> ... it appears that there is no way of finding the general criterion for deciding whether or not a well-formed formula is provable. (We cannot at the moment establish this. Indeed, we have no clue as to how such a proof of undecidability would go.) ... the undecidability is even the *conditio sine qua non* for the contemporary practice of mathematics, using as it does heuristic methods to make any sense.
>
> The very day on which the undecidability does not obtain anymore, mathematics as we now understand it would cease to exist; it would be replaced by an absolutely mechanical prescription (eine absolut mechanische Vorschrift), by means of which anyone could decide the provability or unprovability of any sentence.
>
> Thus, we have to take the position: it is generally undecidable, whether a given well-formed formula is provable or not.

G.H. Hardy made similar comments quoted in [38, p. 85]. However, neither Hardy, von Neumann, nor Gödel had a firm definition of the informal concept of an effectively calculable function.

4 Turing's Paper in 1936

4.1 *What is a Procedure?*

At the beginning of 1936 there was a stalemate in Princeton. Church [2] and Kleene [27] believed in Church's Thesis, but Gödel did not accept it, even though it was based on his own definition [14] of recursive functions. Gödel was looking not merely for a mathematical definition such as recursive functions or Turing machines, but also for a convincing *demonstration* that these captured the intuitive concept of effectively calculable.

In the Nachlass printed in [21, p. 166] Gödel wrote,

> When I first published my paper about undecidable propositions the result could not be pronounced in this generality, because for the notions of mechanical procedure and of formal system no mathematically satisfactory definition had been given at that time. ... The essential point is to define what a procedure is.

Gödel believed that Turing had done so but he was not convinced by Church's argument.

4.2 *Turing's Definition of Effectively Calculable Functions*

Turing's results in 1936 are often presented first as the Turing machine in the order Turing wrote it, and second the definition of effectively calculable function at the end of his paper. However, just as remarkable is Turing's definition of a function calculated by an idealized human being in Sect. 9. Turing [65, p. 249] wrote as follows:

> No attempt has yet been made to show that the computable numbers include all numbers which would naturally be regarded as computable. All arguments which can be given are bound to be, fundamentally, appeals to intuitions, and for this reason rather unsatisfactory mathematically. The real question at issue is 'What are the possible processes which can be carried out in computing a number?' The arguments which I shall use are of three kinds. (a) A direct appeal to intuition. (b) A proof of the equivalence of two definitions (in case the new definition has a greater intuitive appeal). (c) Giving examples of large classes of numbers which are computable.

Turing devoted his attention to (a) rather than to (b) or (c). We shall not repeat Turing's argument for (a) which was clearly presented and which has also been given in Kleene [30] Sect. 70. The key point is that Turing broke up the steps of a procedure into the smallest pieces which could not be further subdivided. Turing's argument was convincing to scholars in 1936 and to most scholars today. When one goes through Turing's analysis of a procedure one is left with something very close to a Turing machine which is designed to carry out these atomic acts.

4.3 *Gödel's Reaction to Turing's Paper*

Kurt Gödel wrote in [15] as reprinted in [21, p. 168]:

> That this really is the correct definition of mechanical computability was established beyond any doubt by Turing.
>
> But I was completely convinced only by Turing's paper.

In his 1964 postscript to [14] Gödel wrote,

> In consequence of later advances, in particular of the fact that, due to A. M. Turing's work, a precise and unquestionably adequate definition of the general concept of formal system can now be given, the existence of undecidable arithmetical propositions and the non-demonstrability of the consistency of a system in the same system can now be proved rigorously for every consistent formal system containing a certain amount of finitary number theory.

He also wrote,

> Turing's work gives an analysis of the concept of mechanical procedure, ... This concept is shown to be equivalent with that of a Turing machine.

Gödel: Gibbs Lecture Yale [17]:

> The greatest improvement was made possible through the precise definition of the concept of finite procedure, ... This concept, ... is equivalent to the concept of a computable function of integers. ...
>
> The most satisfactory way, in my opinion, is that of reducing the concept of finite procedure to that of a machine with a finite number of parts, as has been done by the British mathematician Turing.

4.4 *Church's Reaction to Turing's Paper*

In his review [4] of Turing [65] Church wrote,

> Of the three different notions: computability by a Turing machine, general recursiveness of Herbrand–Gödel–Kleene, and λ-definability, the first has the advantage of making the identification with effectiveness in the ordinary (not explicitly defined) sense evident immediately—i.e., without the necessity of proving preliminary theorems.

4.5 *Kleene's Reaction to Turing's Paper*

Stephen Kleene [33, p. 49] wrote:

> Turing's computability is intrinsically persuasive but λ-definability is not intrinsically persuasive and general recursiveness scarcely so (its author Gödel being at the time not at all persuaded).

In his second book [31, p. 233] Kleene wrote:

> Turing's machine concept arises from a direct effort to analyze computation procedures as we know them intuitively into elementary operations. ...
>
> For this reason, Turing computability suggests the thesis more immediately than the other equivalent notions and so we choose it for our exposition.

5 Church Rejects Post's "Working Hypothesis"

Emil Post [40], independently of Turing, produced a mathematical model very similar to a Turing machine. Post knew of the work by Church [2] but not of Turing's work in 1936. Although Post's mathematical model resembles Turing's, Post's results were considerably weaker in several respects.

1. Post did not attempt to prove that his formalism, called formulation 1, coincided with any other formalism, such as general recursiveness, but merely expressed the expectation that this would turn out to be true, while Turing [67] proved the Turing computable functions equivalent to the λ-definable ones.
2. Post gave no hint of a universal Turing machine.
3. Most important, Post gave no analysis, as did Turing, of the intuitive notion of the effectively calculable functions or why they coincide with those computable in his formal system.
4. Post did not prove the unsolvability of the *Entscheidungsproblem.*

Post offered only as a conjecture that his contemplated "wider and wider formulations" are "logically reducible to formulation 1." This uncertainty was compounded when Post wrote in 1936 that Church's identification in 1936 of effective calculability and recursiveness was merely a *working hypothesis* which is in *"need of continual verification."* This irritated Church who criticized it in his review in [5] of Post [40]. Church wrote:

> The author [Post 1936] proposes a definition of 'finite 1-process' which is similar in formulation, and is equivalent, to computation by a Turing machine (see the preceding review Church [1937].) He does not, however, regard his formulation as certainly to be identified with effectiveness in the ordinary sense, but takes this identification as a 'working hypothesis' in need of continual verification. To this the reviewer would object that effectiveness in the ordinary sense has not been given an exact definition and the working hypothesis in question has not an exact meaning. To define effectiveness as computability by an arbitrary machine, subject to restrictions of finiteness, would seem to be an accurate representation of the ordinary notion, and if this is done the need for a working hypothesis disappears.

Church is clearly saying that the question of whether Turing computability captures the intuitive notion of an effectively calculable function is *not debatable.* It is *not* a working hypothesis, a conjecture, or a thesis. It has been *definitely settled* by Turing's characterization in [65] of what is an effectively calculable function computed by an idealized human being. Gödel and others agreed in the 1930's that there was no debatable issue here. Unfortunately, Kleene [29] and especially in his book [30] moved Turing's work from a demonstrated principle in [65] to the status of a mere debatable hypothesis, a *thesis*.

6 Remarks by Other Authors

In a number of books and articles for the Turing Centennial, and in earlier papers, several prominent authors have made comments about the issues which we address in our present paper. We now address some of these authors one by one. Each provides one more piece in the spectrum of our entire analysis.

7 Gandy and Sieg: Proving Turing's Thesis 1.1

7.1 Gandy

Robin Gandy [11, 12] was one of the first to challenge Kleene's claim that Turing's Thesis could not be proved. Indeed Gandy claimed that Turing [65] had already proved it. Gandy analyzed Turing's argument in detail and wrote [12, pp. 82–84]:

> Turing's analysis does much more than provide an argument for Turing's Thesis, *it proves a theorem*.
>
> Turing's analysis makes no reference whatsoever to calculating machines. Turing machines appear as a result, a codification, of his analysis of calculations by humans.

7.2 Sieg

Wilfried Sieg [47, 48, 52] and elsewhere analyzed Turing's Thesis. He gave an axiomatization for Turing's computorable functions and proved that any function satisfying the axioms is Turing computable. Sieg defined again [52, p. 189] his notion of a computor, a human computing agent who proceeds mechanically. Sieg wrote, "No appeal to a thesis is needed; rather, that appeal has been replaced by the task of recognizing the correctness of the axioms for an intended notion."

8 Feferman: Concrete and Abstract Structures

8.1 Feferman's Review of the History of the Subject

In this volume, Feferman [10] wrote,

> The concepts of recursion and computation were closely intertwined from the beginning of the efforts in the 1930s to obtain a conceptual analysis of the informal notion of *effective calculability*. ... It is generally agreed that the conceptual analysis of effective calculability was first provided most convincingly by Turing [1936–37]. ... Curiously, the subject of effective computation came mostly to be called *Recursion Theory*, even though that is only one of the possible forms of its development.

The footnote states, "Soare in his articles [55, 56] has justifiably made considerable efforts to reconfigure the subject so as to emphasize its roots in computation rather than recursion, ..."

Feferman continues:

> In his influential book, *Introduction to Metamathematics*, Kleene [1952] baptized the statement that every effectively calculable function is general recursive as *Church's Thesis*. He went on to baptize as *Turing's Thesis* the statement that "every function which would naturally be regarded as computable is computable ... by one of his machines ...

Feferman's account of the history closely agrees with ours given above.

8.2 Feferman on Concrete Structures

Feferman [10] states:

> My main concern in this article is to examine proposed theses for computation and recursion on both concrete and abstract structures. By *concrete structures* I mean those given by sets of finite symbolic configurations (e.g., finite strings, trees, graphs, hereditarily finite sets, etc.) together with the appropriate tests and operations for their transformation.
>
> ... *in whatever way one understands CT, there is no calculation without representation*, i.e., it is a necessary ingredient of whatever constitutes effective calculability that one operates only in finite symbolic configurations by means that directly transform such configurations into new ones using appropriate tests along the way.

Turing's Thesis for concrete structures is very similar to the analysis we have given, and this has been considered by Gandy, Sieg, Dershowitz and Gurevich, and others. Feferman then moves to abstract structures which we omit here in order to give a more convincing argument for the concrete case as in TT Thesis 1.1.

8.3 Feferman on Kleene's Naming of CTT

Feferman [10, Sect. 2] agrees with our analysis above of the origin of the terms introduced by Kleene [30] and the ambiguity in Kleene's naming of CTT. Feferman writes at the end of Sect. 2:

> As described in sec. 1 above, it was Kleene [1952] p. 300 *et seq,* who led one to talk of Church's Thesis, Turing's Thesis, and then, ambiguously, of the Church–Turing Thesis for the characterization through these equivalences of the effectively calculable functions. In another influential text, Rogers [1967] pp. 18ff, took the argument by confluence as one of the basic pins for CT and used that to justify informal proofs by Church's Thesis.

9 Gurevich: What is an Algorithm?

Yuri Gurevich [22], *What is an algorithm?*, discusses the concrete notions of algorithm by Sieg and the abstract notions by Moschovakis. In Sect. 2 Gurevich raises the question of whether it is even possible to define all algorithms. In Sect. 2.1 he gives a negative answer, and in Sect. 2.2 he gives a positive answer. This illustrates the difficulty in answering that question and explains why we restrict our discussion in this paper to the computorable functions of Thesis 1.1. Dershowitz and Gurevich [9] also give another approach to proving the thesis.

10 Kripke: On Proving Turing's Thesis

10.1 Kripke on Computorable Functions in Soare

Saul Kripke [38] began:

> Traditionally, many writers, following Kleene [1952], thought of the Church–Turing thesis as unprovable, but having various strong arguments in its favor, including Turing's analysis of human computation. More recently, the beauty, power, and obvious fundamental importance of this analysis—what Turing calls "argument I"–has led some writers to give an almost exclusive emphasis on the argument as *the* justification for the Church–Turing thesis. ...
>
> I resumed thinking about the subject of mathematical arguments for the Church–Turing thesis as a reaction to Soare [1996]. In his admirable survey of the history and philosophy of the subject, he proposed that the subject traditionally called "recursion theory" should be renamed "computability theory," and that correspondingly such terms as "recursively enumerable" should be renamed "computably enumerable." This reform has been generally adopted by writers on the subject. ...
>
> Soare, following Sieg [1994], [1997] and Gandy [1988] gives a careful mathematical and philosophical analysis of Turing's first argument. He proposes to break it into two different steps. One equates the informal notion of effective computability with what he calls "computorability," namely computability by an idealized human computer.

Kripke goes on to discuss various other agreements and differences with Soare [55].

10.2 Kripke's Suggested Proof of the Machine Thesis 1.2

Kripke's main point [38] is to suggest another way of proving the much stronger form of Turing's Thesis 1.2 even for *machine computable* functions using the Gödel Completeness Theorem. Kripke begins by stating what he calls "Hilbert's Thesis":

> namely, that the steps of any mathematical argument can be given in a language based on first order logic (with identity). The present argument can be regarded as either reducing Church's thesis to Hilbert's thesis, or alternatively as simply pointing out a theorem on all computations whose steps can be formalized in a first-order language.

Kripke concluded that his paper does not give an outright proof of Turing's Thesis from the Gödel Completeness Theorem but reduces one hypothesis to another and directly relates the former to the latter. The only open part of this intended proof is to demonstrate that any machine computation can be represented in first order logic. In view of the questions surrounding a characterization of what is an algorithm in Sect. 9 and elsewhere, this seems like a very interesting but formidable program.

Of course, if successful, our present program would be a consequence if we agree that Turing's definition of a computorable function can be represented in first order logic. We are proposing here, along with Gandy and Sieg, that Turing's is clearly the correct representation of the informal notion of effectively calculable, and therefore Turing's Thesis 1.1 has a proof parallel to (rather than as a consequence of) Gödel's Completeness Theorem. Nevertheless, we agree with Kripke that the connection is very important.

11 Turing on Definition Versus Theorem

Turing's last published paper discussed puzzles. Turing wrote of his central assertion (about a function being effectively calculable iff it is computable by a Turing machine) that this assertion lies somewhere between a theorem and a definition.

> In so far as we know a priori what is a puzzle and what is not, the statement is a theorem. In so far as we do not know what puzzles are, the statement is a definition that tells us something about what they are.

In any case, most scholars agree that it is not a thesis. It is not something laid down for continual verification or debate.

Acknowledgements The present paper was partially supported by the Isaac Newton Institute of the University of Cambridge during 2012 and partially by grant # 204186 *Computability Theory and Applications* to Robert Soare funded by the Simons Foundation Program for Mathematics and the Physical Sciences.

References

1. A. Church, An unsolvable problem of elementary number theory, Preliminary Report (abstract). Bull. Am. Math. Soc. **41**, 332–333 (1935)
2. A. Church, An unsolvable problem of elementary number theory. Am. J. Math. **58**, 345–363 (1936)
3. A. Church, A note on the Entscheidungsproblem. J. Symb. Log. **1**, 40–41 (1936). Corrections 101–102
4. A. Church, Review of [65]. J. Symb. Log. **2**(1), 42–43 (1937)
5. A. Church, Review of [40]. J. Symb. Log. **2**(1), 43 (1937)
6. J. Copeland (ed.), *The Essential Turing* (Oxford University Press, Oxford, 2010)
7. J. Copeland, C. Posy, O. Shagrir (eds.), *Computability: Gödel, Church, Turing, and Beyond* (MIT Press, Cambridge, 2013)
8. M. Davis (ed.), *The Undecidable. Basic Papers on Undecidable Propositions, Unsolvable Problems, and Computable Functions* (Raven Press, New York, 1965)
9. N. Dershowitz, Y. Gurevich, A natural axiomatization of computability and proof of Church's thesis. Bull. Symb. Log. **14**(3), 299–350 (2008)
10. S. Feferman, Theses for computation and recursion over concrete and abstract structures, in this volume, pp. 103–124
11. R. Gandy, Church's thesis and principles for mechanisms, in *The Kleene Symposium*, ed. by J. Barwise et al. (North-Holland, Amsterdam, 1980), pp. 123–148
12. R. Gandy, The confluence of ideas in 1936, in [24], pp. 55–111
13. K. Gödel, Über formal unentscheidbare Sätze der Principia Mathematica und verwandter Systeme. I. Monatsch. Math. Phys. **38**, 173–178 (1931). (English trans. in [8, pp. 4–38], in [69, pp. 592–616], and in [19, pp. 145–195])
14. K. Gödel, *On Undecidable Propositions of Formal Mathematical Systems*, ed. by S.C. Kleene, J.B. Rosser. Lectures at the Institute for Advanced Study (Princeton, New Jersey, 1934), 30 pp. (Reprinted in [8, pp. 39–74], and in [19, pp. 346–371])
15. K. Gödel, Undecidable diophantine propositions, in [21, pp. 156–175]
16. K. Gödel, Remarks before the Princeton bicentennial conference of problems in mathematics (1946). (Reprinted in [8, pp. 84–88] and in [20, pp. 150–153])
17. K. Gödel, Some basic theorems on the foundations of mathematics and their implications, in [21, pp. 304–323]
18. K. Gödel, Postscriptum to [13], written in 1946, reprinted in [8, pp. 71–73]
19. K. Gödel, *Collected Works, Volume I: Publications 1929–1936*, ed. by S. Feferman et al. (Oxford University Press, Oxford, 1986)
20. K. Gödel, *Collected Works, Volume II: Publications 1938–1974*, ed. by S. Feferman et al. (Oxford University Press, Oxford, 1990)
21. K. Gödel, *Collected Works, Volume III: Unpublished Essays and Lectures*, ed. by S. Feferman et al. (Oxford University Press, Oxford, 1995)
22. Y. Gurevich, What is an algorithm? (Revised) *Proceedings of the 2011 Studia Logica Conference on Church's Thesis: Logic, Mind and Nature* (2013)
23. Y. Gurevich, Semantics-to-Syntax Analyses of Algorithms, in this volume, pp. 185–204
24. R. Herken (ed.), *The Universal Turing Machine: A Half-Century Survey* (Oxford University Press, Oxford, 1988)
25. D. Hilbert, W. Ackermann, *Grundzüge der theoretischen Logik* (Springer, Berlin, 1928). English translation of 1938 edition (Chelsea, New York, 1950)
26. A. Hodges, *Alan Turing: The Enigma* (Burnett Books and Hutchinson/Simon and Schuster, London/New York, 1983). New Edition, (Vintage, London, 1992)
27. S.C. Kleene, General recursive functions of natural numbers. Math. Ann. **112**, 727–742 (1936)
28. S.C. Kleene, λ-definability and recursiveness. Duke Math. J. **2**, 340–353 (1936)
29. S.C. Kleene, Recursive predicates and quantifiers. Trans. Am. Math. Soc. **53**, 41–73 (1943)

30. S.C. Kleene, *Introduction to Metamathematics* (Van Nostrand, New York, 1952). Ninth reprint (Walters-Noordhoff Publishing Co./North-Holland, Gröningen/Amsterdam, 1988)
31. S.C. Kleene, *Mathematical Logic* (Wiley, New York/London/Sydney, 1967)
32. S.C. Kleene, Origins of recursive function theory. Ann. Hist. Comput. **3**, 52–67 (1981)
33. S.C. Kleene, The theory of recursive functions, approaching its centennial. Bull. Am. Math. Soc. (n.s.) **5**, 43–61 (1981)
34. S.C. Kleene, Algorithms in various contexts, in *Proceedings Symposium on Algorithms in Modern Mathematics and Computer Science* (dedicated to Al-Khowarizimi) (Urgench, Khorezm Region, Uzbek, SSSR, 1979) (Springer, Berlin/Heidelberg/New York, 1981)
35. S.C. Kleene, Reflections on Church's thesis. Notre Dame J. Formal Log. **28**, 490–498 (1987)
36. S.C. Kleene, Turing's analysis of computability, and major applications of it, in [24, pp. 17–54]
37. S.C. Kleene, E.L. Post, The upper semi-lattice of degrees of recursive unsolvability. Ann. Math. **59**, 379–407 (1954)
38. S. Kripke, The Church–Turing "Thesis" as a special corollary of Gödel's completeness theorem, in *Computability: Gödel, Church, Turing, and Beyond,* ed. by J. Copeland, C. Posy, O. Shagrir (MIT Press, Cambridge, 2013)
39. C.S. Peirce, Book 2. Speculative grammar, in *Collected Papers of Charles Sanders Peirce*, ed. by C. Hartshorne, P. Weiss. Elements of Logic, vol. 2 (The Belknap Press of Harvard University Press, Cambridge/London, 1960)
40. E.L. Post, Finite combinatory processes—formulation I. J. Symb. Log. **1**, 103–105 (1936). Reprinted in [8, pp. 288–291]
41. E.L. Post, *Absolutely Unsolvable Problems and Relatively Undecidable Propositions: Account of an Anticipation* (Submitted for publication in 1941). Printed in [8, pp. 340–433]
42. E.L. Post, Formal reductions of the general combinatorial decision problem. Am. J. Math. **65**, 197–215 (1943)
43. E.L. Post, Recursively enumerable sets of positive integers and their decision problems. Bull. Am. Math. Soc. **50**, 284–316 (1944)
44. E.L. Post, Degrees of recursive unsolvability: preliminary report (abstract). Bull. Am. Math. Soc. **54**, 641–642 (1948)
45. H. Rogers Jr., *Theory of Recursive Functions and Effective Computability* (McGraw-Hill, New York, 1967)
46. G.E. Sacks, *Mathematical Logic in the 20th Century* (Singapore University Press, Singapore, 2003). Series on 20th Century Mathematics, vol. 6 (World Scientific Publishing Co., Singapore/New Jersey/London/Hong Kong, 2003)
47. W. Sieg, Mechanical procedures and mathematical experience, in *Mathematics and Mind*, ed. by A. George (Oxford University Press, Oxford, 1994), pp. 71–117
48. W. Sieg, Step by recursive step: Church's analysis of effective calculability. Bull. Symb. Log. **3**, 154–180 (1997)
49. W. Sieg, Gödel on computability. Philos. Math. **14**, 189–207 (2006)
50. W. Sieg, Church without dogma—axioms for computability, in *New Computational Paradigms*, ed. by B. Lowe, A. Sorbi, B. Cooper (Springer, New York, 2008), pp. 139–152
51. W. Sieg, On computability, in *Handbook of the Philosophy of Mathematics*, ed. by A.D. Irvine (Elsevier, Amsterdam, 2009), pp. 535–630
52. W. Sieg, Gödel's philosophical challenge [to Turing], in *Computability: Gödel, Church, Turing, and Beyond,* ed. by J. Copeland, C. Posy, O. Shagrir (MIT Press, Cambridge, 2013), pp. 183–202
53. W. Sieg, J. Byrnes, *K*-graph machines: generalizing Turing's machines and arguments. Preprint (1995)
54. R.I. Soare, *Recursively Enumerable Sets and Degrees: A Study of Computable Functions and Computably Generated Sets* (Springer, Heidelberg, 1987)
55. R.I. Soare, Computability and recursion. Bull. Symb. Log. **2**, 284–321 (1996)
56. R.I. Soare, The history and concept of computability, in *Handbook of Computability Theory*, ed. by E.R. Griffor (North-Holland, Amsterdam, 1999), pp. 3–36

57. R.I. Soare, Computability and incomputability, in *Computation and Logic in the Real World*, ed. by S.B. Cooper, B. Löwe, A. Sorbi. Proceedings of the Third Conference on Computability in Europe, CiE 2007, Siena, Italy, June 18–23, 2007. Lecture Notes in Computer Science, vol. 4497 (Springer, Berlin/Heidelberg, 2007)
58. R.I. Soare, Turing oracle machines, online computing, and three displacements in computability theory. Ann. Pure Appl. Log. **160**, 368–399 (2009)
59. R.I. Soare, An interview with Robert Soare: reflections on Alan Turing, in *Crossroads of the ACM*, vol. 18, no. 3, pp. 15–17, (2012)
60. R.I. Soare, Formalism and intuition in computability theory. Philos. Trans. R. Soc. A **370**, 3277–3304 (2012)
61. R.I. Soare, Turing and the art of classical computability, in *Alan Turing—His Work and Impact*, ed. by B. Cooper, J. Leeuwen (Elsevier, New York, 2013)
62. R.I. Soare, Interactive computing and Turing-Post relativized computability, in *Computability: Gödel, Church, Turing, and Beyond,* ed. by J. Copeland, C. Posy, O. Shagrir (MIT Press, Cambridge, 2013)
63. R.I. Soare, Turing and the discovery of computability, in *Turing's Legacy: Developments from Turing's Ideas in Logic*, ed. by R. Downey. Lecture Notes in Logic (Association for Symbolic Logic/Cambridge University Press, Cambridge, 2013)
64. R.I. Soare, *The Art of Turing Computability: Theory and Applications.* Computability in Europe Series (Springer, Heidelberg, 2014)
65. A.M. Turing, On computable numbers, with an application to the Entscheidungsproblem. Proc. Lond. Math. Soc. Ser. 2 **42**(Parts 3 and 4), 230–265 (1936). Reprinted in [8], pp. 116–154
66. A.M. Turing, A correction. Proc. Lond. Math. Soc. **43**, 544–546 (1937)
67. A.M. Turing, Computability and λ-definability. J. Symb. Log. **2**, 153–163 (1937)
68. A.M. Turing, Systems of logic based on ordinals. Proc. Lond. Math. Soc. **45**, 161–228 (1939). Reprinted in [8, pp. 154–222]
69. J. van Heijenoort (ed.), *From Frege to Gödel, A Sourcebook in Mathematical Logic, 1879–1931* (Harvard University Press, Cambridge, 1967)
70. J. von Neumann, Zur Hilbertschen Beweistheorie. Math. Z. **26**, 1–46, 11–12 (1927)

Incomputability Emergent, and Higher Type Computation

S. Barry Cooper

Abstract Typing of information played an historical role in bringing *consistency* to formulations of set theory and to the foundations of mathematics. Underlying this was the augmentation of language and logical structure with a respect for constructive principles and the corresponding infrastructure of an informational universe. This has important consequences for how we view the computational character of science, the humanities and human creativity. The aim of this article is to make more explicit the anticipations and intuitions of early pioneers such as Alan Turing in informing and making relevant the computability theoretic underpinnings of today's understanding of this. We hope to make clearer the relationship between the typing of information—a framework basic to all of Turing's work—and the computability theoretic character of emergent structure in the real universe. The informational terrain arising is one with comprehensive computational structure, but with theoretical boundaries to those areas accessible to effective exploration in an everyday setting.

Keywords Emergence • Higher type computation • Incomputability • Turing definability

AMS Classifications: 03D10, 68Q1

> I have spent my entire life studying randomness, practicing randomness, hating randomness. The more that time passes, the worse things seem to me, the more scared I get, the more disgusted I am with Mother Nature. The more I think about my subject, the more I see evidence that the world we have in our minds is different from the one playing outside. Every morning the world appears to me more random than it did the day before, and humans seem to be even more fooled by it than they were the previous day. It is becoming unbearable. I find writing these lines painful; I find the world revolting.—Taleb [1, p. 215]

S.B. Cooper (✉)
School of Mathematics, University of Leeds, Leeds LS2 9JT, UK
e-mail: pmt6sbc@leeds.ac.uk

G. Sommaruga, T. Strahm (eds.), *Turing's Revolution*,
DOI 10.1007/978-3-319-22156-4_13

> Some have asked why neuroscience has not yet achieved results as spectacular as those seen in molecular biology over the past four decades. Some have even asked what is the neuroscientific equivalent of the discovery of DNA structure, and whether or not a corresponding neuroscientific fact has been established. There is no such single correspondence ... the equivalent, at the level of mind-producing brain, has to be a large-scale outline of circuit and system designs, involving descriptions at both microstructural and macrostructural levels.—Damasio [2, p. 260]

1 Introduction

When the Earl of Northampton gave rise (so the Oxford English Dictionary tells us) to the first recorded use of the word 'incomputable' back in 1606, the word simply referred to large finite sets being too big to keep track of. With today's radically extended notion of computability, 'incomputable' is something more far reaching. For the Earl of Northampton it meant that his algorithm for counting took more time than he could draw on. It may have meant having an army too big to submit to a role call. For us today, it would mean the counting algorithm eludes us—more like a failure to carry out a role call due to many of the individual foot-soldiers refusing to answer to their names. Worse than incomputability would be randomness. The undisciplined or rebellious soldiers may mill chaotically, impossible to even question in an orderly fashion.

In this brief non-technical introduction, we consider some very basic questions, such as:

- Where does *incomputability* come from? Does it arise from a mere mathematical trick, or from an informational process of common encounter?
- What is its connection with *randomness* and *higher type computation*?
- What do such notions have to do with real information, and how do they relate to the classical Turing model of computation, its derivatives such as cellular automata, and its embodiments?
- How does the mathematics of higher type computation differ from that of the classical model? How do these differences play out in the real world, and what are the practical strategies for dealing with them?
- And—most importantly—how can the mathematics and abstract models help us deal with the computational complexities arising in science and the humanities? Or at least validate and help us understand what we do already?

The aim is to map out some directions in our current thinking about such questions, and to review some of the ways in which the mathematics of computation can rescue us from current confusions, across a whole spectrum of disciplines. We are on the threshold of a new understanding of the causal character of reality, no less than a new causal paradigm replacing that passed down to us from Isaac Newton's time via Laplace's predictive demon. And it is a revolution in thinking about the world which is as much in need of appropriate mathematical underpinnings as was that of 350 years ago. 2004 saw the publication by Cambridge University Press of a collection of 30 contributions from leading thinkers on the subject of *Science and Ultimate Reality: Quantum Theory, Cosmology and Complexity* (edited by John Barrow, Paul Davies and Charles Harper, Jr.). As George Ellis [23] describes (on p. 607) in his contribution on *True complexity and its associated ontology*:

> True complexity involves vast quantities of stored information and hierarchically organized structures that process information in a purposeful manner, particularly through implementation of goal-seeking feedback loops. Through this structure they appear purposeful in their behavior ("teleonomic"). This is what we must look at when we start to extend physical thought to the boundaries, and particularly when we try to draw philosophical conclusions—for example, as regards the nature of existence—from our understanding of the way physics underlies reality. Given this complex structuring, one can ask, "What is real?", that is, "What actually exists?", and "What kinds of causality can occur in these structures?".

What is surprising—reflecting the situation more widely—is that throughout the 740 pages of this fascinating and insightful book, there is no serious consideration of the basic computational structure of information, and its lessons for this extended view of causality. The technical content is largely constrained by a specificity, uninformed by the fundamentals of classical computability, familiar to anybody who has taken an interest in any of the thriving fields dealing with complexity and emergence in the real world. A comprehensive 20-page index contains but one mention of Alan Turing, and just one mention of a Turing machine, and two of 'computable', and all but one of these mentions (one mention of 'computable') appear in David Deutsch's contribution. Unsurprisingly, one looks in vain for the oracle Turing machine model of interactive computation, or—the conceptual 'elephant in the room'—for higher type computation, or definability and its computational content. As one surveys the field, one finds creative thinking and insight in abundance, but reaching us via a generally ad hoc or descriptive expression, with much duplication and circularity of argument.

In Sect. 2 below, we review the definition of the classical Turing machine model; discuss the non-triviality of any embodiment of a particular TM as an actual machine; and discuss the type-reduction implicit in the focus on the logical structure captured in the program, and the sense in which the universal Turing machine is in practice machine, or program, or something else.

In Sect. 3 we describe the powerful and persuasive paradigm of universality historically associated with by the universal Turing machine, and how the paradigm has been translated into areas beyond its origins, via functionalist successes and speculations; back in the mathematical domain, we describe how the type-reduction trick used in the definition of the universal Turing machine, like Kurt Gödel's coding of the elements of first-order Peano arithmetic, enables one to formalise properties of the universal machine taking us to a semantical level not within the purview of the system; and we examine the logical character of this semantical content, revealing it to be built on ingredients more widely occurring, in less abstractly mathematical contexts; in doing this, we need to review the basics of type theory, as a framework within which to bring together mathematics and embodied logical parallels; finally, returning to the quotation from Nassim Taleb's *The Black Swan*, which introduced this article, we clarify the relationship between the deceptively familiar notion of randomness, and the less problematic notion of incomputability which features in this article.

Section 4 is concerned with the theoretical bridge between the mathematics we have discussed, and embodied computation and emergent phenomena; we first examine such mathematically formulated examples as the Mandelbrot set and Turing's computational analogues of morphogenic phenomena, both of which have computer bases and simulations providing persuasive links to more generally encountered emergence; and the opportunity is taken to connect the parallel computability theoretic and definitional ingredients of these examples to historically important intuitions about the wider context; at this point, we take the opportunity to introduce the fundamental notion of definability, and the related language-independent one of invariance. Finally, we put definability in the context of existing proposals from mathematics for higher type computation, explaining how these theoretical models come with rather different characteristics to those of the classical lower type one, which fits well with what we experience in the real world.

During Sect. 5, we take our conceptual and technical toolbox into a range of informative and varied contexts; the aim here is to both exercise our theoretical framework, and to describe both problems and potential for clarification; and to prepare the ground for a return to the mathematics in the final Sect. 6, with the aim of exercising a more informative control over our examples. The main example chosen in Sect. 5—from artificial intelligence and brain science—is selected for its challenges, familiarity and rich literature; in particular, higher type computation is widely identified, via informal discussion, with characteristics of human thought. The overview of Turing definability in Sect. 6 is no more than a first introduction, with references to further reading for the more engaged reader.

2 Incomputability: Unruly Guest at the Computer's Reception

The Earl of Northampton's notion of incomputability was based in our everyday experience of the real world. It seems a simple concept until you try to get beyond its 'common sense' fuzziness. In fact, 'common sense' is far from simple, and its lack of robustness derives from that of the extended mathematical models of *computation* we are driven to apply. Without a model of computation, the notion of incomputability is impossible to pin down, of course. To obtain such a model one needs to accept some limitations on the sort of computational activity one recognizes as feasible—and this is what Alan Turing (see p. 16 of [3] *Alan Turing: His Work and Impact*) did back in the 1930s. See his 'machine' below.

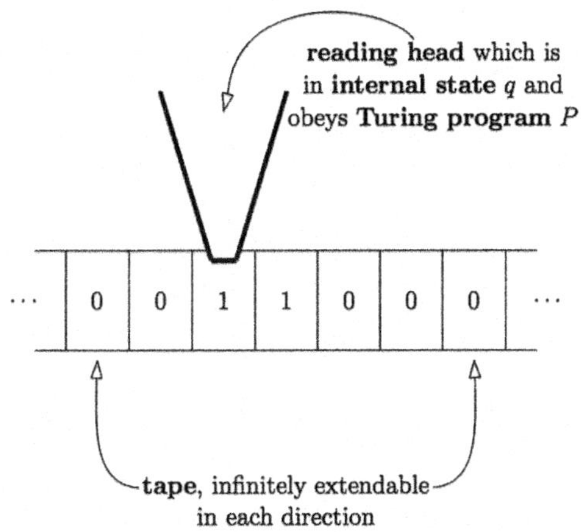

Unlike more mathematical models of computation, such as the recursive functions, or the λ-computable functions, the Turing machine was not so much chosen to fit the mathematician's sophisticated mental world: it was based on the simple computational actions one could ask a human 'computer' to perform with just pencil and paper and a limited—say binary—language. The workspace was a one dimensional tape, subdivided into cells each allowed to record one 'tape symbol', and assumed to be as long as needed. Programming was in terms of simple instructions for a 'reading head'—representing the human 'computer'—to write or erase a symbol (say 1), or move one cell right or left. Programs were consistent finite batches of instructions, with the 'computer' applying whichever instruction applied to the tape symbol she was reading, and the 'state of mind' (or 'internal state') she was in. The inputs and outputs would be words in binary, and could code numbers, for instance. According to the contemporary view (see Cooper [4]), a computation consists of the 'computer' entering an input, and just following instructions until she had to halt due to no appropriate further instruction being provided by the program.

For the young Alan Turing, barely 24 years old, the modeling of computation was very specific, very physically based. It was Turing's mentor Max Newman who was to describe him as 'at heart more of an applied than a pure mathematician'. But it was the disembodiment of computation that the Turing machine was to become famous for. Before Turing, if you wanted a computing machine to do a particular task, the programming was all in the hardware. And that meant you had to laboriously manufacture a new machine for each kind of computational chore. Before the modern era, calculating machines exhibiting varying degrees of programmability were developed (though not always built), including Babbage's Analytical Engine, famously programmed by Ada Lovelace, using notes of Italian mathematician Luigi Menabrea, in the 1840s. But the program was still applied via what were essentially hardware interventions.

Alan Turing's key innovation was the Universal Turing Machine, embodied in the first *stored program* computers during the late 1940s. The universal machine was not only programmable to compute anything computable by a Turing machine—the property of being what we now call *Turing complete*—but it could accept any program as data, then decode, store and implement it, and even modify the program via other stored programs. Not even the program need be embodied as punched cards or similar, the entire computing activity could be governed by abstract logic structure. Essentially, the machine had access to a 'dictionary' which enabled it to decode programs from data. This feature was what was programmed into the universal machine. A machine was not universal by accident. It could be realised as in early US computers via the *von Neumann computer architecture*. It was what we would nowadays identify as the computer system, achieved by a combination of hardware and software features, but modifiable—and installed, as far as the user was concerned, via programming.

And *logically*, the hardware was trivial, and a given Turing machine *was* its program. The program could be coded as a number—called the *index* of the machine—and fed as data into the Universal Machine. Of course, something very strange was happening. It has probably occurred to you that your laptop is physically quite a complicated object: hard to imagine in what sense you could take it and feed it into your desk-top computer! A full mathematical description of it would need to reflect its physicality, and would be far from a simple piece of data. But logically, as a computer, it is a universal Turing machine, and has an *index*, a mere number. We say that the logical view has *reduced* the *type* of the information embodied in the computer, from a type-2 set of reals, say, to a type-0 index. The physical complexities have been *packaged* in a logical structure, digitally coded. But in the next section we will see that suppression of one level complexity comes at the cost of the revelation of new higher order information, based on the enabled ease of observation of collated computational activities.

3 Incomputability, Randomness and Higher Type Computation

The indexing trick arises from being able to describe the *functionality* of the machine—what the machine *does*—via a number. There is a hugely important assumption underlying such a reduction. Looking at another 'computing' object—say a brain, or a stock exchange—how can we be sure we can reduce its functionality to a simple code of its logical structure. The assumption that one can do this in general is a massive leap of faith. At the level of computers we can currently build, it follows Turing's logic, governed by what has come to be called the *Church-Turing Thesis*. The extension of the Church-Turing Thesis to organisms in physics or biology—or economics—is problematic. It is about modeling. It is about comparing structures, mathematical or real, which may elude computable characterization or comparison.

This assumption has given rise to a powerful computing paradigm, aspects of which can be found in many forms. One can think of it as a formal counterpart of the post-Newtonian determinacy familiar to us via Pierre-Simon Laplace's predictive demon. If the functionalist view is as described by Hilary Putnam [5], namely 'the thesis that the computer is the right model for the mind', then one can trace it back to Alan Turing and the beginnings of the computer age. For example, he is quoted as saying to his assistant Donald Bayley while working at Hanslope Park in 1944 that he was 'building a brain'. But the aspect that Putnam develops, with the full sophistication of the philosopher of mind, is that of mental computation being *realizable* on different hardware (see his *Minds and Machines* [5] from 1960). One finds echoes of this outlook in computer science in the fruitful notion of a *virtual machine*. The practical emergence of the idea of virtualizing hardware is traceable back to IBM researchers of the mid-1960s, and is common today across a whole range of software, including Java and Unix.

However, it is the success of the type-reduction which rebounds on us. At the beginning of the last century it was a disregard for the typing of information which led to the set-theoretical paradoxes. A few decades later, it was Turing's deliberate type-reduction, which—as well as anticipating the stored-program computer—'computed' a type-1 object beyond the reach of his computational model.

So what can possibly be incomputable about a computer—an instantiation of a universal Turing machine **U**, say? Let us define the set H to be the set of all pairs (n, s) such that the machine **U** with input n halts at step s of the computation. H is clearly computable—to decide if (n, s) is in H, all one has to do is to give **U** input n, and follow the computation for at most s steps to tell if the machine halts at step s. But say one asks: Does there exist a step s such that (n, s) is in H? The new set can be described more formally as the set H* of numbers n such that $(\exists s)[(n, s) \in H]$ that is, the computable relation $(n, s) \in H$ with an existential quantifier added. And for arbitrary n we cannot decide $(\exists s)[(n, s) \in H]$ computably, since it is quite possible that the computation never halts, and the only way of being sure that happens is to complete an infinite search for a number s such that $(n, s) \in H$, so identifying that $(n, s) \notin H$. This incomputable set of numbers H* is algorithmically connected to H: If we imagine H arrayed in the usual way in 2-dimensional Euclidean space, H* is just the shadow of H under sunlight from directly above the array. We say H* is the *projection* of H onto the n-axis. It can be shown formally that this application of projection, using an existential quantifier, does give an incomputable set.

Notice, the route by which we get incomputability is by observing the behaviour of **U** over its global domain of computation. And remember too that global relations over computationally modeled environments are commonly important to us—such as with changing weather conditions, or macro-economics, or cosmological developments. And this is where the notion of *emergence* comes in—the global patterns can often be clearly observed as emergent patterns in nature or in social environments, with hard to identify global connection to the underlying computational causal context. Like the Halting Set H*, there is a computable basis, the emergent higher type outcome H* may be observable, but H* itself may elude computational prediction or description.

And here we see a computable sampling of what me might term 'big data', revealing enough of the higher order semantics to bring with it the incomputability concealed within. The sampling itself could be done in different ways. Just as in the wider context, the right choice of sampling technique, usually computable, though possibly pseudo-random, may, if we are clever enough and fortunate, bring out the semantics we seek—like Alan Turing applying his Bayesian sampling to decode messages at Bletchley Park.

The halting problem occupies a higher level than the basic logical structure of the operative Turing machine. It forms an entry point to the higher type structure which attracted the attention of Georg Cantor, Bertrand Russell and Kurt Gödel—and provided the type theory which so interested Alan Turing. Here is Gödel's outline of the basics (from *Russell's mathematical logic* [6]):

> By the theory of simple types I mean the doctrine which says that the objects of thought ... are divided into types, namely: individuals, properties of individuals, relations between individuals, properties of such relations, etc. ... , and that sentences of the form: 'a has the property φ ', 'b bears the relation R to c', etc. are meaningless, if a, b, c, R, φ are not of types fitting together. Mixed types (such as classes containing individuals and classes as elements) and therefore also transfinite types (such as the class of all classes of finite types) are excluded. That the theory of simple types suffices for avoiding also the epistemological paradoxes is shown by a closer analysis of these.

This classical typing was that used by Russell (see his 1903 *The Principles of Mathematics*) as a more careful way of observing and structuring set theoretical information, to avoid the paradoxes which had emerged out of the earlier work of Cantor and Frege. The geometry of typed information—numbers, reals, 3-dimensional objects as curves and surfaces, relations between shapes, and so on—gives us a sense of the way in which the mathematics of such a hierarchy can play an embodied role in the real world. Clearly 'big information' occurs in many contexts, not least in the economic or neuro-scientific domains. We also get a sense of how this embodiment has the potential to take reality out of the classical computational domain.

The notion of randomness is sometimes used to describe contexts lacking predictability, as an apparently more intuitive notion than that of incomputability. For instance, Nassim Taleb does not make a distinction between 'incomputable' and 'random' in his best-selling book [1] *The Black Swan*. In fact, he does not mention incomputability.

Unfortunately, randomness turns out to be a much more complicated, and less robust, notion than common sense dictates (see Downey and Hirschfeldt [7]). There is no such thing as randomness as an absolute property, which raises questions regarding assumptions of randomness in physics. *How much* randomness does observation and theory lead us to expect? What are the underlying reasons for that level? Computer scientist Cristian Calude and physicist collaborator Karl Svozil [8] have tried to answer the open question: How random is quantum randomness? As Calude comments, quantum randomness 'passes all reasonable statistical properties of randomness'. But all that the researchers were able to substantiate, building on generally acceptable assumptions about the physics, was incomputability.

On the other hand, the mathematics gives us a very useful model of computability, which fits well with computer science. And it is to the mathematics of the incomputable we turn to clarify the limitations of computational economics.

Hilary Putnam knows all about the mathematics of computability, incomputability and randomness, and from a very different starting point, is traveling the same road as Taleb. Of course, Putnam is still determined to resist the "metaphysical obscurities" that his functionalist vision was intended to rescue us from. As he says in the introduction to *Representation and Reality* [27, p. xi]:

> I may have been the first philosopher to advance the thesis that the computer is the right model for the mind. I gave my form of this doctrine the name "functionalism", and under this name it has become the dominant view – some say the orthodoxy – in contemporary philosophy of mind.

But here he is again (p. 89 of *Is the Causal Structure of the Physical Itself Something Physical?* in *Realism with a Human Face*, [9]), modifying his earlier view in the broader context:

> ... if the physical universe itself is an automaton ... , then it is unintelligible how any particular structure can be singled out as "the" causal structure of the universe. Of course, the universe fulfills structural descriptions – in some way or other it fulfills every structural description that does not call for too high a cardinality on the part of the system being modeled; ... the problem is how any one structure can be singled out as "the" structure of the system.

What is important to observe here is that although Taleb is focusing on economics, and Putnam on the mind, the issues they are grappling with are relevant to both contexts—and any answers, peculiar to one or the other, may be mutually beneficial. There are many ways in which the complexities, interactivity, representations and recursions of the active brain are reflected in those of the contemporary economics.

The aim of this article is to give information, as embodied structure, a role in an overarching basic causality. Within this enriched context, some epistemologically familiar, but physically less often encountered mathematics will play a clarifying role. The mathematics will be used, not according to the classical computing paradigm, to rob information of its 'thingness', but to provide the glue for a rich and surprisingly informative structure of information.

4 Embodiment Beyond the Classical Model of Computation

The Mandelbrot set M (see below) is familiar as a computable object—computable since it is viewable on our computer screen. But mathematically, this viewability is not so simple as it might seem. In fact, the definition of the set shares some important features with the halting set H*.

The Mandelbrot set is got by applying a couple of quantifiers to a simple equation involving members of a sequence of complex numbers. And with closer examination, one of the quantifiers can be eliminated, giving a definition of the points *not* in M via simple rules which we apply an existential quantifier to—just like the definition of the halting set H*. The image on our computer screen is, a computable approximation to M, essentially a type-reducing sampling of M. The computability of M in its full glory, with beautifully complex fractal boundary, is still an open problem.

Courtesy of Niall Douglas

Notice that although the Mandelbrot set is a type-2 object, our type reduction via a computable sampling of M produces a good enough approximation for us to mentally 'compute' with a good sense of the type-2 richness of emergent geometry of M. Effectively, the human brain can carry out a good simulation of type-2 computation with special input M. But such reductions are ad hoc. Our brains extract no such useful information to compute on from H*. There is informative geometry delivered by the definability of M, but not of H*. The practical and nonuniform subtleties of reducing 'big data' to computable information is illustrated by Turing's use of Bayesian sampling in his decoding work at Bletchley Park in the early 1940s (see Mardia and Cooper [10]).

Corresponding to the parallel definitions of M and H* we have parallel computational characteristics. Neither are computable (at least for now), but both H* and the complement of M can be computably enumerated. The everyday correspondence is with simulation, once achieved in analog style via wind-tunnels etc, now very often digitally. The difference between being able to compute a phenomenon, and being able to simulate every detail of it, is that a simulation is never complete, despite the computable unfolding of information.

The Mandelbrot set provides an illuminating link between the pure abstraction of the halting problem, and strikingly embodied examples of emergence in nature. It helps us gather together physically interesting examples of emergence and abstract entities with clear mathematical attributes, to the illumination of both.

Turing's seminal work on morphogenesis in early 1950s Manchester is a particularly striking illustration of the power of mathematical descriptions to capture what seem to be surprising or mysterious developments of form in the real world. The validity of his proposal in the 1952 paper on *The Chemical Basis of Morphogenesis* [11] of an underlying reaction–diffusion basis for a wide range of emergent phenomena has been verified and suitably modified in many contexts. And this only paper of Turing's on the topic before he died has become his most cited paper, ahead of the famous computable numbers and Turing test articles.

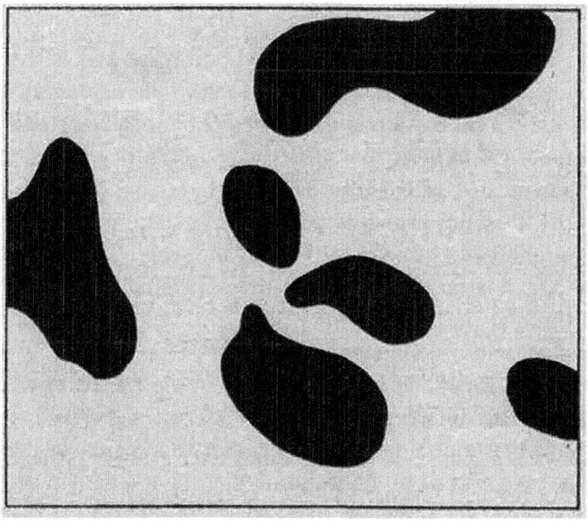

Courtesy of P. N. Furbank

Not only did Turing creatively extract the model, but also provided the differential equations governing its development in various contexts, and in some cases—such as that of the emergence of dappling on a cow's hide—used the newly available computers via Ferranti to simulate the solutions—with remarkably persuasive results.

Of course, Turing's differential equations tended to have computable solutions, even though they logically inhabited the same world as the halting problem or the Mandelbrot set. Today, similar modeling has become standard in many fields, and where analytic solutions are not found to be derivable, computer simulation has become a standard way of imaging emergent patterns, such as 'strange attractors', or global outcomes for economic models—all, according to a broad-brush logical perspective, of a similar level of abstraction (LoA) in relation to the computable base, as the halting set H* is to the universal machine. The well-known sensitivity to initial data exhibited by these simulated emergent outcomes is an indicator of incipient problems with computational prediction.

It becomes clear that in a real situation one may have very real characteristics of the real world describable in some sense, without there being a corresponding computational counterpart. The logical form of Turing's examples points to this possibility, and subsequent experience—such as that described by Nassim Taleb—tends to confirm its actuality. The language may code computational content, but our ability to access computation is limited by the reach of our machines and our brains.

We can take this analysis further, with dramatic consequences. Not only does the language have the power to capture real phenomena in terms of basic underlying causality—just as Turing did for the moving stripes on tropical fish, say, or the stripes on the zebra. The intuition—or *a priori* insight as Wittgenstein would have it in the 1921 *Tractatus Logico-Philosophicus*—is that natural phenomena not only generate descriptions within a causal context, but arise and derive from them in some necessary way. Expressions of this go back to the beginnings of philosophy, finding its well-known expression in the *Principle of Sufficient Reason,* as voiced in Leibniz' 1714 *La Monadologie*.

In the mathematical world, the principle has found expression via the notion of definability. Just as the principle is a basic assumption in science, so is the mathematical ontology of a structure driven by the search for inductive descriptions of individuals and relations in terms of those originally given. The growth of modern logic through the 19th century, and the focus on axiomatic theories in formal versions of natural language, opened the door to projects such as that of the philosopher Hans Reichenbach, who sought in the 1920s to axiomatize special relativity. It fell to Alfred Tarski [12] in the 1930s to clarify the formal definability of truth of a logical statement in an interpretation, and hence of definability in the real world. This gave a basis to the intuition that the mathematics of a given class of structures advanced by showing connections in terms of a commonly shared mathematical language.

There are two important aspects of the mathematics of definability that one might usefully take back into the real world. The first is the realisation that definability sometimes breaks down: in fact, necessarily breaks down, since the type structure rapidly throws up far more relations than the language is capable of describing. And there are more subtle failures of definability, opening the door to different models of a theory, in which important aspects are not uniquely defined—a bit like the state of an atomic particle before a measurement is taken.

The second is that there is a language independent version of definability, useful in an ontological context, since the real universe does not use language to 'define' its structure and laws. If a relation on a structure is left entirely undisturbed by any automorphism of the structure, we say it is *invariant*. The invariance is an expression of the ontological uniqueness of the relation on the structure. If the logic adopted to describe the structure is appropriate, then all definable relations on the structure are invariant.

Drawing things together: The higher order properties of a structure—those which are so important for us to understand in the real world—are the emergent relations of the structure. These large-scale relations are commonly based on reasonably well-understood local structure. Mathematically, we have formalised this in terms of definability—or, as invariance under automorphisms over basic computational structure.

And, to add a final element of the jigsaw: Such definability can be viewed as computation over higher type data. If a structure uniquely pins down a relation, in the sense that there is no alternative presentation of the structure that changes the relation in our observational purview, then we have to conclude that, in some sense, the complexity of the structure has computed that relation. There is, of course, a rich theory of higher type computation (starting with Stephen Kleene, Robin Gandy, Georg Kreisel, Gerald Sacks etc.), which makes explicit how we see its computational infrastructure. John Longley's 2005 survey paper [13] is an excellent guide to the field: and forthcoming is his book with Dag Normann [14] on *Computability at Higher Types*.

5 Living in a World of Higher Type Computation

We cannot expect computation over higher type information to have the reliability or precision of the classical model. And it is at the level of the human brain and its hosting of complex mentality that we can expect this to be specially apparent. Here is Alan Turing, the advocate of computer intelligence, in a talk to the London Mathematical Society, on February 20, 1947 (quoted in Hodges [15], p. 361):

> ... if a machine is expected to be infallible, it cannot also be intelligent. There are several theorems which say almost exactly that.

And here he is again, in the final paragraph of his popular article *Solvable and Unsolvable Problems* in *Penguin Science News* 31, 1954, p. 23 (p. 322 of the *Impact* book [3]), suggesting that human 'common sense' is something additional to algorithmic thinking:

> The results which have been described in this article are mainly of a negative character, setting certain bounds to what we can hope to achieve purely by reasoning. These, and some other results of mathematical logic may be regarded as going some way towards a

demonstration, within mathematics itself, of the inadequacy of 'reason' unsupported by common sense.

One might also look for emergence, in the form of surprisingly emergent outcomes. As mentioned previously, mathematical creativity was something Henri Poincaré was keenly interested in, and Jacques Hadamard in his 1945 book [16] on *The Psychology of Invention in the Mathematical Field* (Princeton University Press) refers to Poincaré's celebrated 1908 lecture at the Société de Psychologie in Paris, where he recounts getting stuck solving a problem related to elliptic functions:

> At first Poincaré attacked [a problem] vainly for a fortnight, attempting to prove there could not be any such function ... [quoting Poincaré]:
> 'Having reached Coutances, we entered an omnibus to go some place or other. At the moment when I put my foot on the step, the idea came to me, without anything in my former thoughts seeming to have paved the way for it ... I did not verify the idea ... I went on with a conversation already commenced, but I felt a perfect certainty.
> On my return to Caen, for conscience sake, I verified the result at my leisure.'

What is striking about this account is not so much the surprising emergence, as the sense we have of there being some definitional ownership taken of the proof, enabling Poincaré to carry home the proof as if packaged in the proof shop. Of course, Poincaré would have known nothing of emergence in 1908 (even though the concept is traceable back to John Stuart Mill's 1843 *System of Logic*), and though chance might be playing a role, an idea echoed by Hadamard in his book.

At the neuroscientific level, we have a huge body of commentary on the 'left brain—right brain' dichotomy, and the evidence of them hosting different kinds of thinking—according to our perspective, different types of computational activity. In his 2009 book [17] *The Master and his Emissary: The Divided Brain and the Making of the Western World,* Iain McGilchrist describes how:

> The world of the left hemisphere, dependent on denotative language and abstraction, yields clarity and power to manipulate things that are known, fixed, static, isolated, decontextualised, explicit, disembodied, general in nature, but ultimately lifeless. The right hemisphere by contrast, yields a world of individual, changing, evolving, interconnected, implicit, incarnate, living beings within the context of the lived world, but in the nature of things never fully graspable, always imperfectly known—and to this world it exists in a relationship of care. The knowledge that is mediated by the left hemisphere is knowledge within a closed system. It has the advantage of perfection, but such perfection is bought ultimately at the price of emptiness, of self-reference. It can mediate knowledge only in terms of a mechanical rearrangement of other things already known. It can never really 'break out' to know anything new, because its knowledge is of its own representations only. Where the thing itself is present to the right hemisphere, it is only 're-presented' by the left hemisphere, now become an idea of a thing. Where the right hemisphere is conscious of the Other, whatever it may be, the left hemisphere's consciousness is of itself.

This description of the functionality of the brain fits surprisingly well our developing awareness of the differences between type-1 and type-2 computation; shadowing the differences between the algorithmic, as embodied in today's computer, and the more open, creative thinking that the human brain appears capable of.

An historically persistent question concerns not so much the computational nature of mentality, as the *relationship* between the physical host, with its left and right hemispheres, and the mentality we appreciate in abstract form, and attach such significance to. In this sense, our computational paradigm in the making has been knocking at the door for nearly 400 years. Philosophically, the question has become framed in terms of the supervenience of mentality on the physical. As Jaegwon Kim [18] succinctly describes it (*Mind in a Physical World*, MIT Press, 1998, pp. 14–15), supervenience:

> ... represents the idea that mentality is at bottom physically based, and that there is no free-floating mentality unanchored in the physical nature of objects and events in which it is manifested.

More formally, the Stanford Encyclopedia of Philosophy puts it:

> A set of properties **A** supervenes upon another set **B** just in case no two things can differ with respect to **A**-properties without also differing with respect to their **B**-properties.

Kim asks various questions, which make it clear we are in need of an updated model, and one which delivers a coherent overview of the interactions of these domains which are so basic to our everyday lives. The sort of questions one is driven to face up to include: How can mentality have a computational role in a world that is fundamentally physical? A highly reductive physicalist view, in which the effective role of such phenomena as consciousness become illusory, has distinguished protagonists. For instance, in *Turing's 'Strange Inversion of Reasoning'* (*Alan Turing: His Work and Impact*, [3], pp. 569–573), Daniel Dennett finds that a gradualist Darwinian model, and parallel thinking of Turing concerning intelligent machines, a useful antidote to those who 'think that a Cartesian *res cogitans*, a thinking thing, cannot be constructed out of Turing's building blocks'. And what about *overdetermination*—the problem of phenomena having both mental and physical causes? In a book whose title points to the quest for a convincing *Physicalism, or Something Near Enough* (Princeton, 2005), Jaegwon Kim [19] nicely expresses the problem we have physically connecting mentality in a way that gives it the level of autonomy we observe:

> ... the problem of mental causation is solvable only if mentality is physically reducible; however, phenomenal consciousness resists physical reduction, putting its causal efficacy in peril.

More brain-like computational models, *connectionist* models or *neural nets*, go back to the 1940s (Warren McCulloch and Walter Pitts, 1943, and Alan Turing's 'unorganised machines', 1948). These have come a long way since then, to a point where we can read Paul Smolensky (*On the proper treatment of connectionism*, in *Behavioral and Brain Sciences*, 11, 1988, pp. 1–74) saying:

> There is a reasonable chance that connectionist models will lead to the development of new somewhat-general-purpose self-programming, massively parallel analog computers, and a new theory of analog parallel computation: they may possibly even challenge the strong construal of Church's Thesis as the claim that the class of well-defined computations is exhausted by those of Turing machines.

The intuition is that the model may support an emergent form of incomputability, the interactivity supporting a more clearly embodied example than that derivable via the Turing machine model. Is that all?

Such functionality does not impress Steven Pinker (for instance) for whom 'neural networks alone cannot do the job'. Pinker—echoing Floridi's [20] 'levels of abstraction' in the mental context—identifies [21] in *How the Mind Works*, a 'kind of mental fecundity called recursion', with an example:

> We humans can take an entire proposition and give it a role in some larger proposition. Then we can take the larger proposition and embed it in a still-larger one. Not only did the baby eat the slug, but the father saw the baby eat the slug, and I wonder whether the father saw the baby eat the slug, the father knows that I wonder whether he saw the baby eat the slug, and I can guess that the father knows that I wonder whether he saw the baby eat the slug, and so on.

It falls to Antonio Damasio [22] to get us engaged with a more detailed description of the basic physicality of the emergent and 'recursive' character of observed human creative thinking (*The Feeling of What Happens*, 1999, p. 170):

> As the brain forms images of an object—such as a face, a melody, a toothache, the memory of an event—and as the images of the object affect the state of the organism, yet another level of brain structure creates a swift nonverbal account of the events that are taking place in the varied brain regions activated as a consequence of the object-organism interaction. The mapping of the object-related consequences occurs in first-order neural maps representing the proto-self and object; the account of the causal relationship between object and organism can only be captured in second-order neural maps. ... one might say that the swift, second-order nonverbal account narrates a story:
> that of the organism caught in the act of representing its own changing state as it goes about representing something else.

The physicalism is still conceptually challenging, but contains the sort of ingredients which echo the higher type computation we identify within the messiness and strange effectiveness of right-hemisphere functionality. At the same time, one detects a complexity of interaction involving representation, sampling and looping between types/levels of abstraction, that is characteristic of what George Ellis [23] calls "true complexity".

This is a complexity of scenario reinforced by experience of trying to build intelligent machines. The lesson of the neural net insufficiency is that the basic logical structure contains the seeds of computation beyond the Turing barrier, but that the embodiment needs to be capable of hosting mechanisms for type reduction, employing the sort of representational facility Damasio observes in the brain, and enabling the kind of recursions that Pinker demands. In artificial intelligence the purely logical approach has encountered unsurmountable difficulties. While some have been driven to an ad hoc empirical approach, following the recipe for machine-human interaction and learning anticipated by Turing. Once again, we see a community showing the effects of paradigmic confusion. From Rodney Brooks in *Nature* in 2001 we have:

> ... neither AI nor Alife has produced artifacts that could be confused with a living organism for more than an instant.

While Marvin Minsky is quoted at Boston University in May 2003 as proclaiming that:

AI has been brain-dead since the 1970s.

We are feeling our way. Type-2 information cannot be fully represented in general in a form we can present to our computing machines. It is embodied in the real world, and as such becomes input to computations as processes. The data-machine duality may be present, but the universality is lost. And the 'machine'—say a particular human organism—may be computed within a world we find hard to simulate. This is not to say we may make huge progress in understanding and simulating aspects of our world, but the process may be slow. We are participants, not masters. There are no masters.

At the same time, the mathematics tells us that there is a radical difference between type-1 and type-2 computation, in which the Faustian pact with type-2 is at the expense of certainty. There is a hugely challenging and unpredictable control task, as anyone working with people already knows. At the practical level, this means an AI developed with care within a learning context, with specific tasks in mind. Of course, another lesson is that algorithmic and 'intelligent' thinking is an alliance, that part of the control process is keeping this clear. The left brain-right brain paradigm is fundamental. Why else would the brain have evolved as it did?

All this does not just apply to human intelligence. It applies to any environment to which the human organism brings its 'levels of abstraction', including instances of artistic, social and economic contexts. In general, these will not be subject to detailed control or prediction. Though we may come to understand and simulate many aspects of the world in its computational activity.

6 Turing's Mathematical Host

If we want to model and understand causal structure, we need a computational model which gives due weight to information and the computational underpinnings of language—a model based in the science where we have a degree of trust, and a sound footing. Of course, in the scientific context, the basic route to representation of information is the real numbers.

In Turing's 1939 paper [24] on *Systems of Logic Based on Ordinals* (Proc. London Math. Soc. (2) 45 (1939), 161–228) he grapples with incompleteness and incomputability. Faced with the incomputability of important relations, Turing introduced the notion of an *oracle Turing machine*, in order to compare *degrees* of incomputability. At a stroke, this gave us a structuring of information according to its *relative* computability. The oracle machine computes just like the standard machine described earlier, but has an extra facility for occasionally asking a question about a finite part of a given real number provided as the *oracle*—the oracle being a repository of auxiliary information to be used during any given computation.

This idea gave us a model of how we compute using data given to us from unknown sources. It not only gave us a way of comparing computability of reals. It could also be used to model real world causality, where the mathematical law for computing applications of the law from one state to another was computable. This was a framework within which Newtonian computable laws fitted nicely, at the basic level. Looked at globally, one obtained a model of computable content of structures, based on *partial computable (p.c.) functionals* over the real numbers. So any physical context, characterised by computable relationships between pairs of reals representing particular physical states, could be presented as an instantiation of part of the resulting mathematical structure—often termed the *Turing universe*. Strangely, despite Turing's later interest in interactive computation, he never returned to the oracle machine.

In 1948, Emil Post mathematically prepared the Turing universe for further investigation by grouping reals together into equivalence classes (called *degrees of unsolvability*) of inter-computable reals, the degrees ordered by the induced ordering got from relative computation. The resulting structure, over the years, was discovered to have a very rich infrastructure. Mathematically, what was seen as a high degree of pathology provided the means for defining a rich array of relations and individuals. The complexity, of course, was just what one might expect from a mathematical structure capable of reflecting causal structure from different physical contexts. The definable relations formed a mathematical counterpart of the richness of emergent phenomena in the real world.

By the 1960s, the theory of this basic structure—now termed the *Turing degrees*—was becoming a very active area of pure research, one of formidable technical complexity. And this brings us to what has become known as Hartley Rogers' Programme, originating with a paper [25] of his on *Some problems of definability in recursive function theory*, based on an invited talk at the 1965 Logic Colloquium in Leicester in 1965. The fundamental problem implicit in the paper and talk, was that of characterising the Turing definable/invariant relations over the structure of the Turing degrees. The intuition was that these definable relations are key to pinning down how higher order relations on the real world can appear to be computed. While the breakdown of definability could point to ontological and epistemological ambiguities in the real world—such as confused mental states, or uncertainty at the quantum level in physics. For further information on a range of such topics, see Cooper's book [4] on *Computability Theory*, or his paper [26] with Piergiorgio Odifreddi on *Incomputability in Nature*.

The topic of physical computation, the Turing universe and emergent relations is treated in more detail elsewhere. The general message is that in the real world, one may describe global relations in terms of local structure, so capturing the computational basis of large-scale phenomena. And that mathematically this can be formalized as definability—or more generally invariance under automorphisms—over structure based on oracle computations. If nothing else, the model can give an idea of the causal complexity of so much in the modern world.

References

1. N. Taleb, *The Black Swan: The Impact of the Highly Improbable* (Allen Lane, London, 2007)
2. A. Damasio, *Descartes' Error: Emotion, Reason and the Human Brain* (G.P. Putnam's Sons, New York, 1994)
3. S.B. Cooper, J. van Leeuwen (eds.), *Alan Turing: His Work and Impact* (Elsevier, Amsterdam, 2013)
4. S.B. Cooper, *Computability Theory* (Chapman & Hall/CRC, London, 2004)
5. H. Putnam, *Minds and Machines*. Reprinted in Putnam, *Mind, Language and Reality. Philosophical Papers,* vol. 2, (Cambridge University Press, 1975)
6. K. Gödel, Russell's mathematical logic, in *The Philosophy of Bertrand Russell*, ed. by P.A. Schilpp (Northwestern University, Evanston and Chicago, 1944), pp. 123–153
7. R.G. Downey, D.R. Hirschfeldt, *Algorithmic Randomness and Complexity* (Springer, New York, 2010)
8. C.S. Calude, K. Svozil, Quantum randomness and value indefiniteness. Adv. Sci. Lett. **1**(2), 165–168 (2008)
9. H. Putnam, *Realism with a Human Face* (Harvard University Press, Cambridge, 1990)
10. K.V. Mardia, S.B. Cooper, Alan Turing and enigmatic statistics. Bull. Brasilian Section Int. Soc. Bayesian Anal. **5**(2), 2–7 (2012)
11. A.M. Turing, The chemical basis of morphogenesis. Philos. Trans. R. Soc. Lond. Ser, B **237**, 37–72 (1952)
12. A. Tarski, *On Definable Sets of Real Numbers*; translation of Tarski (1931) by J.H. Woodger. In A. Tarski, *Logic, Semantics, Metamathematics*, 2nd edn, ed. by J. Corcoran (Hackett, Indianapolis, 1983), pp. 110–142
13. J. Longley, Notions of computability at higher types I, in *Logic Colloquium 2000*, ed. by R. Cori, A. Razborov, S. Todorcevic, C. Wood (A K Peters, Natick, 2005)
14. J. Longley, D. Normann, *Computability at Higher Types*. Theory and Applications of Computability (Springer, Heidelberg, New York)
15. A. Hodges, *Alan Turing: The Enigma* (Vintage, New York, 1992); reprinted in Centennial edition, Princeton University Press, 2012
16. J. Hadamard, *The Psychology of Invention in the Mathematical Field* (Princeton University Press, Princeton, 1945)
17. I. McGilchrist, *The Master and His Emissary: The Divided Brain and the Making of the Western World* (Yale University Press, New Haven, 2009)
18. J. Kim, *Mind in a Physical World* (MIT Press, Cambridge, 1998)
19. J. Kim, *Physicalism, or Something Near Enough* (Princeton University Press, Princeton, 2005)
20. L. Floridi, *The Philosophy of Information* (Oxford University Press, Oxford, 2011)
21. S. Pinker, *How the Mind Works* (W.W. Norton, New York, 1997)
22. A. Damasio, *The Feeling of What Happens: Body and Emotion in the Making of Consciousness* (Harcourt Brace, New York, 1999)
23. G.F.R. Ellis, True complexity and its associated ontology, in *Science and Ultimate Reality: Quantum Theory, Cosmology and Complexity*, ed. by J.D. Barrow, P.C.W. Davies, C.L. Harper, Jr. (eds.) (Cambridge University Press, Cambridge, 2004)
24. A.M. Turing, Systems of logic based on ordinals. Proc. London Math. Soc. **45**(2), 161–228 (1939); reprinted in Cooper and van Leeuwen, pp. 151–197
25. H. Rogers, Jr., Some problems of definability in recursive function theory, in *Sets, Models and Recursion Theory* (Proceedings of Summer School Mathematical Logic and Tenth Logic Colloquium, Leicester, 1965) (North-Holland, Amsterdam, 1967), pp. 183–201
26. S.B. Cooper, P. Odifreddi, Incomputability in nature, in *Computability and Models: Perspectives East and West*, ed. by S.B. Cooper, S.S. Goncharov (Plenum, New York, 2003), pp. 137–160
27. H. Putnam, Why functionalism didn't work, in *Representation and Reality* (Putnam, MIT Press, 1988), pp. 73–89

CPSIA information can be obtained
at www.ICGtesting.com
Printed in the USA
LVOW02*1154310116
473068LV00001B/3/P

9 783319 221557